Design Analysis in Rock Mechanics

Design Analysis in Rock Mechanics

Contributors

Mahdi Rasouli Maleki, Mohammad Mahyar et al.

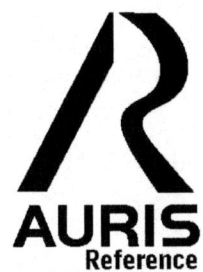

www.aurisreference.com

Design Analysis in Rock Mechanics

Contributors: Mahdi Rasouli Maleki, Mohammad Mahyar et al.

Published by Auris Reference Limited
www.aurisreference.com

United Kingdom

Copyright 2016
Printed in 2017 for Sale in the Indian Subcontinent

The information in this book has been obtained from highly regarded resources. The copyrights for individual articles remain with the authors, as indicated. All chapters are distributed under the terms of the Creative Commons Attribution License, which permit unrestricted use, distribution, and reproduction in any medium, provided the original author and source are credited.

Notice

Contributors, whose names have been given on the book cover, are not associated with the Publisher. The editors and the Publisher have attempted to trace the copyright holders of all material reproduced in this publication and apologise to copyright holders if permission has not been obtained. If any copyright holder has not been acknowledged, please write to us so we may rectify.

Reasonable efforts have been made to publish reliable data. The views articulated in the chapters are those of the individual contributors, and not necessarily those of the editors or the Publisher. Editors and/or the Publisher are not responsible for the accuracy of the information in the published chapters or consequences from their use. The Publisher accepts no responsibility for any damage or grievance to individual(s) or property arising out of the use of any material(s), instruction(s), methods or thoughts in the book.

Design Analysis in Rock Mechanics

ISBN: 978-1-78154-906-3

British Library Cataloguing in Publication Data
A CIP record for this book is available from the British Library

Printed in the United Kingdom

Exclusively distributed by CBS Publishers & Distributors Pvt. Ltd.

Sales & Distribution Rights only for India, Pakistan, Bangladesh, Sri Lanka, Nepal and Bhutan. This book is not to be sold outside these territories.

Contents

List of Abbreviations .... *vii*
List of Contributors .... *ix*
Preface .... *xiii*

Chapter 1   **Design of Overall Slope Angle and Analysis of Rock Slope Stability of Chadormalu Mine Using Empirical and Numerical Methods** 1
Mahdi Rasouli Maleki, Mohammad Mahyar, Kambiz Meshkabadi

Chapter 2   **Numerical Study on Crack Propagation in Brittle Jointed Rock Mass Influenced by Fracture Water Pressure** 11
Yong Li, Hao Zhou, Weishen Zhu, Shucai Li and Jian Liu

Chapter 3   **Study on the Constitutive Model for Jointed Rock Mass** 29
Qiang Xu, Jianyun Chen, Jing Li, Chunfeng Zhao, Chenyang Yuan

Chapter 4   **Study on Calculation of Rock Pressure for Ultra-Shallow Tunnel in Poor Surrounding Rock and its Tunneling Procedure** 59
Xiaojun Zhou, Jinghe Wang, Bentao Lin

Chapter 5   **A Generalized Plasticity-Based Model for Sandstone Considering Time-Dependent Behavior and Wetting Deterioration** 83
Meng-Chia Weng

Chapter 6   **Application of Geostatistical Models for Estimating Spatial Variability of Rock Depth** 115
Pijush Samui, Thallak G. Sitharam

Chapter 7   **Proposal of a New Parameter for the Weathering Characterization of Carbonate Flysch-Like Rock Masses: The Potential Degradation Index (PDI)** 137
M. Cano, R. Tomás

Chapter 8   **Rock Mass Hydraulic Conductivity Estimated by Two Empirical Models** 175
Shih-Meng Hsu, Hung-Chieh Lo, Shue-Yeong Chi and Cheng-Yu Ku

Chapter 9	Application of Base Force Element Method on Complementary Energy Principle to Rock Mechanics Problems	209
	Yijiang Peng, Qing Guo, Zhaofeng Zhang, and Yanyan Shan	
Chapter 10	Three-Dimensional Numerical Model of Hydraulic Fracturing in Fractured Rock Masses	255
	B. Damjanac, C. Detournay, P.A. Cundall and Varun	
Chapter 11	Roughness Research of Center Profile Curve on Rock Fracture Surface Based on Statistical Method	269
	Xuezai Pan, Zhigang Feng, Guoxing Dai, and Hongguang Liu	
	Citations	285
	Index	287

List of Abbreviations

ASTM	American Society for Testing and Materials
BEM	Boundary Element Method
BFEM	Base Force Element Method
BPM	Bonded Particle Model
BQ	Basic Quality
DDM	Displacement Discontinuity Method
DEM	Discrete Element Method
DFN	Discrete Fracture Network
DI	Depth Index
FE	Finite Element
FEM	Finite Element Method
FOS	Factor of Safety
GCD	Gouge Content Designation
GIS	Geographic Information System
GRL	Ground Reduced Levels
GSI	Geological Strength Index
HF	Hydraulic Fracture
HPFM	Heat Pulse Flowmeter
ISRM	International Society for Rock Mechanics
JRC	Joint Roughness Coefficients
LPI	Lithology Permeability Index
MTS	Mechanics Test Systems
NFR	Naturally Fractured Reservoirs
PDI	Potential Degradation Index
PFC	Particle Flow Code
RAC	Recycled Aggregate Concrete
RMR	Rock Mass Rating
RQD	Rock Quality Designation
SDT	Slake Durability Test
SJM	Smooth Joint Model
SRM	Synthetic Rock Mass
SRMR	Slope Rock Mass Rating

List of Contributors

Mahdi Rasouli Maleki
Engineering Geology & Rock Mechanic Department, Tunnel Consulting Engineers, Tehran, Iran

Mohammad Mahyar
Mining Engineering, Tunnel Consulting Engineers, Tehran, Iran

Kambiz Meshkabadi
Lecturer of Civil Engineering Department, Islamic Azad University of Ahar, Iran

Yong Li
Geotechnical & Structural Engineering Research Center, Shandong University, Jinan 250061, Shandong, China

Hao Zhou
Geotechnical & Structural Engineering Research Center, Shandong University, Jinan 250061, Shandong, China

Weishen Zhu
Geotechnical & Structural Engineering Research Center, Shandong University, Jinan 250061, Shandong, China

Shucai Li
Geotechnical & Structural Engineering Research Center, Shandong University, Jinan 250061, Shandong, China

Jian Liu
School of Civil Engineering, Shandong University, Jinan 250061, Shandong, China

Qiang Xu
School of Civil and Hydraulic Eng., Dalian University of Technology, Dalian, China

Jianyun Chen
State Key Lab.of Coastal and Offshore Eng., Dalian University of Technology, Dalian, China
School of Civil and Hydraulic Eng., Dalian University of Technology, Dalian, China

Jing Li
School of Civil and Hydraulic Eng., Dalian University of Technology, Dalian, China

Chunfeng Zhao
School of Civil and Hydraulic Eng., Dalian University of Technology, Dalian, China

Chenyang Yuan
School of Civil and Hydraulic Eng., Dalian University of Technology, Dalian, China

Xiaojun Zhou
Key Laboratory of Transportation Tunnel Engineering of Ministry of Education, School of Civil Engineering, Southwest Jiaotong University, Chengdu 610031, China

Jinghe Wang
Key Laboratory of Transportation Tunnel Engineering of Ministry of Education, School of Civil Engineering, Southwest Jiaotong University, Chengdu 610031, China

Bentao Lin
The 2nd Institute of Civil and Architecture Engineering, China Railway Eryuan Engineering Corporation Ltd., Chengdu 610031, China

Meng-Chia Weng
Department of Civil and Environmental Engineering, National University of Kaohsiung, 700, Kaohsiung University Rd, Kaohsiung 81148, Taiwan, ROC

Pijush Samui
Centre for Disaster Mitigation and Management, VIT University, Vellore, India

Thallak G. Sitharam
Department of Civil Engineering, Indian Institute of Science, Bangalore, India

M. Cano
Departamento de Ingenierı́a Civil, Escuela Politécnica Superior, Universidad de Alicante, 03080 Alicante, Spain

R. Tomás
Departamento de Ingeniería Civil, Escuela Politécnica Superior, Universidad de Alicante, 03080 Alicante, Spain

Shih-Meng Hsu
Sinotech Engineering Consultants, Inc

Hung-Chieh Lo
Sinotech Engineering Consultants, Inc

Shue-Yeong Chi
Sinotech Engineering Consultants, Inc

Cheng-Yu Ku
National Taiwan Ocean University, Taiwan

Yijiang Peng
The Key Laboratory of Urban Security and Disaster Engineering, Ministry of Education, Beijing University of Technology, Beijing 100124, China

Qing Guo
The Key Laboratory of Urban Security and Disaster Engineering, Ministry of Education, Beijing University of Technology, Beijing 100124, China

Zhaofeng Zhang
The Key Laboratory of Urban Security and Disaster Engineering, Ministry of Education, Beijing University of Technology, Beijing 100124, China

Yanyan Shan
The Key Laboratory of Urban Security and Disaster Engineering, Ministry of Education, Beijing University of Technology, Beijing 100124, China

B. Damjanac
Itasca Consulting Group, Inc., Minneapolis, Minnesota, USA

C. Detournay
Itasca Consulting Group, Inc., Minneapolis, Minnesota, USA

P.A. Cundall
Itasca Consulting Group, Inc., Minneapolis, Minnesota, USA

Varun
Itasca Consulting Group, Inc., Minneapolis, Minnesota, USA

Xuezai Pan
School of Mathematics, Nanjing Normal University, Taizhou College, Taizhou, China
Faculty of Science, Jiangsu University, Zhenjiang, China

Zhigang Feng
State Key Laboratory of Coal Resources and Safe Mining, China University of Mining and Technology, Beijing, China
Faculty of Science, Jiangsu University, Zhenjiang, China

Guoxing Dai
Faculty of Science, Jiangsu University, Zhenjiang, China

Hongguang Liu
Faculty of Civil Engineering and Mechanics, Jiangsu University, Zhenjiang, China

Preface

Rock mechanics is a theoretical and applied science of the mechanical behavior of rock and rock masses; compared to geology, it is that branch of mechanics concerned with the response of rock and rock masses to the force fields of their physical environment. The text *Design Analysis in Rock Mechanics* treats the basics of rock mechanics in a clear and straightforward manner and discusses important design problems in terms of the mechanics of materials. The purpose of first chapter is to determine the bench slope angle and overall slope of the west wall in Chadormalu mine in points susceptible to rupture. In second chapter, to simulate the fissure development of jointed rock mass under fissure water pressure, we propose a novel numerical model on the basis of secondary development in Lagrangian analysis of continua (FLAC3D), which is an explicit finite difference method (FDM). Study on the constitutive model for jointed rock mass has been presented in third chapter. The aim of fourth chapter is to find out a simple way to calculate the asymmetric rock pressure for design of tunnel lining in super-shallow surrounding rock. Based on the concept of generalized plasticity, fifth chapter proposes a constitutive model to describe the time-dependent behavior and wetting deterioration of sandstone. In sixth chapter, a semivariogram model has been developed along with the kriging model for the reduced level of the rock in the subsurface of Bangalore. The aim of seventh chapter is to propose a method for characterizing the weathering behavior of carbonate lithologies that outcrop in heterogeneous Flysch-like slopes. Eighth chapter proposes two empirical models to estimate hydraulic conductivity of fractured rock mass. In ninth chapter, the base force element method (BFEM) on complementary energy principle is used to analyze the engineering problems of rock mechanics. Three-dimensional numerical model of hydraulic fracturing in fractured rock masses has been focused in tenth chapter. Roughness research of center profile curve on rock fracture surface based on statistical method has been presented in last chapter.

Chapter 1

DESIGN OF OVERALL SLOPE ANGLE AND ANALYSIS OF ROCK SLOPE STABILITY OF CHADORMALU MINE USING EMPIRICAL AND NUMERICAL METHODS

Mahdi Rasouli Maleki[1], Mohammad Mahyar[2], Kambiz Meshkabadi[3]

[1]Engineering Geology & Rock Mechanic Department, Tunnel Consulting Engineers, Tehran, Iran

[2]Mining Engineering, Tunnel Consulting Engineers, Tehran, Iran

[3]Lecturer of Civil Engineering Department, Islamic Azad University of Ahar, Iran

ABSTRACT

In engineering projects associated with rock mechanic science like open pit mines, assessment and slope stability of mine walls is one of the important performance in generate of these structures. Estimating and knowledge of stable slope angle is one of main parts that should be occurring to special attention in open pit mines studies phase. Considering the importance of economic costs in mining issues, the need for appropriate design slope angle that can cause an adverse minimize project costs and throws the other hand, the stability conditions in the safe walls of the mine life will provide essential and seems obvious. Therefore, in this study to determine the optimal slope angle of overall and bench of west wall of the Chadormalu ore iron mine, has been trying, first, done field studies on the discontinuity of western wall, engineering classification and geomechanical properties of rock masses of wall, then assess the amount of optimal slope angle using empirical method. Finally, in order to ensure stability and accuracy of the wall slope angle based on the obtained (empirical method) tries to analysis is amount of Factor of Safety (FOS), displacements and mean stress condition atwalls calculated from drilling use Phase2D powerful software.

INTRODUCTION

The purpose of this study is to determine the bench slope angle and overall slope of the west wall in Chadormalu mine in points susceptible to rupture. To do so, the survey tries to; first, detect sensitive points by current empirical methods. Then it determines the bench and overall angle of slope.

In order to be sure about the results validity obtained by the empirical methods, the study attempts to analyze stability and determine the slope safety factor and the wall displacements using finite element method and powerful Phase2D software.

POSITION AND GEOLOGY OF THE WEST WALL OF THE MINE

Chadormalu iron ore mine is located in central Iran and in northern slope of Chah-Mohammad grey mountains in southern margin of Saghand salt marsh about 180 km from north-east of Yazd and 300 km from south of Tabas desert. According to the geology studies performed in this region, it was cleared that Chadormalu fault between the plain and high lands is the major factor of ore creation and mineralization in the region formed in two forms of northern and southern anomaly. Also, petrography studies on the mine rocks shows that major rocks in Chadormalu mine area are Metasomatite, Albitite, Diorite, Magnetite and Hematite[4,5]. It should be mentioned that, performance of different faults in this area makes the mine rocks tobe severely tectonizedand provide suitable conditions fordifferent ruptures of the wall. The western wall of this mine is made up of igneous rocks and various metamorphic rocks such as Diorite, Albitite and Metasomatite[4]. Generally speaking, **Figure 2** shows the geological profile perpendicular in the western wall of the mine together with its lithological combination.

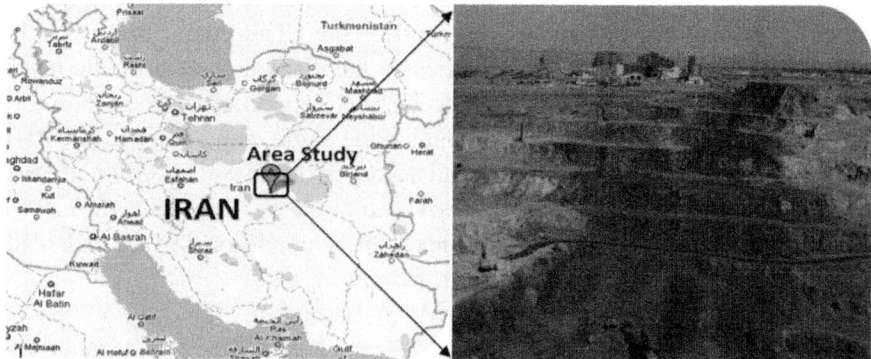

Figure 1: Location of area study on the Iran map.

Profile on the midcourse of block 11 & Block 12

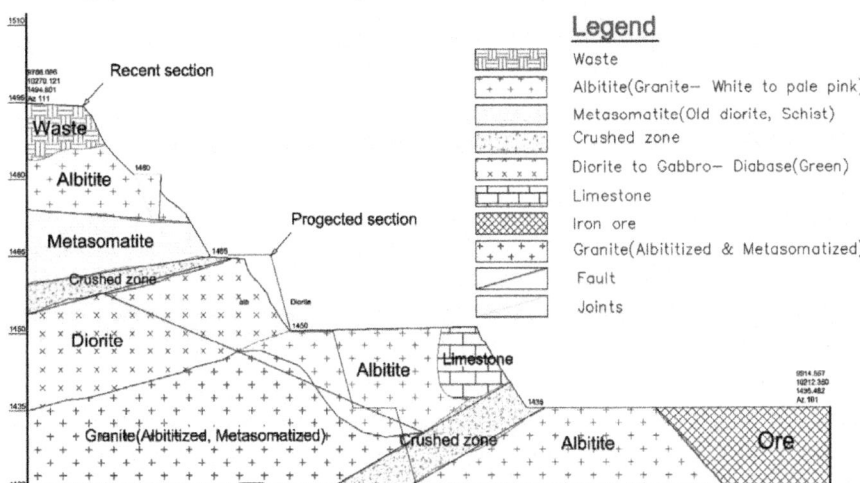

Figure 2: The geological profile perpendicular on the west wall of Chador-malu iron ore mine.

In this study, in order to save time and costs, three blocks (No. B-10, B-11, and B-13) in instability intensity wear detected as the conclusion of the geological surveys on the mine western wall, than engineering surveys and joint studies ware done on each of these blocks [5].

CLASSIFICATION OF ROCK MASS AND DETERMINATION OF ENGINEERING PARAMETERS

As the main purpose of engineering projects is to use classification systems to determine geomechanical characteristics of rocks by simple methods, this study tries to do joint studies on present discontinuities; and then it classifies the rocks enclosed in each of the blocks using Rock Mass Rating (RMR), Geological Strength Index (GSI) and Slope Rock Mass Rating (SRMR) classification systems [3]. **Table 1** indicates the results of the classification of the rocks of the western wall in Chadormalu iron ore mine.

Results obtained from this table indicate that the quality of the rocks in the west wall area in Chadormalu mine is poor due to breakings and development of lots of joints and fractures.

As in engineering works especially in analysis of rock slopes, the purpose is to classify rocks to estimate and measure their engineering and geomechanical features correctly, this study uses results of rocks classification and Roclab software [6] for each of the rock pieces enclosed in block B-11 to detect those parameters that have been introduced by empirical methods established by researches throughout the world. **Table 2** shows the most important calculated engineering parameters which are used in Phase2D software [7].

DETERMINATION OF BENCH SLOPE AND OVERALL SLOPE

According to definition of slope geometry, it is said that height, width, and angle of bench slope are the most significant geometrical parameters of slopes and steep surfaces where any alternation each of these features can put a direct effect on the slope stability. On the basis of these words, therefore, one can admit that optimum determination of these geometrical features of a slope in preventing rupture is one of most important parts of rock and soil slopes analysis, so that importance of this issue in open pit mine activities and road cuttings are observable and underst-andable [2].

This study tries to apply not only ranking system, but also other empirical methods in determina-tion of bench slope and overall steep of the slope in order to promote the obtained results safety factor. Therefore, value of the rocks of each block, one can determine slope steep angle by the empirical methods obtained from Rock Mass Rating (RMR), Mine Rock Mass Rating (MRMR) and Slope Rock Mass Rating (SRMR) values. Results on overall angle and safe bench slope angle for each block are shown in Tables 3 and 4 respectively. Also, **Table 5** indicates features and final geometry of the west wall in Chadormalu mine.

Table 1: Results of the classification of the west wall Chadormalu mine according to various classification systems

System	NO. Block	Metasomatite	Albitite	Diorite	Fault	Crushed zone	Average
RMR	B - 10	28	-	-	25	23	25
	B - 11	26	26	-	-	21	24
	B - 12	39	39	38	-	-	39
GSI	B - 10	23	-	-	20	18	20
	B - 11	21	21	-	-	16	19
	B - 12	34	34	33	-	-	34
SRMR	B - 10	52	-	-	45	44	47
	B - 11	52	49	-	-	44	48
	B - 12	65	62	60	-	-	63

Table 2: The most important engineering parameters of each rock groups

Material	Albitite	Metasomatite	Granite	Diorite	Crushed zone
Unit weight (MN/m^3)	0.024	0.028	0.024	0.028	0.026
Compressive Strength of Rock mass (MPa)	2.29	2.52	3.24	3.31	1.83
Young's modulus (MPa)	1632	1691	3901.5	3270	1268
Poisson's ratio	0.26	0.23	0.23	0.24	0.26
Tensile strength (MPa)	0	0	0	0	0
Peak friction angle (degrees)	27.1	26	56.1	33.5	22.2
Peak cohesion (MPa)	0.67	0.68	0.21	1.01	0.54
Dilation Angle (degrees)	0	0	0	0	0
Residual Friction Angle (degrees)	27.1	26	56.1	33.5	22.2
Residual Cohesion (MPa)	0.67	0.68	0.21	1.01	0.54

STABILITY ANALYSES OF THE MINE WEST SLOPE

Introduction

Today, there are several methods for slope stability analysis, each has its own advantages and disadvantages. Numerical analyses methods are the most common ones that are used for rock and soil slope analyses. One of the software's that can analyze rock slope stability in a numerical way is powerful software called Phase2D. This software was used in the present study for analyses of stability in the west wall of Chadormalu mine.

Hypotheses of Analyses

In this research, it is supposed the all analyses have been done in conditions prior to excavation of berm 1435 and for both static and dynamic states whit 0.31 g earthquake acceleration.

Analyses and Results of the Slope Stability

Modeling and bordering the concerned slope in finite element Phase2D software and running the program, result of stability in the west slope of the mine were examined. The results show that safety factor of the slope designed under static and dynamic conditions will be 3.39 m and 2.26 m, respectively (**Table 6**). Also, displacement due to berms excavation is 1.5 cm for static state and 1.6 cm for dynamic one. Figures 3 and 4 represents the outcome model of Phase2D software (Factor of safety, total displacement and mean stress status) in condition prior to excavation of bench 1435 and **Figure 5** shows the shear-strain changes for both static and dynamic states, respectively.

Table 3: Safe overall slope angle obtained by Bieniawski (1989)[1] method

NO. Block	Metasomatite	Albitite	Diorite	Fault	Crushed zone	Average
B - 10	45.6	-	-	41.7	38.7	42.0
B - 11	42.6	42.6	-	-	35.0	40.1
B - 12	57.4	57.4	56.5	-	-	57.1

Table 4: Safe bench slope angle obtained by Slope Rock Mass Rating (SRMR)

NO. Block	Metasomatite	Albitite	Diorite	Fault	Crushed zone	Average
B - 10	69	-	-	66	65	66
B - 11	69	67	-	-	65	67
B - 12	74	73	72	-	-	73

CONCLUSIONS

Results obtained from this study confirm that; in dynamic condition, to obtain a safety factor upper than 2.2, the bench slope angle and its overall slope angle should be 70 degree and 44 degree. Also, according to the features of the rock masses in this area, numerical analysis results indicate that; under such a condition, displacement due to bench excavation will be led than 1.5 cm.

Table 5: Features and final geometry obtained for bench for the west wall in Chadormalu mine

Slope Parameters	Value
High (m)	15
Width of bream (m)	8.5
Bench Angle (degree)	70
Inter-ramp Angle (degree)	47
Overall Angle (degree)	44

Table 6: Slope safety factor and displacement due to slopes excavation

Parameters	Static	Dynamic
Safety of Factor (SRF)	3.39	2.26
Horizontal displacement (m)	0.015	0.016

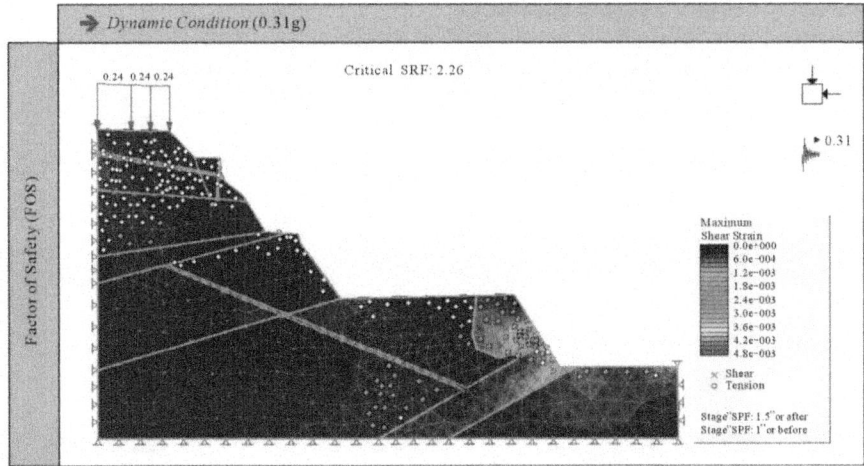

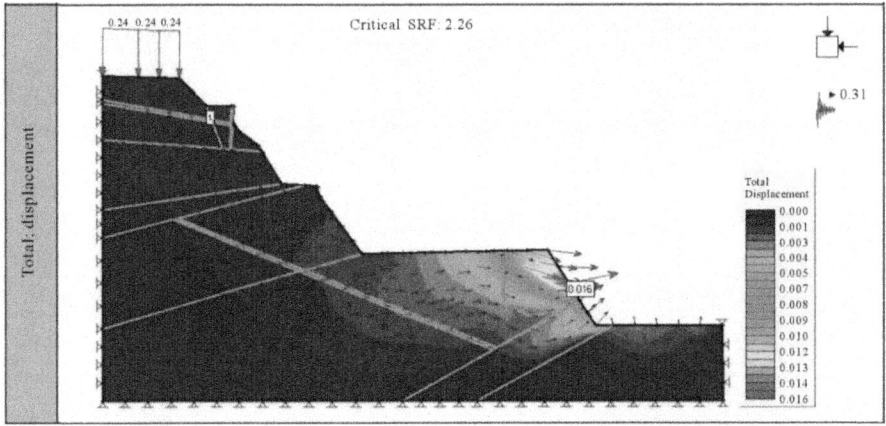

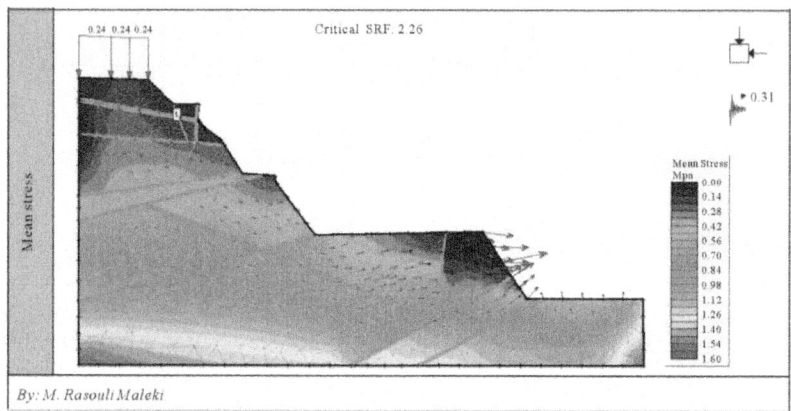

Figure 3: Output model of Phase2D software, factor of safety (FOS), total displacements and mean stress for static state.

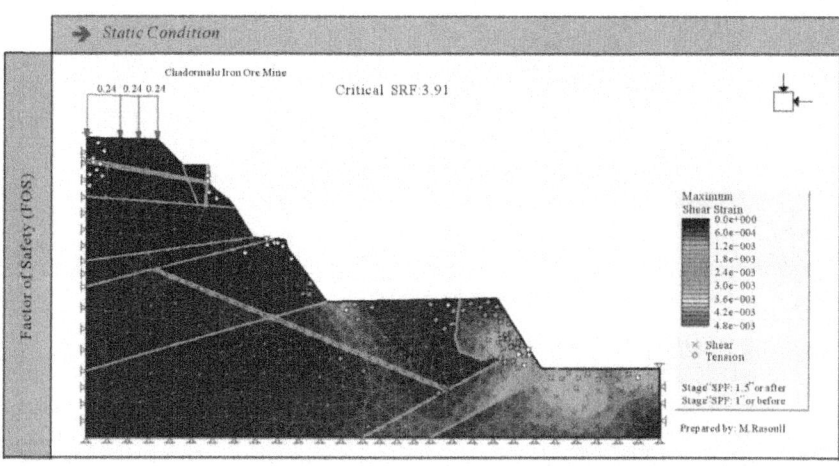

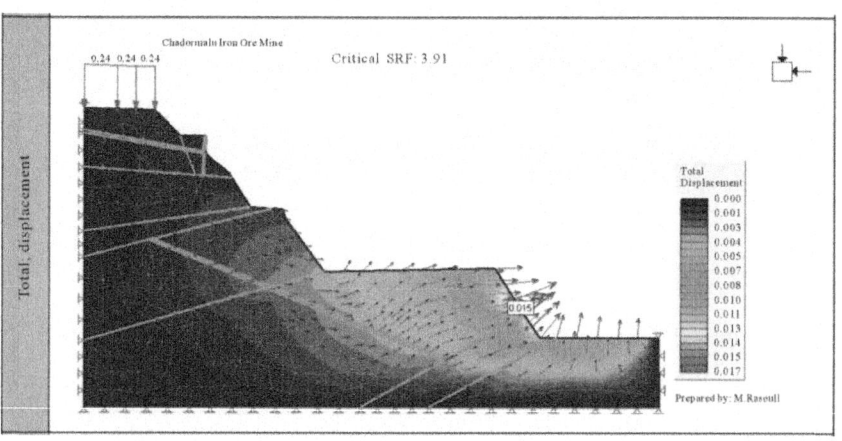

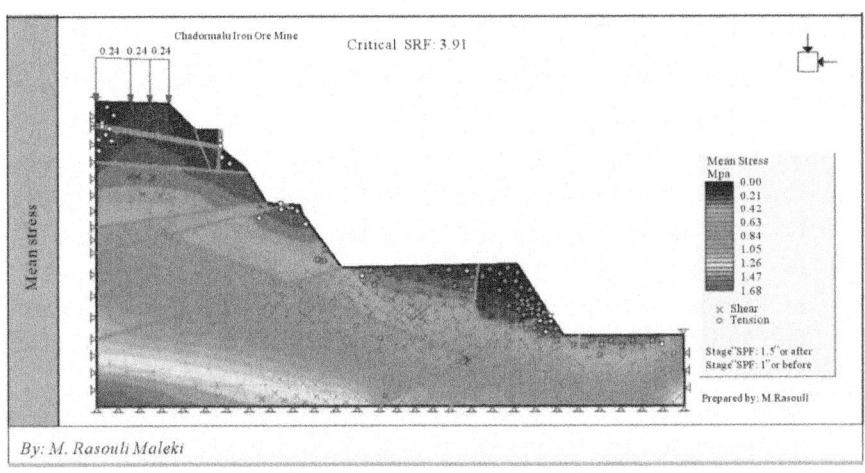

Figure 4: Output model of Phase2D software, factor of safety (FOS), total displacements and mean stress for dynamic state.

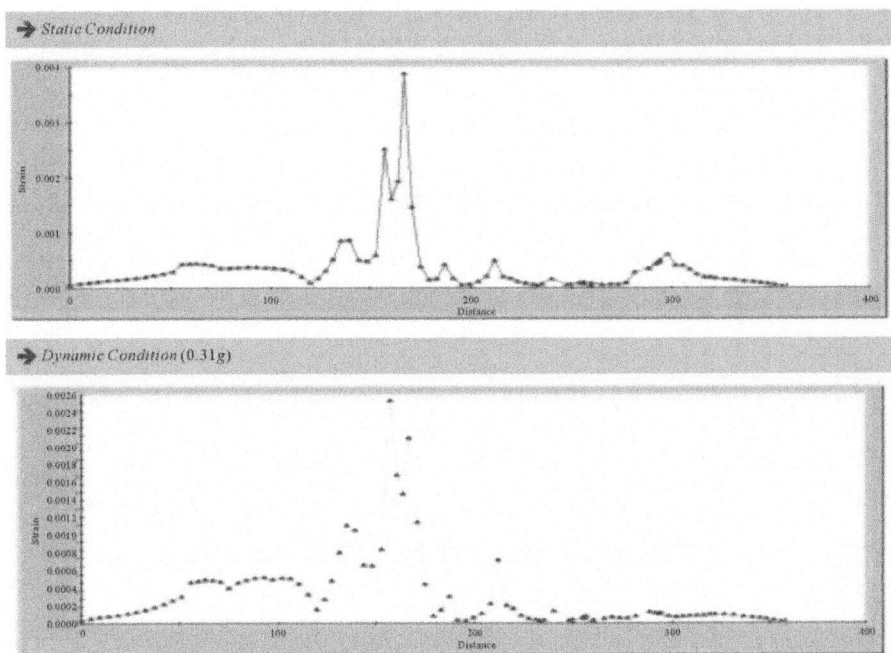

Figure 5: Shear-strain changes for static and dynamic states.

REFERENCES

1. Z. T. Bieniawski, "Engineering Rock Mass Classifications," Wiley, New York, 1989, pp. 5-249.
2. M. Rasouli, "Study of the Engineering Geological Problems of the Havasan Dam, with Emphasis on ClayFilled Joints in the Right Abutment," International Journal of Rock Mechanics and Rock Engineering, 2011, pp. 1-16. doi:10.1007/s00603-011-0165-2
3. M. Rasouli and M. Mahyar, "Assessment of Dominant Type of Failures in the Cutting of Transit Road Iran - Armenia Based on SMR Classification System," 4th National Conference on Rock Mechanics, Iran, 2011.
4. M. Rasouli and M. Mahyar, "Assessment and comparison of occurring probability of rock failures based on empirical and kinematical methods," 4th National Conference on Rock Mechanics, Tehran, May 2011.
5. M. Rasouli, "Assessment of Occurrence Probability for Planar & Wedge Failures under Dynamic & Static Conditions in Abutments of a Double Arch Concrete Dam (Case Study)," 4th International Conference on Geotechnical Engineering and Soil Mechanics, Tehran, 2-3 November 2010.
6. Roc Science, Rock Mass Strength Analysis Using the Hoek-Brown failure criterion, 2005.
7. Roc Science, Two-Dimensional Finite Element Slope Stability Analysis, 2005

Chapter 2

NUMERICAL STUDY ON CRACK PROPAGATION IN BRITTLE JOINTED ROCK MASS INFLUENCED BY FRACTURE WATER PRESSURE

Yong Li [1], Hao Zhou [1], Weishen Zhu [1], Shucai Li [1] and Jian Liu [2]

[1]Geotechnical & Structural Engineering Research Center, Shandong University, Jinan 250061, Shandong, China

[2]School of Civil Engineering, Shandong University, Jinan 250061, Shandong, China

ABSTRACT

The initiation, propagation, coalescence and failure mode of brittle jointed rock mass influenced by fissure water pressure have always been studied as a hot issue in the society of rock mechanics and engineering. In order to analyze the damage evolution process of jointed rock mass under fracture water pressure, a novel numerical model on the basis of secondary development in fast Lagrangian analysis of continua (FLAC3D) is proposed to simulate the fracture development of jointed rock mass under fracture water pressure. To validate the feasibility of this numerical model, the failure process of a numerical specimen under uniaxial compression containing pre-existing fissures is simulated and compared with the results obtained from the lab experiments, and they are found to be in good agreement. Meanwhile, the propagation of cracks, variations of stress and strain, peak strength and crack initiation principles are further analyzed. It is concluded that the fissure water has a significant reducing effect on the strength and stability of the jointed rock mass.

INTRODUCTION

To a great extent, it is the nearly ubiquitous presence of fractures that makes the mechanical behavior of rock masses different from that of most engineering materials. These fractures have a controlling influence on the mechanical behavior of rock masses, since existing fractures provide planes of weakness on which further deformation can more readily occur. Fractures also often provide the major conduits through which fluids can flow [1]. As the Chinese economy gradually grows, the Chinese government will begin to construct numerous huge engineering projects, like hydropower stations, mining, tunnels, large-scale underground caverns for energy storage, *etc.* Therefore, the related problems in jointed rock mass will be encountered in the future [2]. A series of cracking processes finally control the overall behavior of the rock, which have prompted extensive experimental studies of pre-cracked specimens of different materials, including rock-like brittle/semi-brittle materials and natural rocks: glass [3], molded gypsum [4], sand-stone-like material [5], granite [6], marble [7], *etc.* Numerous numerical methods have also been used to simulate the fracture development. These methods could be divided into two types: continuous and discontinuous numerical methods. Tang *et al.* [8] developed some numerical methods to simulate the initiation and coalescence of flaws in rock-like materials, including the finite element method (FEM), boundary element method (BEM) and displacement discontinuity method (DDM), and Tang [9] also proposed a new numerical code named RFPA2D (Rock Failure Process Analysis) to simulate the propagation and coalescence of cracks in a rock bridge area. In addition, the discrete element method (DEM) is also used to simulate the mechanical behavior of rock-like materials [10,11,12,13]. The above research was not entirely conducted under the conditions of fissure water pressure. Fang and Harrison [14,15] adopted a degradation model to simulate the brittle failure in heterogeneous rocks. Xie *et al.* [16] proposed a micromechanical analysis of damage and related inelastic deformation in saturated porous quasi-brittle materials in 2012. Then, Zhu *et al.* [17] gave a deep discussion about two dissipative processes in microcracks. Bikong *et al.*[18] proposed a micro-macro model for the time-dependent behavior of clayey rocks in 2015. Richardson *et al.* [19] presented a method for simulating quasi-static crack propagation in 2D, which combines XFEM with a simple integration technique and a very general algorithm for cutting triangulated domains. In this paper, to simulate the fissure development of jointed rock mass under fissure water pressure, we propose a novel numerical model on the basis of secondary development in Lagrangian analysis of continua (FLAC3D) [20], which is an explicit finite difference method (FDM). Finally, the numerical model is used to study the fissure development of rock specimens.

AN ELASTIC-BRITTLE CONSTITUTIVE MODEL AND HYDRO-MECHANICAL COUPLING

An Elastic-Brittle Constitutive Model

As is known to us all, the nonlinear stress-strain relationship of brittle materials, like rock, concrete *etc.*, does not result from plastic deformations. It is caused by the initiation, propagation and coalescence of the micro-cracks in heterogeneous materials. Therefore, it is appropriate to adopt an elastic-damage model to describe the micro-mechanical properties of brittle materials. The behavior of the rock element undergoing failure, as used in the analysis of the behavior of a rock specimens [21,22], may be simplified to either elastic-brittle, elastic-strain softening (a combination of brittle and ductile) or elastic-ductile (plastic) mechanisms, as shown in Figure 1.

The above elastic-plastic model and strain-softening model could not effectively simulate the failure development of rock materials; even some microscopic problems are difficult to be solved due to the large plastic zone appearing in the crack tips. According to the curves of elastic-brittle stress-strain relations, a piecewise function could be used to express the whole process of the stress-strain relations.

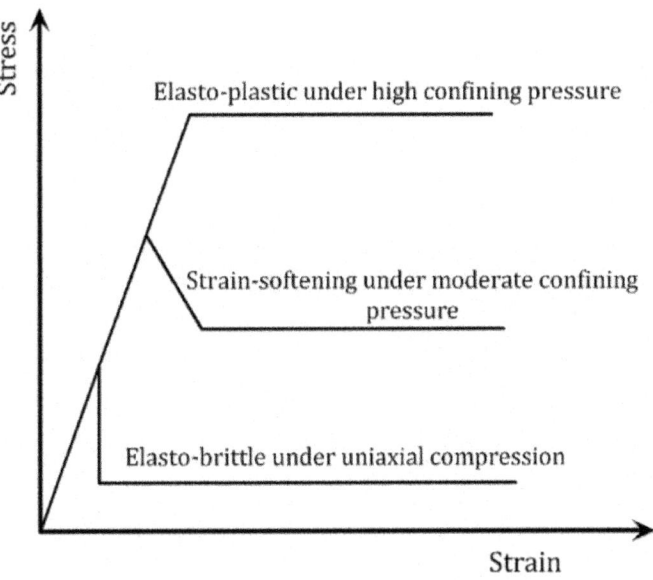

Figure 1. Simplified stress-strain relations of rock elements under different confining pressures within a stressed rock body.

As for the post-failure elements, the mechanical properties must be degraded, and the stress field must be redistributed. Consequently, tensile failure occurs, and cracks are initiated. This model could be effectively used to simulate the complex fissure development in heterogeneous materials. In this elastic-brittle damage model, the specimens under uniaxial tensile loads still have residual strength after they undergo yield strength. This model could be expressed as the following equations:

$$D = \begin{cases} 0, \varepsilon < \varepsilon_{t0} \\ 1 - \dfrac{\sigma_i}{\varepsilon \cdot E_0}, \varepsilon_{t0} \leq \varepsilon \leq \varepsilon_{tu} \\ 1, \varepsilon > \varepsilon_{tu} \end{cases} \quad (1)$$

$$\sigma_i = \eta \cdot \sigma_t \quad (2)$$

where σ_i is the residual strength; ε_{t0} is the initial damage threshold; ε_{tu} is the limit of tensile strain; η is the residual strength coefficient; D represents the damage variable; and σ_t is the uniaxial tensile strength.

According to Mazars' method [23], the tensile strain ε in Equation (1) could be substituted by an equivalent strain $\bar{\varepsilon}$ in three-dimensional conditions:

$$D = \begin{cases} 0, \bar{\varepsilon} < \varepsilon_{t0} \\ 1 - \dfrac{\sigma_i}{\bar{\varepsilon} \cdot E_0}, \varepsilon_{t0} \leq \bar{\varepsilon} \leq \varepsilon_{tu} \\ 1, \bar{\varepsilon} > \varepsilon_{tu} \end{cases} \quad (3)$$

where $\bar{\varepsilon} = \sqrt{(\varepsilon_1)^2 + (\varepsilon_2)^2 + (\varepsilon_3)^2}$

Based on elastic damage mechanics, the stress-strain relations of the constitutive model could be described as the following equations:

$$\sigma_{ij} = \begin{cases} 2G \cdot \varepsilon_{ij} + \lambda \cdot \delta_{ij} \cdot \varepsilon_{kk}, \bar{\varepsilon} < \varepsilon_{t0} \\ \dfrac{\sigma_i}{\bar{\varepsilon} \cdot E_0}(2G \cdot \varepsilon_{ij} + \lambda \cdot \delta_{ij} \cdot \varepsilon_{kk}), \varepsilon_{t0} \leq \bar{\varepsilon} \leq \varepsilon_{tu} \\ 0, \bar{\varepsilon} > \varepsilon_{tu} \end{cases} \quad (4)$$

where $G = \dfrac{E_0}{2(1+v)}$; $\lambda = \dfrac{E_0 \cdot v}{(1+v)\cdot(1-v)}$; and $\delta_{ij} = \begin{cases} 1, i = j \\ 0, i \neq j \end{cases}$.

The damage evolution equations of shear failure are expressed as below:

$$D = \begin{cases} 0, \varepsilon_1 < \varepsilon_{c0} \\ 1 - \dfrac{\sigma_{rc}}{\varepsilon_1 \cdot E_0}, \varepsilon_1 \geq \varepsilon_{c0} \end{cases} \quad (5)$$

where σ_{rc} is the residual strength of shear damage; and ε_{c0} is the strain threshold of shear damage.

Finally, when an element is experiencing shear failure, the equations of the constitutive model could be expressed as below:

$$\sigma_{ij} = \begin{cases} 2G \cdot \varepsilon_{ij} + \lambda \cdot \delta_{ij} \cdot \varepsilon_{kk}, \varepsilon_1 < \varepsilon_{c0} \\ \dfrac{\sigma_{rc}}{\varepsilon_1 \cdot E_0}(2G \cdot \varepsilon_{ij} + \lambda \cdot \delta_{ij} \cdot \varepsilon_{kk}), \varepsilon_{c0} \leq \varepsilon_1 \leq \varepsilon_{ut} \\ 0, \bar{\varepsilon} > \varepsilon_{ut} \end{cases} \quad (6)$$

Hydro-Mechanical Coupling of Jointed Rock Mass

The hydro-mechanical coupling of jointed rock mass is realized by the stress equilibrium equation and the continuous seepage equation. The stress equilibrium equation is usually expressed by the principle of virtual work. This means that the virtual work difference of body forces and plane forces at any time is zero:

$$\int \delta\varepsilon^T \, d\sigma \, dV - \int \delta u^T \, df \, dV - \int \delta u^T \, dt \, dS = 0 \quad (7)$$

where $\delta\varepsilon$ is the virtual strain; δu is the virtual displacement; t is the plane force; and f is the body force.

When porous media is considered, the expression of the Biot effective stress is:

$$\sigma' = \sigma - \alpha\bar{p} \quad (8)$$

where σ' is the effective stress; σ is the total stress; α is Biot's coefficient; and $\bar{p}$ is the average stress of the fluid. Biot's coefficient would evolve with the damage process, but it is very difficult to obtain its variation principle during the coupling process. According to the research results of Walsh [24] and Zhao [25], Biot's coefficient is between zero and one.

The constitutive model could be expressed by the strain increment:

$$d\sigma' = D_{ep}(d\varepsilon - d\varepsilon_l) \quad (9)$$

where D_{ep} represents an elastic-plastic matrix; and dεl is the particle compression induced by pore flow. Here, this is calculated as the following equation:

$$\varepsilon = \sigma'(1-D)E_0 \qquad (10)$$

The continuous seepage equation is expressed as below based on a hypothesis of the Darcy flow:

$$S_w\left[m^T - \frac{m^T D_{ep}}{3K_s}\right]\frac{d\varepsilon}{dt} - \nabla^T\left[k_0 k_r\left(\frac{\nabla p_w}{p_w} - g\right)\right]$$
$$+ \left\{\zeta n + n\frac{S_w}{k_w} + S_w\left[\frac{1-n}{3K_s} - \frac{m^T D_{ep} m}{(3K_s)^2}\right](S_w + p_w\zeta)\right\}\frac{dp_w}{dt} = 0 \qquad (11)$$

where S_w is the degree of saturation; p_w is the pore water pressure; ζ=dswdpw; k_0 is the initial permeability coefficient tensor; k_r is the permeability coefficient; g is the gravity acceleration vector; n is the porosity; and k_w is the bulk modulus of water. The above equations provide the theoretical fundamentals in hydro-mechanical coupling of jointed rock mass.

IMPLEMENTATION OF THE ELASTIC-BRITTLE COUPLING MODEL IN FLAC3D

A survey of commercially available codes shows that the program fast Lagrangian analysis of continua (FLAC3D), produced by Itasca Consulting Group [20], uses an explicit finite difference scheme for the analysis of problems in engineering mechanics. FLAC3D implements an explicit time marching scheme to solve Newton's second law to describe material deformation and embodies a number of basic constitutive models for use in the analysis of the mechanical behavior of geo-materials. Based on these, users can incorporate their own constitutive models by writing a function using a built-in programming language, which is called the FISH language. This provides an easy way to enhance the program, and hence, solve complex problems in rock mechanics and rock engineering. Thus, FLAC has been adopted for the implementation of the elastic-brittle coupling model. Figure 2 shows the procedure for the implementation of the elastic-brittle coupling model in FLAC3D.

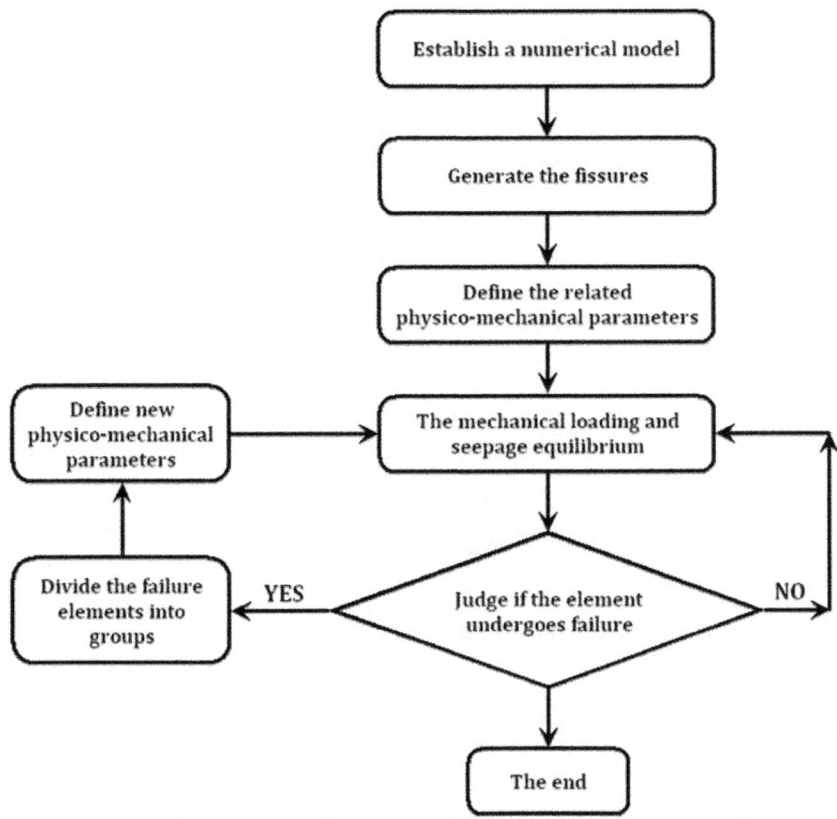

Figure 2. The procedure for the implementation of the elastic-brittle coupling model in Lagrangian analysis of continua (FLAC3D).

NUMERICAL SIMULATIONS ON THE SPECIMENS CONTAINING PRECAST FISSURES

A Two-Dimensional Numerical Simulation

The dimensions of the length, height and thickness in the two-dimensional numerical model shown in Figure 3 are 50 mm, 100 mm and 1 mm, respectively. This numerical model is divided into 7524 elements. It contains two types of media, the intact rock mass and precast fissures. The two parallel fissures are located in the center of the model. The vertical distance between them is 16 mm; the length of the fissure is 18 mm; the thickness is 1 mm; and the dip angle is 45°. The whole numerical model is freely meshed by hexahedral elements. The related physico-mechanical parameters are shown in Table 1.

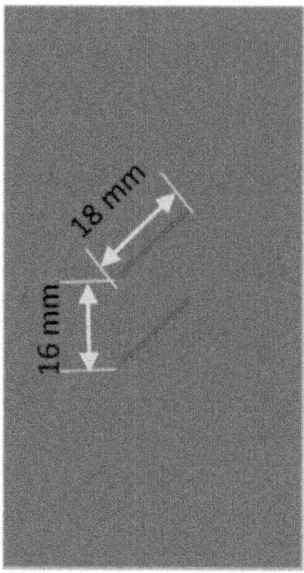

Figure 3. The two-dimensional numerical model.

Table 1. Physico-mechanical parameters of intact rock mass and precast fissure

Rock types	Elastic modulus (GPa)	Poisson's ratio	Tensile strength (MPa)	Cohesion (MPa)	Friction angle (°)	Dilatancy angle (°)
Intact rock mass	45.0	0.25	0.9	1.6	40	0
Precast fissure	1.5	0.35	0.5	0.8	20	0

Next, the elasto-plastic model, the strain-softening model and the elastic-brittle model are also used to simulate the fracture development under uniaxial compression. Figure 4 shows the numerical simulation results. Figure 4a is obtained by the elasto-plastic model. Although the failure occurs near the fissure tips and large-area plastic zones appear, the development of the secondary cracks could not be better observed. Figure 4b is obtained by the strain softening model. Although the plastic zone becomes smaller, it has the same difficulty as Figure 4a. Figure 4c is obtained by the elastic-brittle model. We find that the simulation results are absolutely different with those obtained by the above two models. We observe the development of secondary cracks, and no large-area plastic zones appear, which is extremely close to the results obtained in the laboratory testing specimens [26].

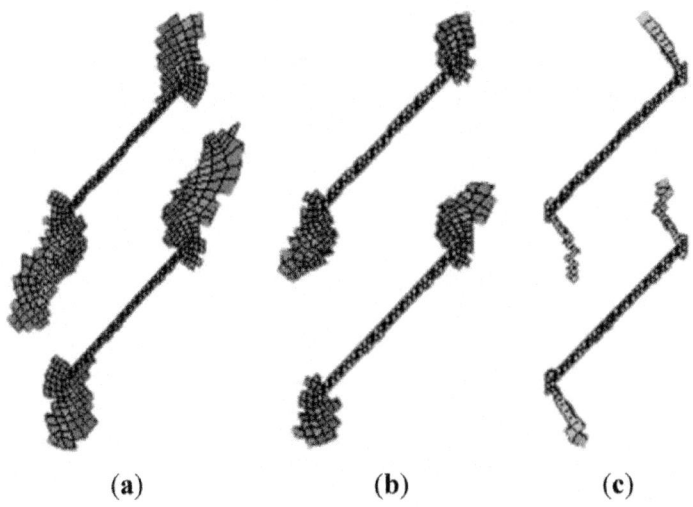

(a)　　　　　　　　(b)　　　　　　　　(c)

Figure 4. Numerical simulation results using elasto-plastic, strain softening and elastic brittle models. (**a**) Plastic zones obtained by elasto-plastic model; (**b**) Plastic zones obtained by strain-softening model; (**c**) Plastic zones obtained by elastic-brittle model.

From the numerical results, it is concluded that the elastic-brittle model is more appropriate to simulate the fracture development of brittle geo-materials.

Three-Dimensional Numerical Simulations

The three-dimensional numerical model adopts a cuboid model, as shown in Figure 5, and the dimensions of the length, width and height are 50 mm, 50 mm and 100 mm, respectively. Two elliptic mica sheets are used to simulate the double fissures, which is more appropriate than the metal sheets in mechanical behavior. The long axis, short axis and thickness of the elliptic fissure are 18 mm, 15 mm and 1 mm. The fissure planes have an inclination angle of 45° to the horizontal plane, and the vertical distance between them is 16 mm. The rolling constraint is fixed to the top and bottom surfaces. In order to clearly observe the fissure development, super fine meshes are generated, and the number of elements is 270,603, as shown in Figure 6. The uniaxial loading is applied on the top and bottom surfaces. The related physico-mechanical parameters are also shown in Table 1.

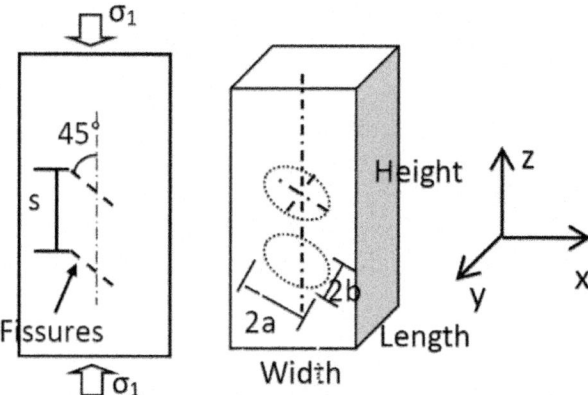

Figure 5. Location of double parallel fissures.

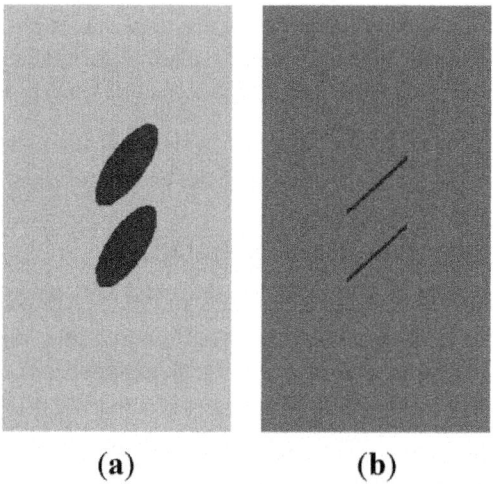

Figure 6. The three-dimensional model. (a) A side view; (b) A front view.

When the elastic brittle numerical model is used to perform the numerical simulations on the fissure development in the specimens, the following essentials should be noticed. As for the elements experiencing shear failures, their residual shear strength should be reduced to five percent of the original shear strength, and for the elements experiencing tensile failures, their tensile strength and cohesion should be reduced to only one percent of the original values. However, whatever failures the elements experience, the friction angle should be kept invariant, and the bulk and shear modulus should be degraded to the same order of magnitude.

Case Study I: Numerical Analysis on the Double-Fissured Specimen under Uniaxial Compression without Fissure Water Pressure

In this case, the double-fissured specimen under uniaxial compression without fissure water pressure is numerically simulated based on the elastic brittle model. Figure 7 shows the fissure development and the profile of the secondary cracks.

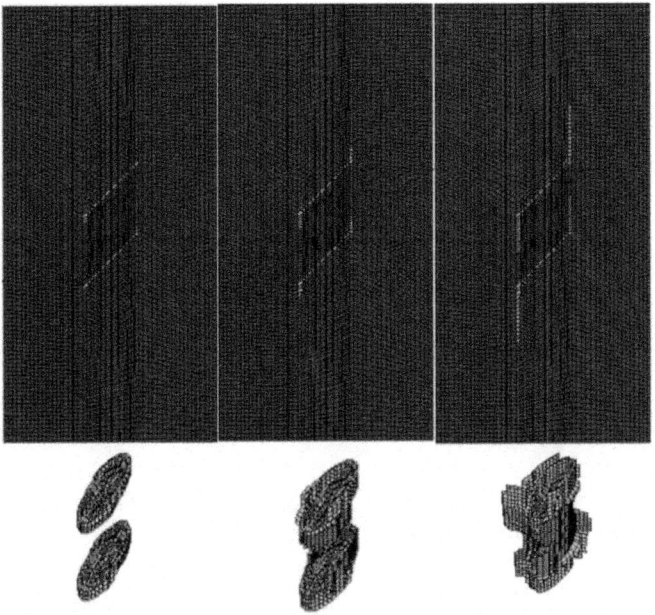

Figure 7. The fissure development and profiles of secondary cracks.

During the beginning of uniaxial compressive loading, we could observe that small encapsulated failure planes initiate near the long-axis tips of the elliptical fissure sheet; the secondary cracks start to extend along the loading direction, and the wing cracks are formed. Although the secondary cracks initiate and extend gradually, they have no coalescence areas, and the rock mass between the two fissure sheets remains intact at this stage. As the loading continues to increase to 44.7% of the peak compressive strength, a major failure zone induced by the propagation of secondary cracks is formed in the rock bridge. Afterwards, the failure planes start to extend along the fissure edges until the final failure occurs. The peak compressive strength is 58.2 MPa.

The complete stress-strain curve is drawn in Figure 8. In the beginning of loading, the specimen is in the linear elastic stage. When the curves approach the peak values, the axial and transvers strains increase faster, and the specimen appears dilatancy. When the peak strength appears, the stress decreases rapidly. The whole process behaves with the typical characteristics of brittle materials.

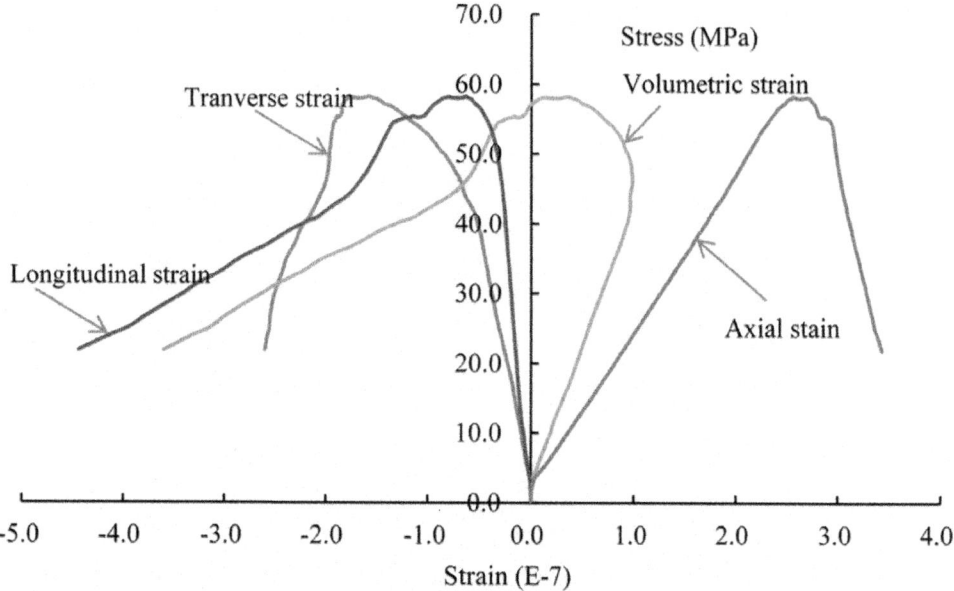

Figure 8. The complete stress-strain curve of the double-fissured specimen under uniaxial compression without fissure water pressure.

Case Study II: Numerical Analysis on the Double-Fissured Specimen under Uniaxial Compression Considering Fissure Water Pressure

In this case, the double-fissured specimen under uniaxial compression considering fissure water pressures is numerically simulated based on the elastic brittle coupling model. According to the actual laboratory experiments, two representative fissure water pressures are considered. One is 3.5 percent of the uniaxial peak strength (3.5% σ_a), and the other is 3.5% σ_a. Figure 9 shows the fissure development and profiles of secondary cracks under fissure water pressure (3.5% σ_a). Figure 10 is the complete stress-strain curve under fissure water pressure (3.5% σ_a). Figure 11 and Figure 12 show the numerical results under fissure water pressure (7% σ_a).

The following results could be concluded from the numerical simulations:

- When the fissure water pressure is equal to 3.5% σ_a, the final uniaxial peak strength is 59.0 MPa. At the beginning of loading, we could find that the fissure development has a similar principle. As the loading continues to increase to 41.3% of the peak compressive strength, a major failure zone induced by the propagation of secondary cracks is formed in the rock bridge. Therefore, it is concluded that the fissure water pressure (3.5% σ_a) has a slight impact on the fissure development.

- As the fissure water pressure is increased to 7% σ_a its impact could not be neglected. The final uniaxial peak strength has obviously been decreased to 42.2 MPa. The cracks initiate and extend while the uniaxial load is 2.11 MPa. As the loading continues to increase, the cracks extend rapidly, because the fissure water pressure intensifies the tensile effects at the fissure tips. When the uniaxial loading is 27.85 MPa, a major failure zone induced by the propagation of secondary cracks is formed in the rock bridge. Afterwards, the failure planes start to extend along the fissure edges until the final failure occurs.

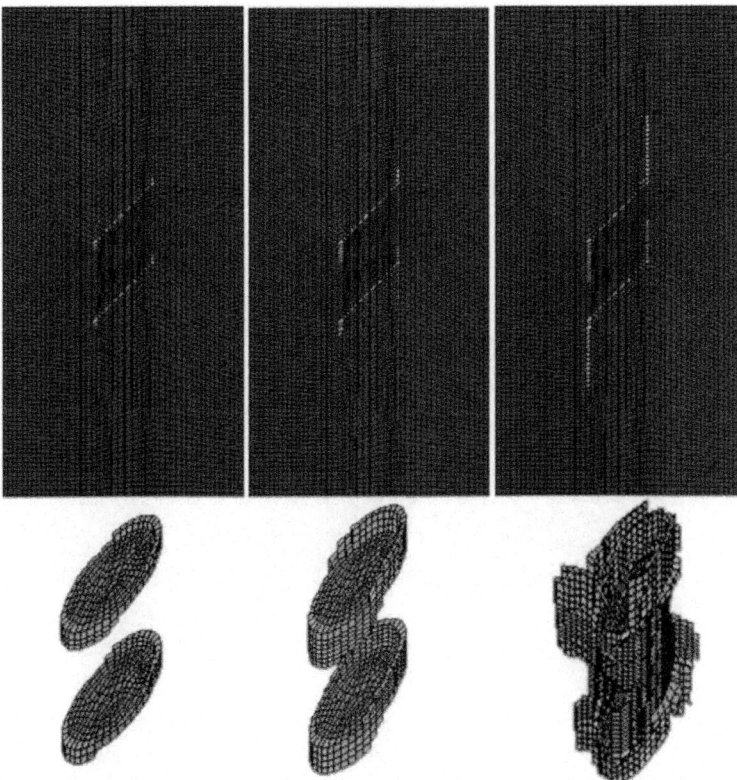

Figure 9. Fissure development and profiles of secondary cracks under fissure water pressure (3.5% σ_a).

24   Design Analysis in Rock Mechanics

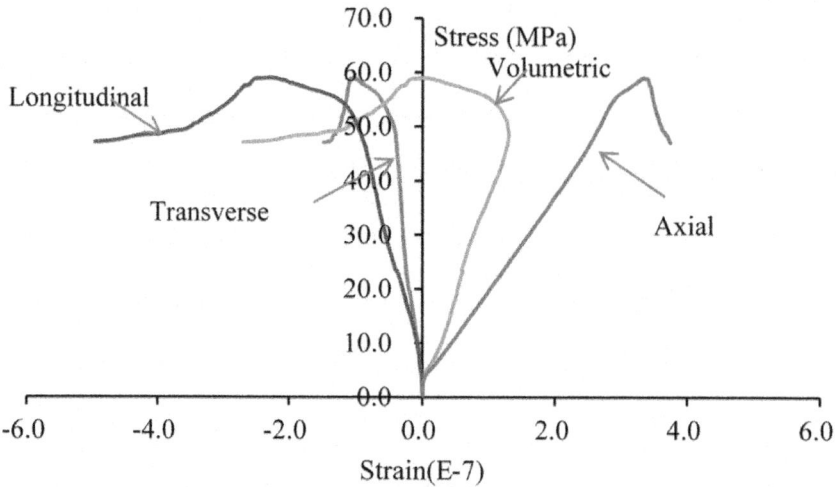

Figure 10. The complete stress-strain curve of the double-fissured specimen under uniaxial compression under fissure water pressure (3.5% σ_a).

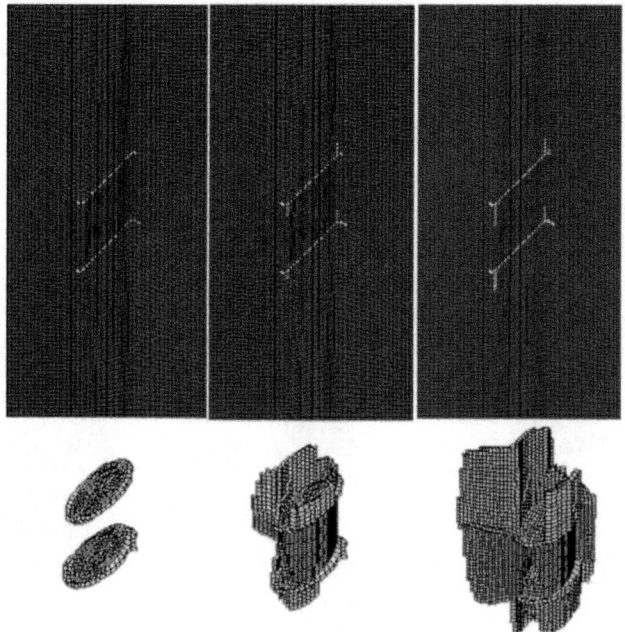

Figure 11. Fissure development and profiles of secondary cracks under fissure water pressure (7% σ_a).

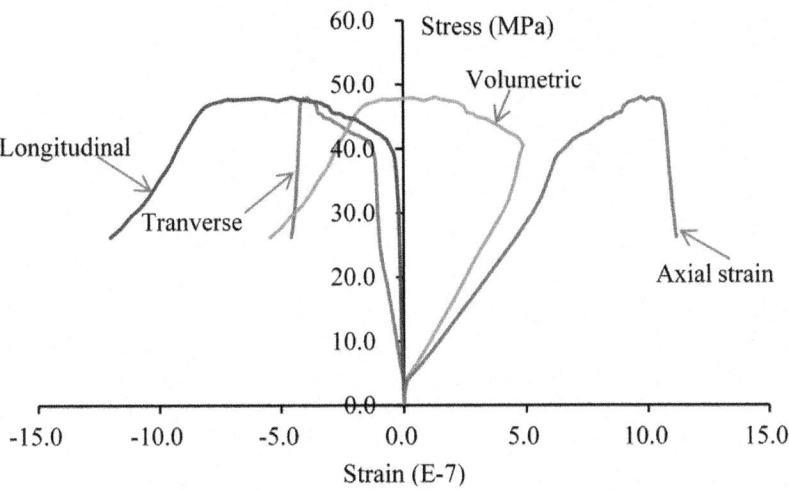

Figure 12. The complete stress-strain curve of the double-fissured specimen under uniaxial compression under fissure water pressure (7% σ_a).

CONCLUSIONS

- An elastic brittle coupling constitutive model on the basis of secondary development in FLAC3D is proposed to simulate the fracture development of jointed rock mass under fracture water pressure. The two-dimensional numerical results are found to be in good agreement with the laboratory results.
- The fissure water pressure has a significant impact on the peak strength of pre-cracked rock specimen. When the value is small, the peak strength may be increased a little. However, when it increases to a bigger value, the peak strength would be decreased rapidly, and a large-area failure zone would appear in the rock bridge. The failure planes start to extend along the fissure edges until the final failure occurs.

ACKNOWLEDGMENTS

The work was supported by the National Science and Technology Support Program of China (2015BAB07B05), the Natural Science Foundation of Shandong Province (BS2012NJ006) and the Specialized Research Fund for the Doctoral Program for Higher Education (No. 20110131120034). We would also like to express our sincere gratitude to the editor and the two anonymous reviewers for their valuable contributions to this paper.

AUTHOR CONTRIBUTIONS

This present research article is based on the research work of Yong Li and Hao Zhou. During the research work, Weishen Zhu, Shucai Li and Jian Liu have provided substantial contributions to the theoretical innovation and experimental guidance. Hao Zhou performed a lot of work in laboratory experiments and Yong Li conducted the numerical simulation and analyzed the related data. Finally, Yong Li finished the whole writing of this paper.

CONFLICTS OF INTEREST

The authors of the paper declare that there is no conflict of interest regarding the publican of this paper. The authors do not have a direct financial relation with the commercial identity that might lead to a conflict of interest for any of the authors.

REFERENCES

1. Jaeger, J.C.; Cook, N.G.W. *Fundamentals of Rock Mechanics*, 3rd ed.; Chapman and Hall: London, UK, 1979.
2. Li, Y.; Zhu, W.S.; Fu, J.W.; Guo, Y.H.; Qi, Y.P. A damage rheology model applied to analysis of splitting failure in underground caverns of Jinping I hydropower station. *Int. J. Rock Mech. Min. Sci.* **2014**, *71*, 224–234.
3. Hoek, E.; Bieniawski, Z.T. Brittle fracture propagation in rock under compression. *Int. J. Fract.* **1965**, *1*, 137–155.
4. Bobet, A. The initiation of secondary cracks in compression. *Eng. Fract. Mech.* **2000**, *66*, 187–219.
5. Mughieda, O.; Alzo'ubi, A.K. Fracture mechanisms of offset rock joints—A laboratory investigation. *Geotech. Geolog. Eng.* **2004**, *22*, 545–562.
6. Miller, J.T.; Einstein, H.H. Crack coalescence tests on granite. In Proceedings of 42nd US Rock Mechanics Symposium, San Francisco, CA, USA, 28 June–2 July 2008. ARMA-08-162.
7. Li, Y.P.; Chen, L.Z.; Wang, Y.H. Experimental research on pre-cracked marble under compression. *Int. J. Solids Struct.* **2005**, *42*, 2505–2516.
8. Tang, C.A.; Lin, P.; Wong, R.H.C.; Chau, K.T. Analysis of crack coalescence in rock-like materials containing three flaws-part II: Numerical approach. *In. J. Rock Mech. Min. Sci.* **2001**, *38*, 925–939.
9. Tang, C.A.; Kou, S.Q. Crack propagation and coalescence in brittle materials under compression. *Eng. Fract. Mech.* **1998**, *61*, 311–324.

10. Lee, H.; Jeon, S. An experimental and numerical study of fracture coalescence in pre-cracked specimens under uniaxial compression. *Int. J. Solids Struct.* **2011**, *48*, 979–999.

11. Yang, S.Q.; Huang, Y.H.; Jing, H.W.; Liu, X.R. Discrete element modeling on fracture coalescence behavior of red sandstone containing two unparallel fissures under uniaxial compression. *Eng. Geol.* **2014**, *178*, 28–48.

12. Manouchehrian, A.; Sharifzadeh, M.; Marji, M.F.; Gholamnejad, J. A bonded particle model for analysis of the flaw orientation effect on crack propagation mechanism in brittle materials under compression. *Arch. Civ. Mech. Eng.* **2014**, *14*, 40–52.

13. Mughieda, O.; Omar, M.T. Stress analysis for rock mass failure with offset joints. *Geotech. Geol. Eng.* **2008**, *26*, 543–552.

14. Fang, Z.; Harrison, J.P. Development of a local degradation approach to the modeling of brittle fracture in heterogeneous rocks. *Int. J. Rock Mech. Min. Sci.* **2002**, *39*, 443–457.

15. Fang, Z.; Harrison, J.P. Application of a local degradation model to the analysis of brittle fracture of laboratory scale rock specimens under triaxial conditions. *Int. J. Rock Mech. Min. Sci.* **2002**, *39*, 459–476.

16. Xie, N.; Zhu, Q.Z.; Shao, J.F.; Xu, L.H. Micromechanical analysis of damage in saturated quasi brittle materials. *Int. J. Solids Struct.* **2012**, *49*, 919–928.

17. Zhu, Q.Z.; Shao, J.F. A refined micromechanical damage–friction model with strength prediction for rock-like materials under compression. *Int. J. Solids Struct.* **2015**, *60–61*, 75–83.

18. Bikong, C.; Hoxha, D.; Shao, J.F. A micro-macro model for time-dependent behavior of clayey rocks due to anisotropic propagation of microcracks. *Int. J. Plast.* **2015**, *69*, 73–88.

19. Richardson, C.L.; Hegemann, J.; Sifakis, E.; Hellrung, J.; Teran, J.M. An XFEM method for modeling geometrically elaborate crack propagation in brittle materials. *Int. J. Numer. Methods Eng.* **2011**, *88*, 1042–1065.

20. *Fast Lagrangian Analysis of Continua in 3-Dimensions*; version 5.0, manual; Itasca Consulting Group Inc: Itasca, MS, USA, 2012.

21. Fang, Z. A Local Degradation Approach to The Numerical Analysis of Brittle Fracture in Heterogeneous Rocks. Ph.D. Thesis, University of London, London, UK, 6 May 2001.

22. Fang, Z.; Harrison, J.P. A mechanical degradation index for rock. *Int. J. Rock Mech. Min. Sci.* **2001**, *38*, 1193–1199.

23. Mazars, J.; Pijaudier-Cabot, G. Continuum damage theory-application to concrete. *J. Eng. Mech. ASCE* **1989**, *115*, 345–365.
24. Walsh, J.B. Effect of pore pressure and confining pressure on fracture permeability. *Int. J. Rock Mech. Min. Sci. Geom. Abstr.* **1981**, *18*, 429–435.
25. Zhao, Y.S.; Hu, Y.Q.; Zhao, B.H.; Yang, D. Coupled mathematical model for solid deformation and gas seepage of rock matrix-fractured media and its applications. *J. China Coal Soc.* **2003**, *28*, 41–45.
26. Guo, Y.S.; Wong, R.H.C.; Zhu, W.S.; Chau, K.T.; Li, S.C. Study on fracture pattern of open surface-flaw in gabbro. *Chin. J. Rock Mech. Eng.* **2007**, *26*, 525–531.

Chapter 3

STUDY ON THE CONSTITUTIVE MODEL FOR JOINTED ROCK MASS

Qiang Xu[2], Jianyun Chen[1,2], Jing Li[2], Chunfeng Zhao[2], Chenyang Yuan[2]

[1] State Key Lab.of Coastal and Offshore Eng., Dalian University of Technology, Dalian, China
[2] School of Civil and Hydraulic Eng., Dalian University of Technology, Dalian, China

ABSTRACT

A new elasto-plastic constitutive model for jointed rock mass, which can consider the persistence ratio in different visual angle and anisotropic increase of plastic strain, is proposed. The proposed the yield strength criterion, which is anisotropic, is not only related to friction angle and cohesion of jointed rock masses at the visual angle but also related to the intersection angle between the visual angle and the directions of the principal stresses. Some numerical examples are given to analyze and verify the proposed constitutive model. The results show the proposed constitutive model has high precision to calculate displacement, stress and plastic strain and can be applied in engineering analysis.

INTRODUCTION

In the rock engineering, joints have significant effect on the stress-strain relationship of jointed rock mass. Generally speaking, there are two categories of approaches: the first method is that joint element is utilized to simulate jointed rock mass. The other method is that special constitutive model is utilized to simulate jointed rock mass.

Constitutive models for jointed rock masses are important for numerical modeling of the behavior of jointed rocks. Many constitutive models for rock joints, based on both empirical and theoretical approach, such as are summarized in [1]. The behavior of the joints is dependent on their sizes, because the scale dependence of surface roughness of the joints whose thresholds are a scaling parameter [2–4]. Some researchers took study on landslide problems and the dynamic frictional processes of the joints using theories of dynamic chaos and catastrophe for an analysis of the interactions between the fracture surfaces regarding friction, fracture stiffness and elastic materials for the jointed rocks [5]. Some researchers utilized joint factor to simulate jointed rock mass based on the finite element method [6]. Some researchers proposed the model for the equivalent elastic parameters of jointed rock mass [7, 8]. Some researchers performed the modeling of dynamic rock fracture sliding using the state variable friction models. In the model, the shear stresses are the functions of both the sliding history and velocity. And the model represented the evolution of rate effects and the path-dependence of the frictional properties [9]. Some researchers utilized representative volume method to analyze nonlinear characteristics of one-way joint and the interaction of two-way orthogonal joints [10]. Some researchers reported new 3D constitutive models for rough rock fractures based on experimentally determined relations between the contact areas under normal loads and asperity inclination angles [11, 12]. Some researchers established the model to calculate the physical parameters of jointed rock mass [13]. Some researchers established softening model for multi-joints [14].

In this paper, the studies on elasto-plastic constitutive model for jointed rock mass are made. The influences of joints on the jointed rock mass are analyzed. Based on these studies, a constitutive model for jointed rock mass, which can consider anisotropic strength of jointed rock mass and anisotropic increase of plastic strain, is constructed. And then the numerical examples are performed to analyze and verify the proposed constitutive model.

THE CONSTITUTIVE MODEL FOR JOINTED ROCK MASS

The Construction of Constitutive Model

Morh-Coulomb model is well-known model in geotechnical engineering application, including in rock engineering modelling and design. The basic concepts of the Mohr-Coulomb model suggest that the behaviors of a rock material are made up of two parts: a constant cohesion and a friction coefficient. And it can be described as

$$\tau_s = \sigma_n tg\varphi + c \tag{1}$$

where τ_s is the shear strength, σ_n is the normal stress, c is the cohesion, φ is friction angle. The parameters of this model are only two, and it is widely used due to the simple expression. But this model is based on the isotropy theory. It can only describe the isotropic material. And jointed rock mass is anisotropic material. The classical Morh-Coulomb model cannot describe the behaviors of jointed rock. So it need to be improved due to its limitations.

Figure 1 shows rock bridges exist in jointed rock masses because of the non-persistent nature of joints. In order to calculate the decrease of strength of jointed rock masses in different directions, It defines the mechanical persistence ratio of rock mass as that the ratio of joint network on the shear failure path when jointed rock mass is sheared to damaged state along a certain direction[15]. Figure 2 shows that the mechanical persistence ratio k is calculated by

$$k_{\beta_0} = \frac{\sum JL}{\sum JL + \sum RBR} \tag{2}$$

where JL and RBR are the projection length of joints and rock bridges in the shear failure path respectively. β_0 is the visual angle, which can be used to express the direction of joints.

Figure 1. The joint network of rocks.

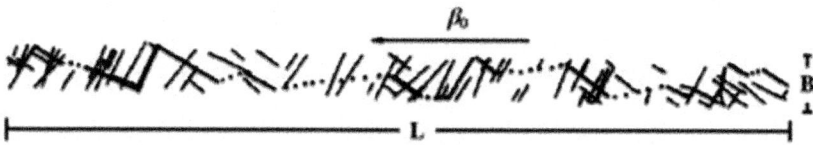

Figure 2. The failure path of joints and rock bridges.

It defines cohesion c_{β_0} and friction angle φ_{β_0} of jointed rock masses in direction of β_0 [16,17] as

$$c_{\beta_0} = (1-k)c_r + kc_j \tag{3}$$

$$\tan\varphi_{\beta_0} = (1-k)\tan\varphi_r + k\tan\varphi_j \tag{4}$$

where c_r and φ_r are the cohesion and friction angle of rock bridges, respectively. c_j and φ_j are the cohesion and friction angle of joints.

Thus, based on Mohr-Coulomb model, the yield strength criterion f can be given by

$$f = |\tau| + \sigma tg\varphi_{\beta_0} - c_{\beta_0} \tag{5}$$

where τ and σ are the shear stress and normal stress in direction of β_0, respectively.

The Mohr-Coulomb model is based on plotting Mohr's circle for states of stress at failure in the plane of the maximum and minimum principal stresses. According to Figure 3, we have

$$\sigma = \frac{1}{2}(\sigma_3 + \sigma_1) + \frac{1}{2}(\sigma_3 - \sigma_1)\cos 2\beta \quad (6)$$

$$|\tau| = \frac{1}{2}(\sigma_1 - \sigma_3)\sin 2\beta \quad (7)$$

where σ_1 and σ_3 are the maximum and minimum principal stresses, respectively. β is the intersection angle between β_0 and the directions of the maximum principal stresses σ_1.

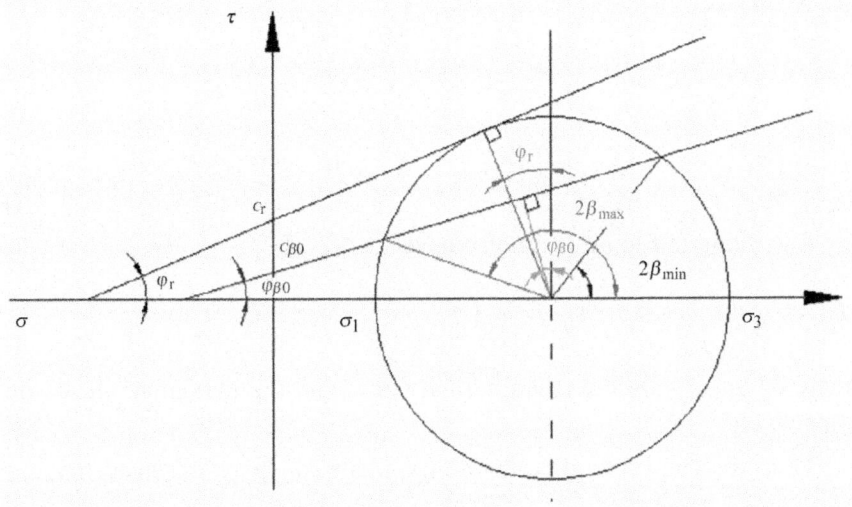

Figure 3. Mohr circle of stress for jointed rock masses.

Thus, the yield strength criterion f in plane can be rewritten as

$$f = \begin{cases} \frac{1}{2}(\sigma_1 - \sigma_3)\sin 2\beta + \left(\frac{1}{2}(\sigma_3 + \sigma_1) + \frac{1}{2}(\sigma_3 - \sigma_1)\cos 2\beta\right) tg\varphi_{\beta_0} - c_{\beta_0} \\ , when \beta_{\min} \leq \beta \leq \beta_{\max} \\ \frac{1}{2}(\sigma_1 - \sigma_3)\cos\varphi_r + \left(\frac{1}{2}(\sigma_3 + \sigma_1) + \frac{1}{2}(\sigma_3 - \sigma_1)\sin\varphi_r\right) tg\varphi_r - c_r \\ , when \beta < \beta_{\min} \text{ or } \beta > \beta_{\max} \end{cases} \quad (8)$$

in which

$$2\beta_{\min} = \varphi_{\beta_0} + \sin^{-1}\left(\left(1 + \frac{(c_{\beta_0} ctg\varphi_{\beta_0} - \sigma_1)(1 - \sin\varphi_r)}{-\sigma_1 \sin\varphi_r + c_r \cos\varphi_r}\right)\sin\varphi_{\beta_0}\right) \quad (9)$$

$$2\beta_{max} = \pi + 2\varphi_{\beta_0} - 2\beta_{min} \tag{10}$$

From (8), through calculating $df/d\beta = 0$, we can obtain the least angle β_L in β when $\beta_{min} \leq \beta \leq \beta_{max}$ and we have

$$\beta_L = 45° + \frac{\varphi_{\beta_0}}{2} \tag{11}$$

And the minimum value of σ_1 and σ_3 obey the function f_{min} when $\beta = \beta_L$, and we have

$$f_{min} = \frac{1}{2}(\sigma_{1,min} - \sigma_{3,min})\cos\varphi_{\beta_0}$$
$$+ \left(\frac{1}{2}(\sigma_{3,min} + \sigma_{1,min}) + \frac{1}{2}(\sigma_{3,min} - \sigma_{1,min})\sin\varphi_{\beta_0}\right)tg\varphi_{\beta_0} - c_{\beta_0} \tag{12}$$

According to Figure 3, we also have

$$\sigma_m = \frac{1}{2}(\sigma_1 + \sigma_3), \tau_m = \frac{1}{2}(\sigma_1 - \sigma_3) \tag{13}$$

where σ_m and τ_m are the mean normal stress and the maximum shear stress, respectively.

Thus, the yield strength criterion f in plane can be rewritten as

$$f = \begin{cases} \tau_m(\sin 2\beta + tg\varphi_{\beta_0}\cos 2\beta) + \sigma_m tg\varphi_{\beta_0} - c_{\beta_0}, & \text{when } \beta_{min} \leq \beta \leq \beta_{max} \\ \tau_m \sec\varphi_r + \sigma_m tg\varphi_r - c_r, & \text{when } \beta < \beta_{min} \text{ or } \beta > \beta_{max} \end{cases} \tag{14}$$

The yield strength criterion f in plane is extended to three-dimensional yield strength criterion and we have

$$f = \begin{cases} R_{mc,\beta_0}q - p\tan\varphi_{\beta_0} - c_{\beta_0}, & \text{when } \beta_{min} \leq \beta \leq \beta_{max} \\ R_{mc,r}q - p\tan\varphi_r - c_r, & \text{when } \beta < \beta_{min} \text{ or } \beta > \beta_{max} \end{cases} \tag{15}$$

in which

$$R_{mc,\beta_0} = \frac{1}{\sqrt{3}\cos\varphi_{\beta_0}}\sin\left(\theta + \frac{\pi}{3}\right) + \frac{1}{3}\cos\left(\theta + \frac{\pi}{3}\right)\tan\varphi_{\beta_0} \tag{16}$$

$$R_{mc,r} = \frac{1}{\sqrt{3}\cos\varphi_r}\sin\left(\theta + \frac{\pi}{3}\right) + \frac{1}{3}\cos\left(\theta + \frac{\pi}{3}\right)\tan\varphi_r \tag{17}$$

$$p = I_1/3, q = \sqrt{3}\sqrt{J_2}, \cos(3\theta) = (J_3/q)^3 \tag{18}$$

where I_1, J_2 and J_3 are the first invariant of stress tensor, the second and third invariant of deviatoric stress tensor, respectively.

In plasticity theory, the strain increment can be decomposed into two parts

$$d\boldsymbol{\epsilon} = d\boldsymbol{\epsilon}^e + d\boldsymbol{\epsilon}^p \tag{19}$$

where $d\boldsymbol{\varepsilon}$ is the incremental strain tensor; $d\boldsymbol{\varepsilon}^e$ and $d\boldsymbol{\varepsilon}^p$ are the incremental elastic and plastic strain tensor, respectively.

The stress—strain relationship is expressed as

$$d\boldsymbol{\sigma}' = \boldsymbol{D}^{ep} : d\boldsymbol{\epsilon} \tag{20}$$

where $d\boldsymbol{\sigma}'$ is the incremental stress tensor; $\boldsymbol{D}^{ep}$ is the elasto-plastic stiffness tensor.

The elasto-plastic stiffness tensor is expressed as:

$$\boldsymbol{D}^{ep} = \boldsymbol{D}^e - \frac{\boldsymbol{D}^e : \boldsymbol{n}_g : \boldsymbol{n}^T : \boldsymbol{D}^e}{\boldsymbol{n}^T : \boldsymbol{D}^e : \boldsymbol{n}_g} \tag{21}$$

in which

$$\boldsymbol{n}_g = \frac{\partial g}{\partial \sigma}, \boldsymbol{n} = \frac{\partial f}{\partial \sigma} \tag{22}$$

where σ is the stress tensor; $\boldsymbol{D}^e$ is the elasto stiffness tensor; $\boldsymbol{n}$ and $\boldsymbol{n}_g$ are the loading and flow direction vectors, respectively; f and g are the yield and plastic potential functions, respectively. And in the model, the plastic potential function g is adopted as the same as the yield function f.

The fluidity variable Λ can be expressed as

$$\Lambda = \frac{\boldsymbol{n}^T : \boldsymbol{D}^e : d\boldsymbol{\epsilon}}{\boldsymbol{n}^T : \boldsymbol{D}^e : \boldsymbol{n}_g} \tag{23}$$

The distinction between loading and unloading directions is described through the following criteria:

$$\Lambda > 0 \text{ (loading)} \quad \Lambda \leq 0 \text{ (unloading)} \tag{24}$$

Because the plastic strain will also increase in the process of reloading, the incremental plastic strain is

$$d\boldsymbol{\epsilon}^p = \langle \Lambda \rangle \boldsymbol{n}_g \tag{25}$$

where the symbol $\langle \rangle$ is defined as $\langle \Lambda \rangle = \Lambda$ for $\Lambda > 0$ and $\langle \Lambda \rangle = 0$ for $\Lambda \leq 0$. It shows that the plastic strain will increase if jointed rock mass is in the state of loading. In other word, we have

$$\begin{cases} d\epsilon^p \neq 0, & \text{in the plastic state when } \Lambda > 0 (\text{loading}) \\ d\epsilon^p = 0, & \text{in the elastic state when } \Lambda \leq 0 (\text{unloading}) \end{cases} \quad (26)$$

Numerical Implementation

The integral algorithm based on fully implicit backward Euler return mapping algorithm is adopted to calculate the updated stresses. The convergence rule is adopted according to the difference of updated stresses less than tolerance. Figure 4 shows the iterative steps of proposed constitutive model.

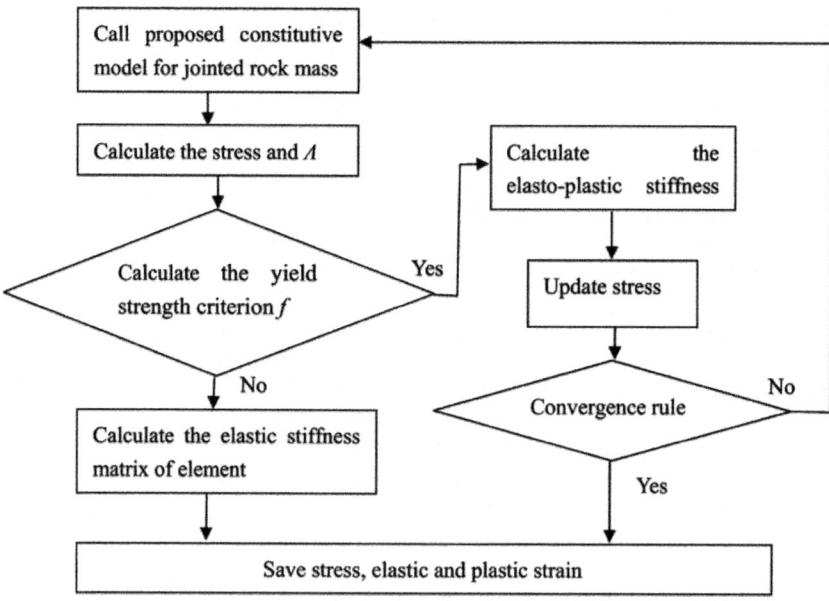

Figure 4. The flow chart of iterative steps of proposed constitutive model.

NUMERICAL EXAMPLES AND RESULTS

The Numerical Example for the Strength of Jointed Rock Mass

In order to analyze the strength of jointed rock mass calculated by proposed elasto-plastic constitutive model, the tests for numerical simulating jointed rock mass are taken. The elastic modulus E and Poisson ratio v of jointed rock mass are 4.00GPa and 0.25, respectively. Table 1 shows the strength parameters of the joint surface and the rock bridge. Figure 5 shows the persistence ratio, friction coefficient and cohesion of jointed rock mass at the visual angle β_0.

Table 1. The strength parameters of the joint surface and the rock bridge

Material	Friction coefficient f	Cohesion c(MPa)
Joints	0.7	0.2
Rock	1.7	2.0

doi:10.1371/journal.pone.0121850.t001

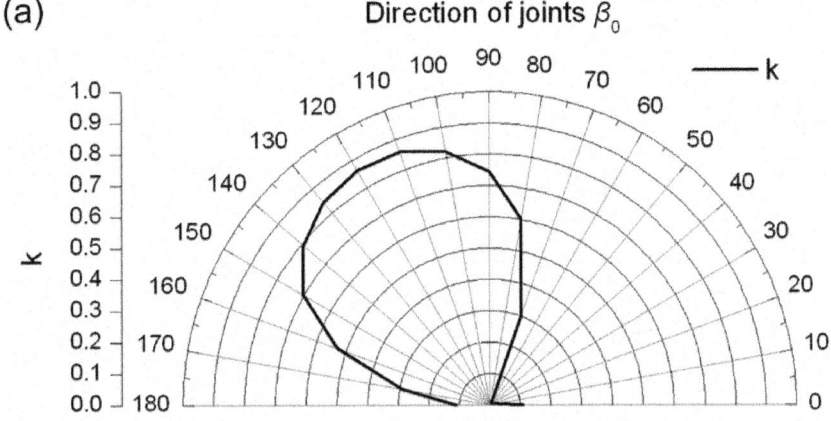

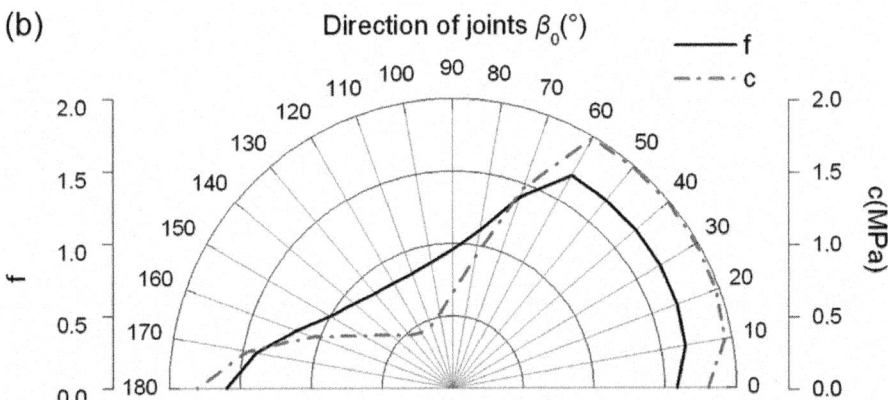

Figure 5. The rose diagrams of the persistence ratio, friction coefficient and cohesion((a) The rose diagrams of the persistence ratio of jointed rock mass; (b) The rose diagrams of friction coefficient and cohesion of jointed rock mass).

Through observing the results of Figures 5–8, they show that the yield strength criterion f in plane of jointed rock mass is not only related to the friction angle $\varphi_{\beta 0}$ and cohesion $c_{\beta 0}$ of jointed rock masses in direction of β_0 (the visual angle). The yield strength criterion f is also related to β (the intersection angle between the visual angle and the directions of the maximum principal stresses). The yield strength criterion f has the relation of $\varphi_{\beta 0}$ and $c_{\beta 0}$ only when $\beta_{min} \leq \beta \leq \beta_{max}$. The relation of β and β_0 is also important to the yield strength criterion f. The different relation of β and β_0 leads to different yield strength criterion f. In some special relation of β and β_0, such as Figure 8 (c), the friction angle $\varphi_{\beta 0}$ and cohesion $c_{\beta 0}$ has no use for the yield strength criterion f. In other word, the persistence ratio k has no use for the yield strength criterion f in some special condition.

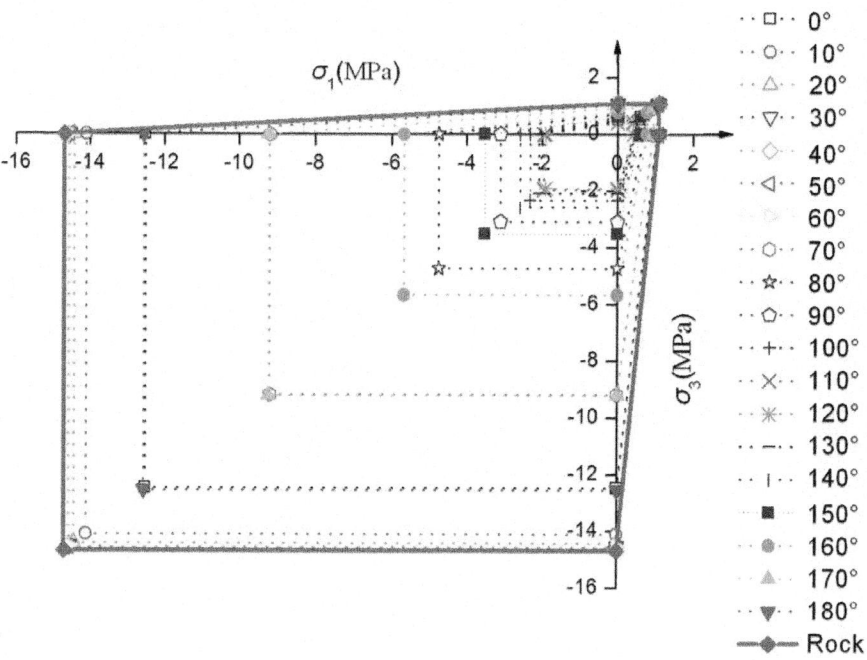

Figure 6. The relation between strength in σ_1-σ_3 **plane and the visual angle** β_0.

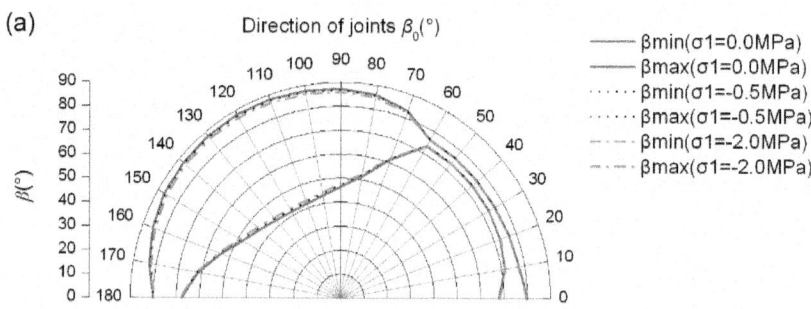

Study on the Constitutive Model for Jointed Rock Mass 39

(b)

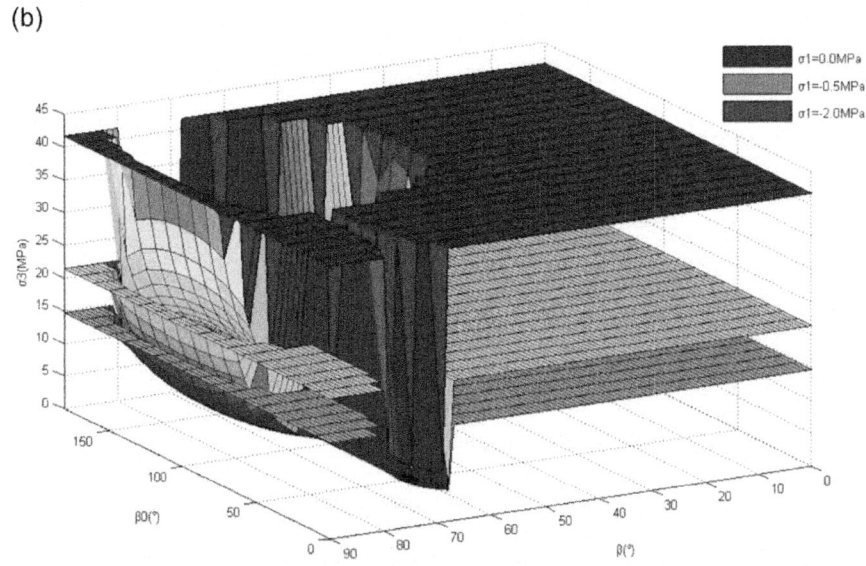

Figure 7. The strength of σ_3 when σ_1 is given((a) The change of β_{min} and β_{max} with different visual angle β_0; (b) The change of strength of σ_3 with different β_0 andβ).

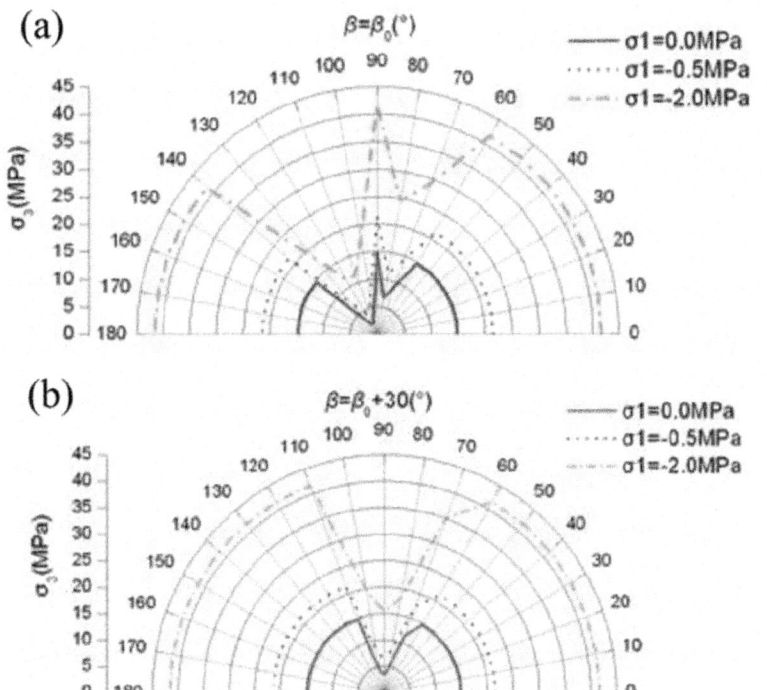

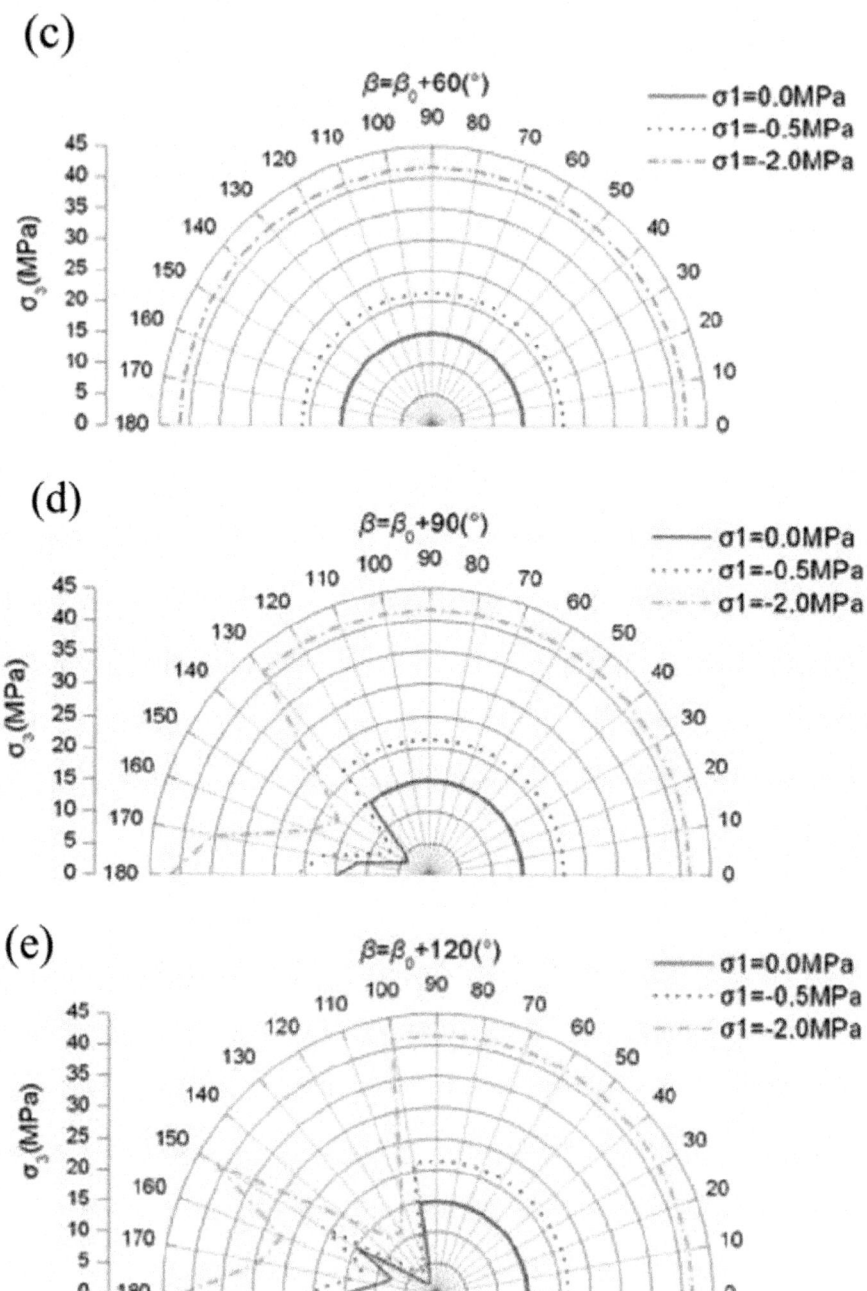

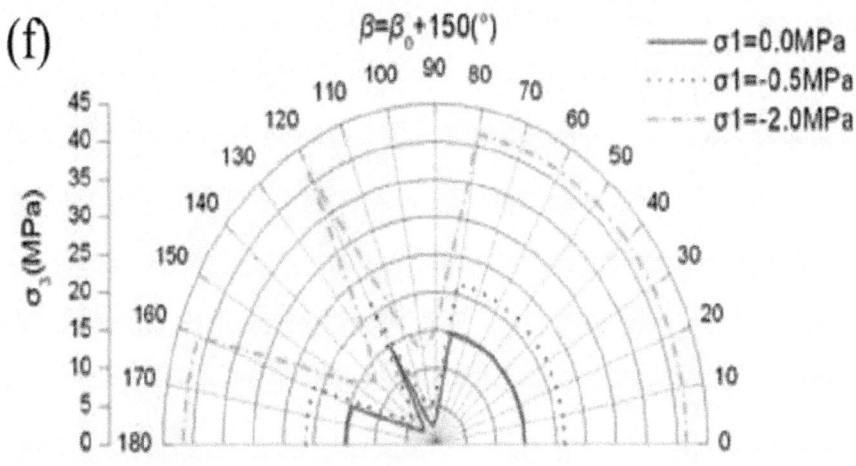

Figure 8. The strength of σ_3 when the relation of β_0 and β is given((a) The strength of σ_3 when $\beta = \beta_0$; (b) The strength of σ_3 when $\beta = \beta_0+30°$; (c) The strength of σ_3 when $\beta = \beta_0+60°$; (d) The strength of σ_3 when $\beta = \beta_0+90°$; (e) The strength of σ_3 when $\beta = \beta_0+120°$; (f) The strength of σ_3 when $\beta = \beta_0+150°$).

Jointed rock direct shear experiment and numerical simulation by proposed model

To verify proposed constitutive model, the compared results of proposed model and experiment are given. The rock mass samples containing joints are 0.3m×0.3m. The visual angles β_0 of joints of two rock mass samples are 0° and 30°, respectively. And the positions of joints are shown in Figure 9. The boundary conditions and numerical model are shown in Figure 10. The normal stress is 1.0MPa, and the shear displacements of experiment are taken to 5mm. The parameters of the joints and the rock mass samples are shown in Table 2.

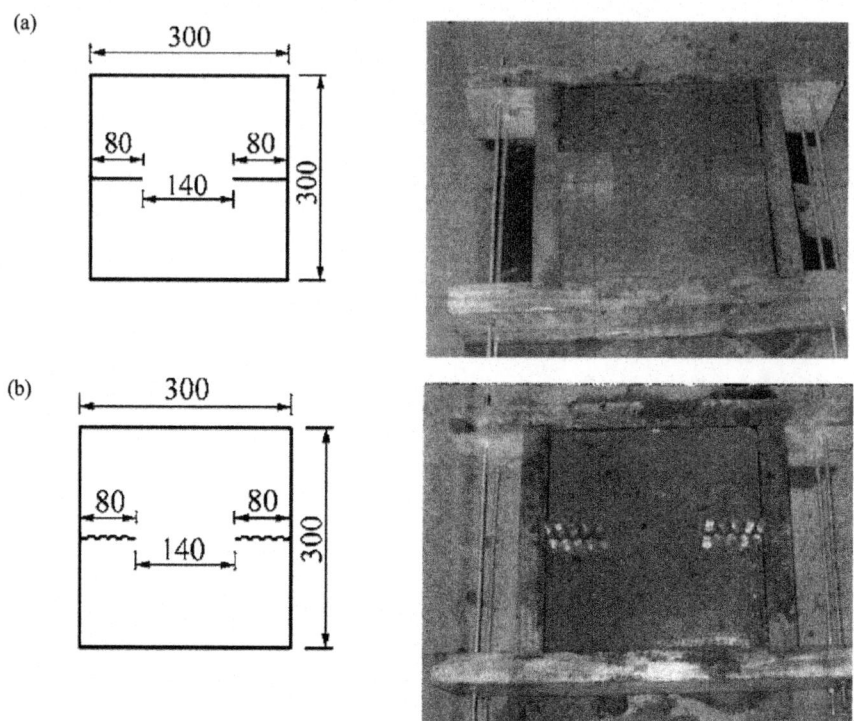

Figure 9. The rock mass samples containing joints [18] ((a) The first rock mass sample containing joints (the visual angles β_0 **of joints = 0°**) ;(b) The second rock mass sample containing joints (the visual angles β_0 **of joints = 30°**)).

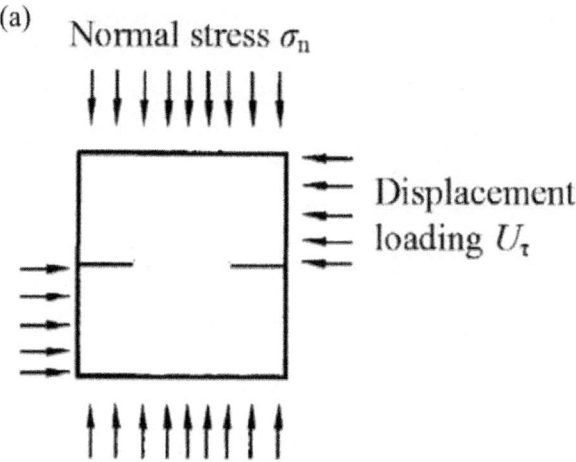

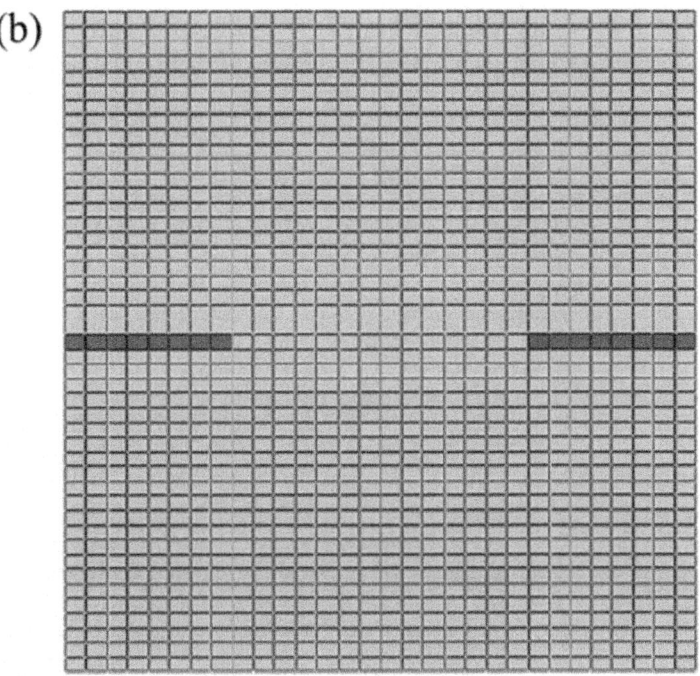

Figure 10. The boundary conditions and numerical model ((a) The boundary conditions of jointed rock direct shear experiment ;(b) The numerical model calculated by proposed constitutive model).

Table 2. The parameters of the joints and the rock

Material	The elastic modulus (GPa)	Poisson ratio	Friction coefficient f	Cohesion c(MPa)
Joints	0.50	0.16	0.25	0.10
Rock	3.70	0.16	0.89	3.93

doi:10.1371/journal.pone.0121850.t002

Through observing the results of Figure 11 and Figure 12, they show that the results of failure mode of rock mass by experiment and numerical simulation by proposed model are similar. And the curves of shear stress-displacement of experiment and numerical simulation by proposed model are close. These results verify the proposed constitutive model. And it shows proposed model can describe the behavior of jointed rock well.

(a)
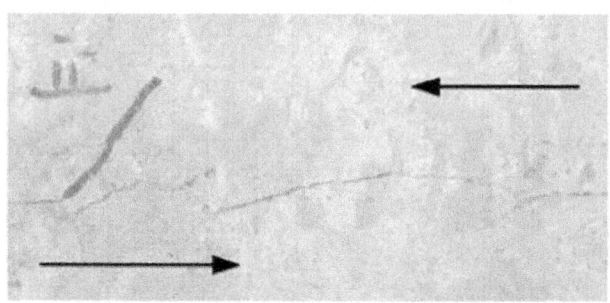

(b) SDV13
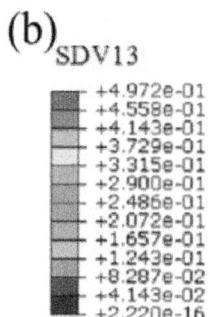

(c)
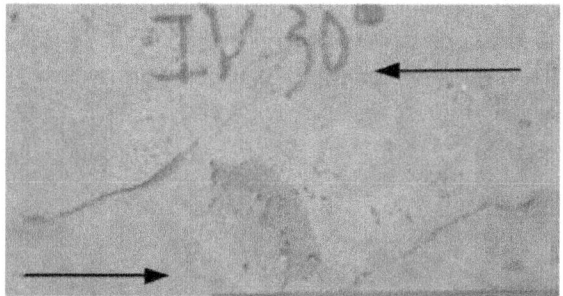

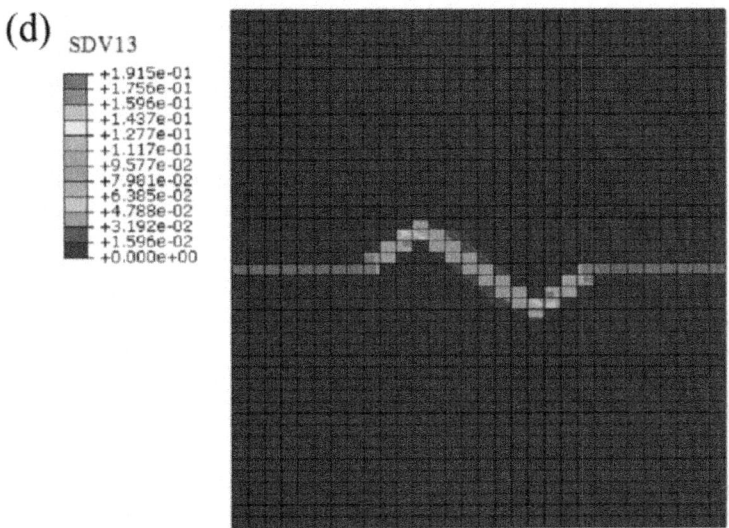

Figure 11. The results of experiment and numerical simulation((a) The failure mode of first rock mass sample in jointed rock direct shear experiment (the visual angles β_0 of joints = 0°);(b) The contour for equivalent plastic deviator strain calculated by proposed constitutive model(the visual angles β_0 of joints = 0°);(c) The failure mode of of second rock mass sample in jointed rock direct shear experiment (the visual angles β_0 of joints = 30°); (d)The contour for equivalent plastic deviator strain calculated by proposed constitutive model(the visual angles β_0 of joints = 30°))

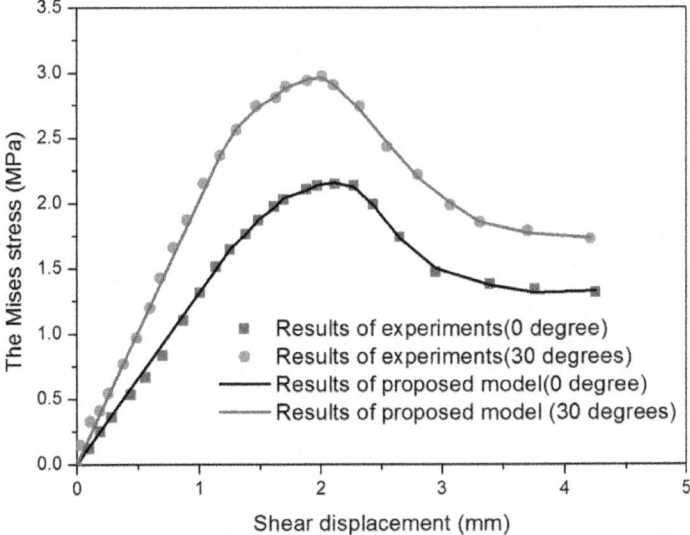

Figure 12. Curves of shear stress-displacement of experiment and numerical simulation

The numerical Examples for a Rectangle Foundation of Jointed Rock Mass

Figure 13 shows the plane stress model (finite element (FE) with 4-nodes) for a rectangle foundation of jointed rock mass (length = 120m, depth = 10m) subjected to uniform load p = 2GPa. The visual angle β_0 = 100°. There are two kinds of materials in the rectangle foundation. The blue regions are calculated by linear elastic constitutive model. The green region is calculated by proposed constitutive model and ubiquitous-joint constitutive model[19] in commercial software Abaqus, respectively. Table 3 shows the physical parameters of the rectangle foundation.

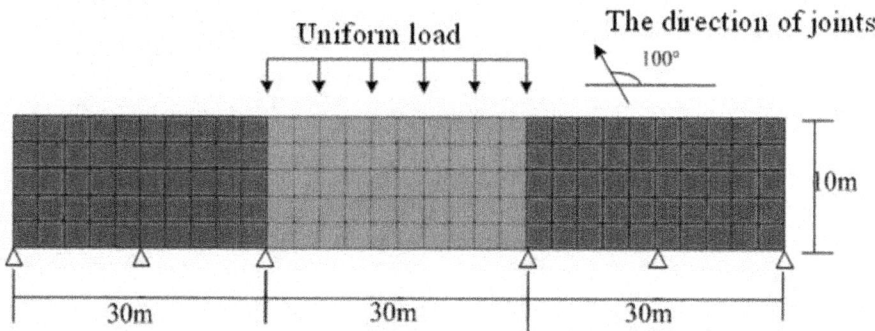

Figure 13. The model for FE analyses with difference materials

Table 3. The parameters of the rectangle foundation

	Elastic modulus(GPa)	Poisson ratio	Friction angle(°)	Cohesion(MPa)
Linear elastic material	25	0.3	-	-
The parameters of jointed rock mass in the visual angle β_0	25	0.3	35	0.27

doi:10.1371/journal.pone.0121850.t003

Through observing the results of Figure 14 and Figure 15, they show that the results of displacement and stress calculated by proposed constitutive model are close to that calculated by ubiquitous-joint constitutive model in commercial software Abaqus, which has been verified. The maximum relative errors of results of displacement and Mises stress calculated by proposed model are 1.77% and 15.25%, respectively.

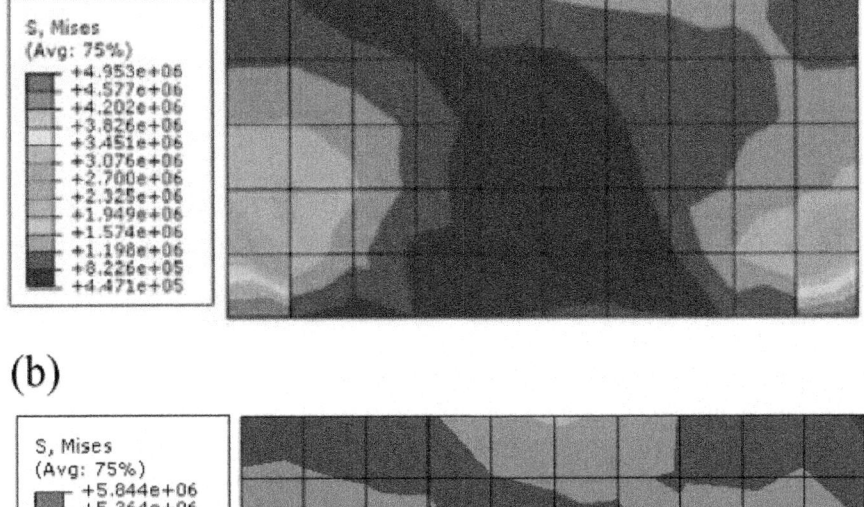

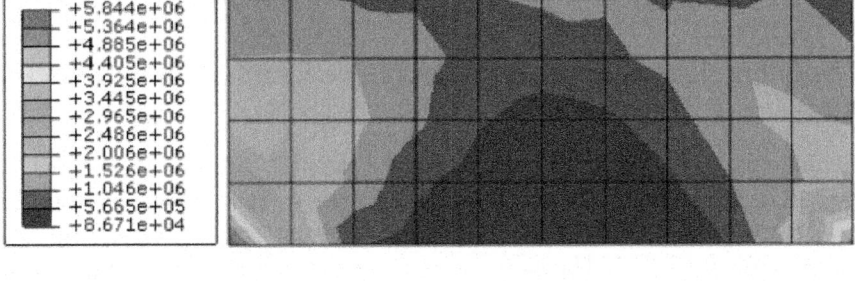

Figure 14. The Mises stress contour of the green region of the rectangle foundation calculated by different constitutive models(Pa) ((a)The Mises stress contour calculated by proposed constitutive model; (b)The Mises stress contour calculated by ubiquitous-joint model).

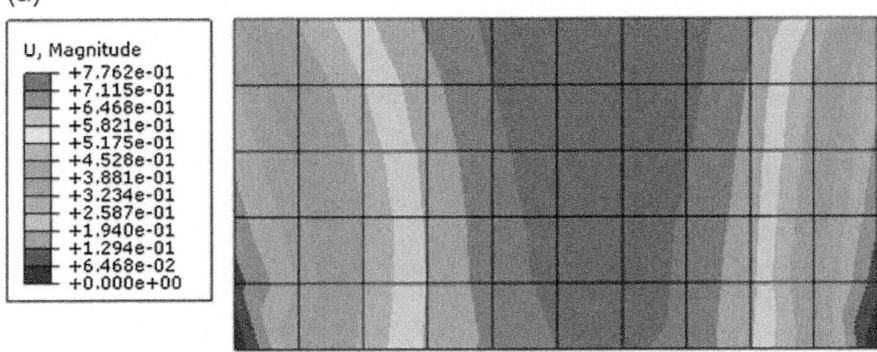

(b)

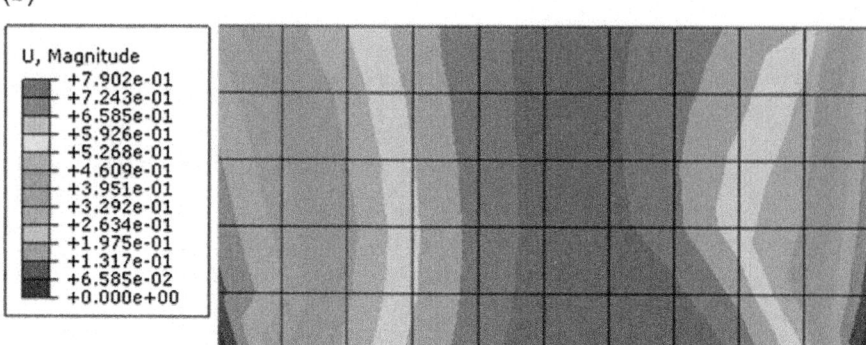

Figure 15. The displacement contour of the green region of the rectangle foundation calculated by different constitutive models(m) ((a)The contour for magnitude of displacement calculated by proposed constitutive model; (b) The contour for magnitude of displacement contour calculated by ubiquitous-joint model).

The Numerical Example for a Rectangle Beam of Jointed Rock Mass

Figure 16 shows the plane strain model (FE with 4-nodes) for a rectangle beam (length = 4m, width = 2m) of jointed rock mass subjected to uniform load p = 0.45MPa. Both sides of beam are restraint against displacement. The visual angle $\beta_0 = 120°$. The beam is calculated by proposed constitutive model and ubiquitous-joint constitutive model in commercial software Abaqus, respectively. Table 4 shows the physical parameters of the rectangle beam.

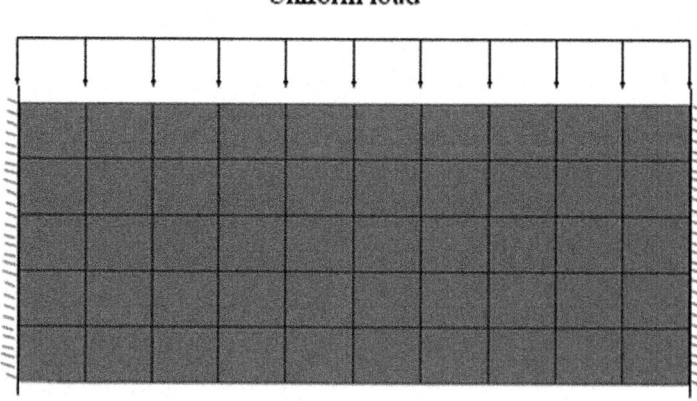

Figure 16. The FE model of the beam.

Table 4. The parameters of the rectangle beam

	Elastic modulus(GPa)	Poisson ratio	Friction coefficient f	Cohesion(MPa)
Rock	25	0.3	1.0	1.18
The parameters of jointed rock mass in the visual angle β_0	25	0.3	0.74	0.33

doi:10.1371/journal.pone.0121850.t004

Through observing the results of Figures 17–19, they show that the results of displacement, stress and plastic strain calculated by proposed constitutive model are also close to that calculated by ubiquitous-joint constitutive model in commercial software Abaqus. The maximum relative errors of results of displacement, stress and plastic strain calculated by proposed model are 8.31%, 1.08% and 19.71%, respectively. The results show the proposed constitutive model has some precision and verify the proposed constitutive model.

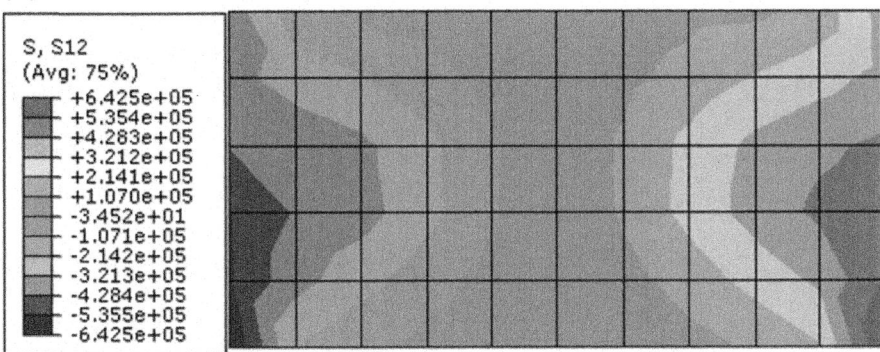

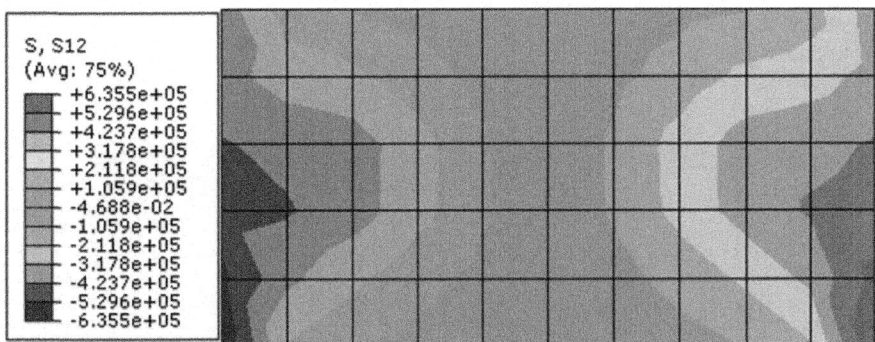

Figure 17. The shear stress contour of the rectangle beam calculated by different constitutive models(Pa) ((a) The shear stress contour calculated by proposed constitutive model; (b) The shear stress contour calculated by ubiquitous-joint model).

50 Design Analysis in Rock Mechanics

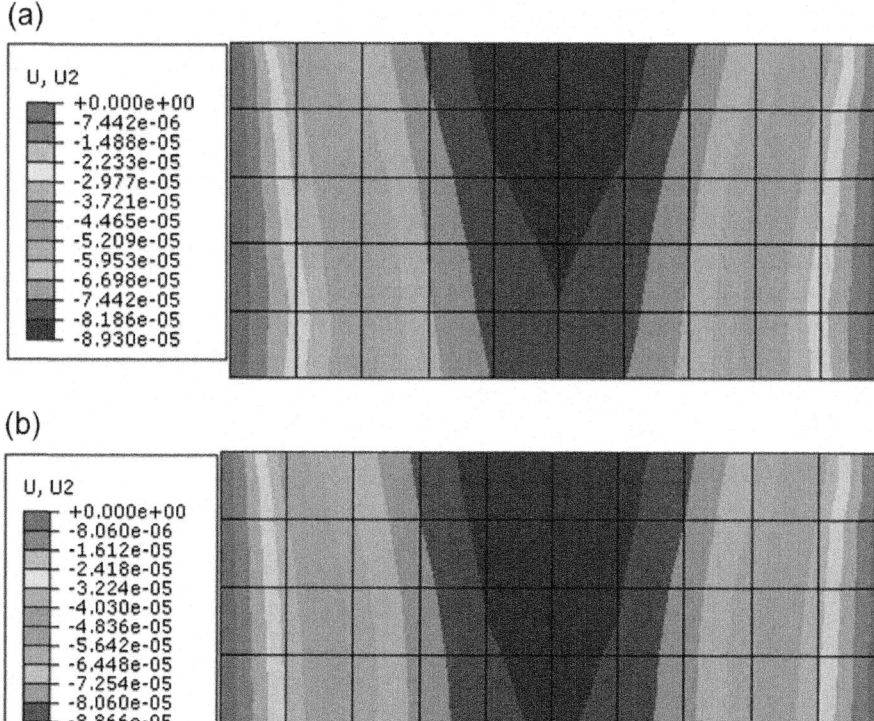

Figure 18. The vertical displacement contour of the rectangle beam calculated by different constitutive models(m) ((a) The vertical displacement contour calculated by proposed constitutive model; (b) The vertical displacement contour calculated by ubiquitous-joint model)

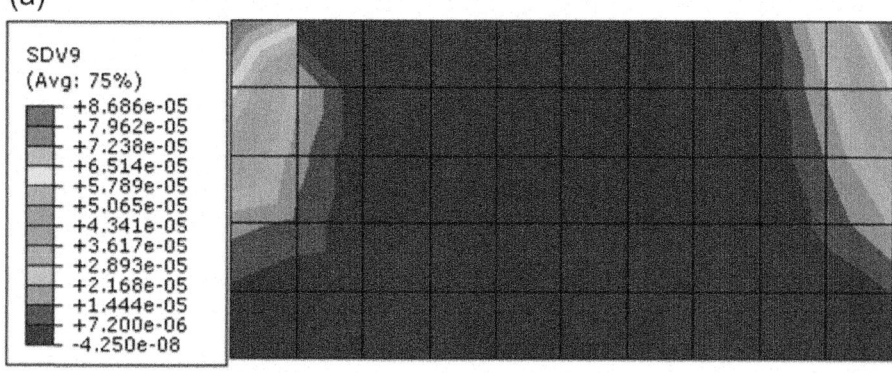

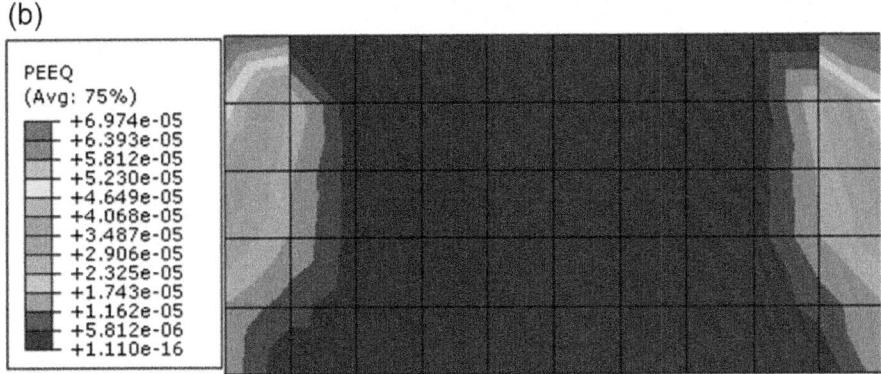

Figure 19. The contour for equivalent plastic deviator strain of the rectangle beam calculated by different constitutive models((a) The contour for equivalent plastic deviator strain calculated by proposed constitutive model;(b) The contour for equivalent plastic deviator strain calculated by ubiquitous-joint model).

The Numerical Example for Slope

Figure 20 shows the plane stress model (FE with 4-nodes) for a slope of jointed rock mass subjected to gravity. The elastic modulus and Poisson ratio of jointed rock mass are 4.00GPa and 0.30, respectively. Figure 21 shows two kinds of jointed rock mass, whose persistence ratios are different, are used to calculate. Table 5 shows the parameters of slope.

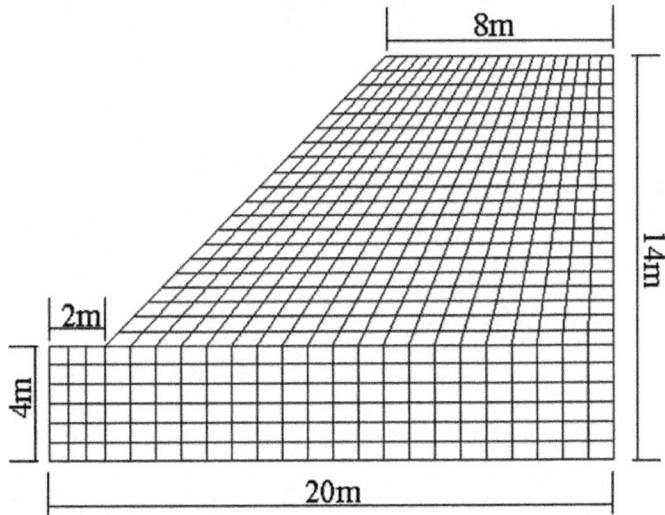

Figure 20. The model of non-homogeneous rock slope.

52 Design Analysis in Rock Mechanics

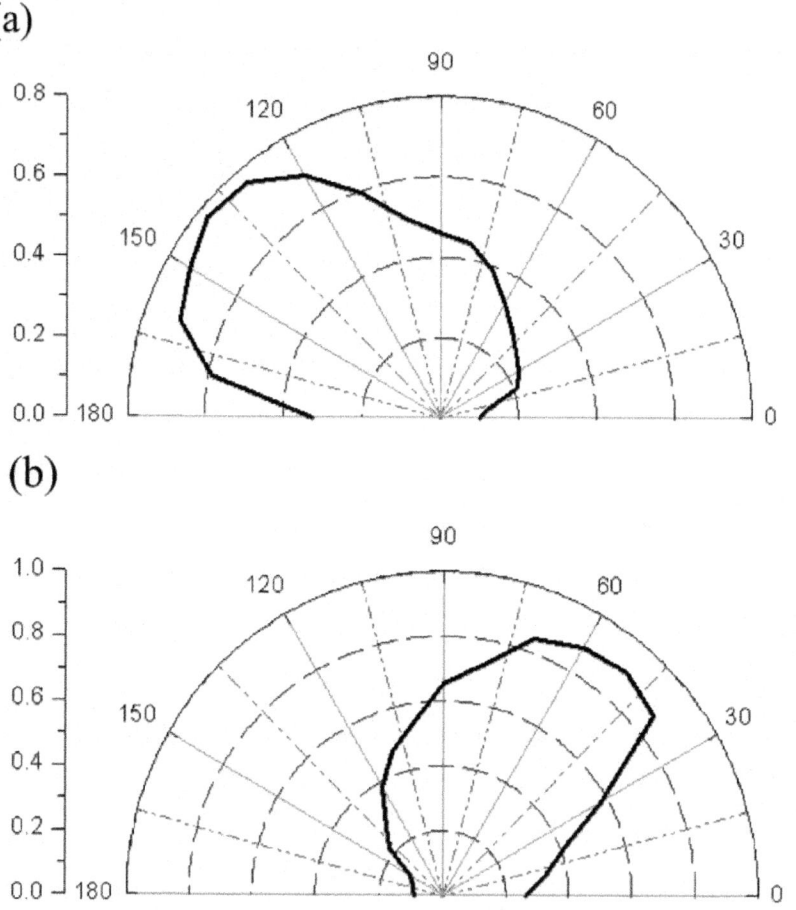

Figure 21. The rose diagrams of the persistence ratio k **of jointed rock mass for the different** slopes((a) The rose diagrams of the persistence ratio k **of jointed rock mass for the first slope; (b) The rose diagrams of the persistence ratio** k **of jointed rock mass for the second slope).**

Table 5. The parameters of slope

Material	Friction coefficient f	Cohesion c(MPa)
Joints	0.2	0.1
Rock	1.9	1.1

doi:10.1371/journal.pone.0121850.t005

Through observing the results of Figure 22 and Figure 23, they show that equivalent plastic strain is easy to develop along the direction, which has higher persistence ratio k. It is unfavorable to anti-slide stability if the visual angle β_0, which has higher persistence ratio k, is similar to the angle of rock slope. And it shows the proposed constitutive model can consider the persistence ratio k in different visual angle β_0.

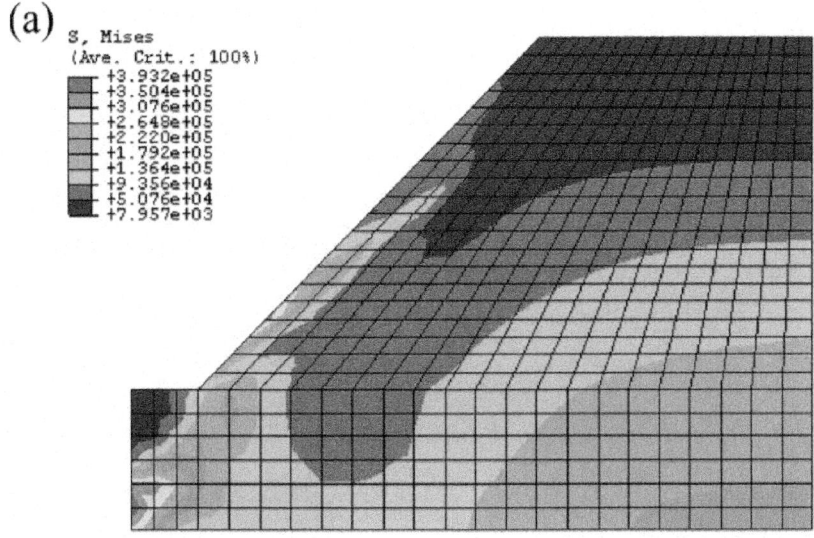

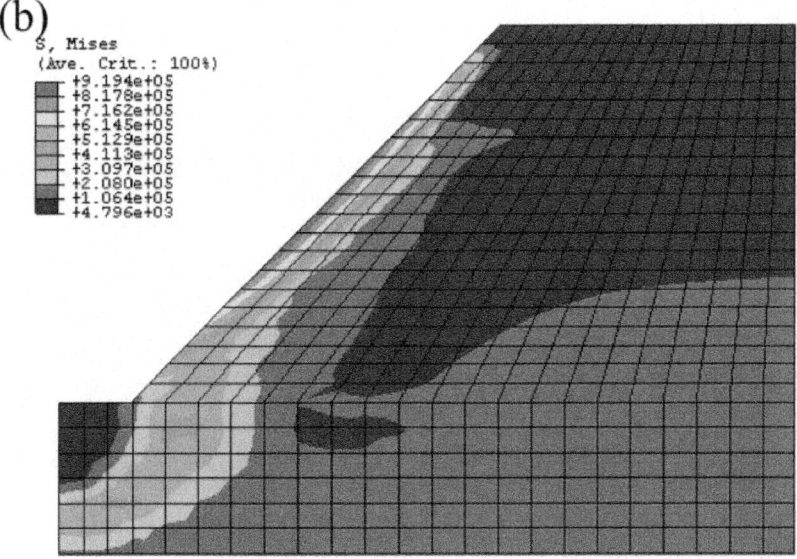

Figure 22. The Mises stress contour calculated by proposed constitutive model for the different slopes((a) The Mises stress contour calculated by proposed constitutive model for the first slope; (b) The Mises stress contour calculated by proposed constitutive model for the second slope).

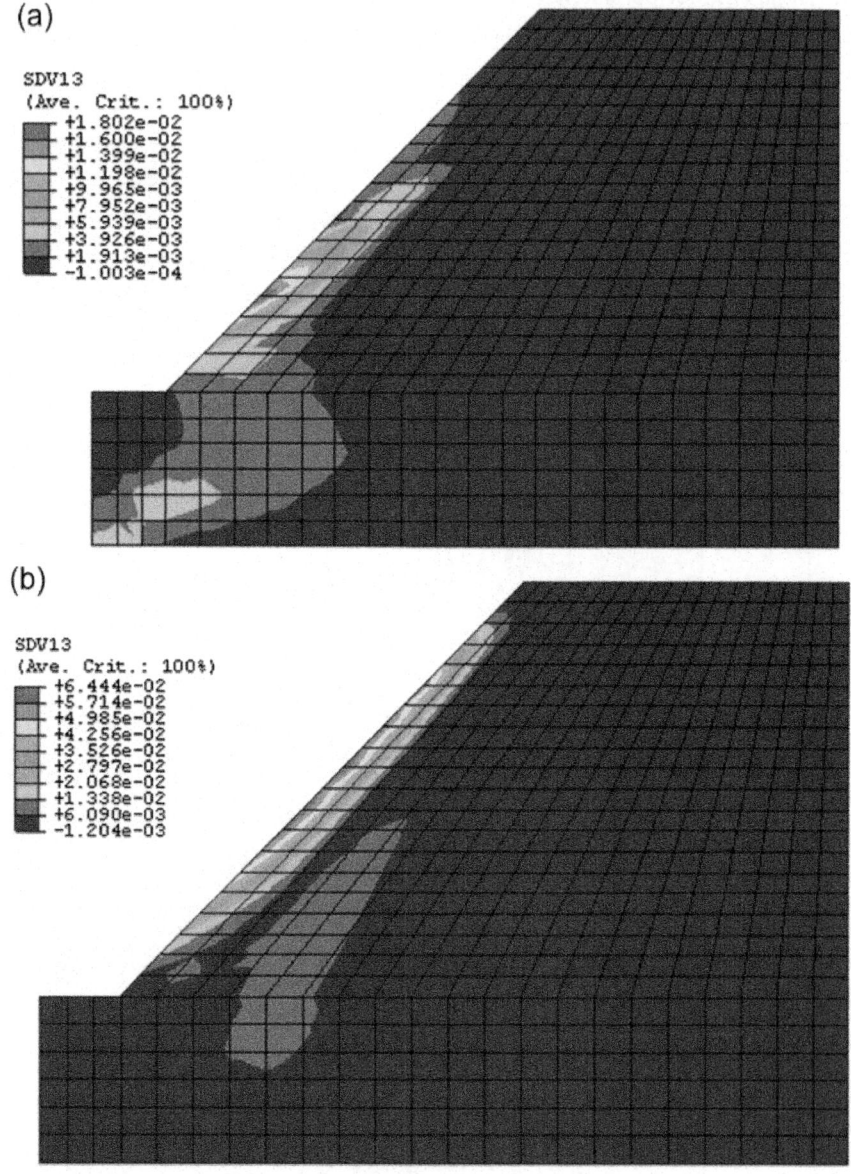

Figure 23. The contour for equivalent plastic deviator strain calculated by proposed constitutive model for the different slopes((a) The contour for equivalent plastic deviator strain calculated by proposed constitutive model for the first slope; (b) The contour for equivalent plastic deviator strain calculated by proposed constitutive model for the second slope).

CONCLUSION

A constitutive model for jointed rock mass, which can consider the persistence ratio in different visual angle, is proposed. The proposed the yield strength criterion f is not only related to friction angle and cohesion of jointed rock masses at the visual angle but also related to the intersection angle between the visual angle and the directions of the maximum principal stresses. From above analysis, it shows that the yield strength criterion f in proposed constitutive model is not only related to $\varphi_{\beta 0}$ and $c_{\beta 0}$ but also related to β. The yield strength criterion f has the relation of k (in other word, $\varphi_{\beta 0}$ and $c_{\beta 0}$) only when $\beta_{min} \leq \beta \leq \beta_{max}$. The relation of β and β_0 is also important to the yield strength criterion f. The different relation of β and β_0 leads to different yield strength criterion f. The proposed constitutive model can consider the persistence ratio k in different visual angle β_0. The proposed constitutive model has precision to calculate displacement, stress and plastic strain. The results show the proposed constitutive model has precision to calculate displacement, stress and plastic strain. However, the proposed constitutive model also has limitations. The model can describe the anisotropic strength of jointed rock, but cannot describe the anisotropic behaviors of elastic modulus and Poisson ratio. And the anisotropic strength of proposed model is also homogenized results. It cannot describe the localization phenomena of jointed rock precisely. And these problems need more research.

ACKNOWLEDGMENTS

This study was supported by the State Key Development Program for Basic Research of China (No.2013CB035905), the National Natural Science Foundation of China (Grant No.51138001, 51178081) and Fundamental Research Funds for the Central Universities(DUT14QY10).

AUTHOR CONTRIBUTIONS

Conceived and designed the experiments: QX JC. Performed the experiments: QX CZ CY. Analyzed the data: QX JL. Contributed reagents/materials/analysis tools: QX JC. Wrote the paper: QX.

REFERENCES

1. Jing L (2003) A review of techniques, advances and outstanding issues in numerical modelling for rock mechanics and rock engineering. International Journal of Rock Mechanics and Mining Sciences 40(3):283–353. doi: 10.1016/s1365-1609(03)00013-3

2. Lanaro F (2000) A random field model for surface roughness and aperture of rock fractures. International Journal of Rock Mechanics and Mining Sciences 37(8):1195–1210. doi: 10.1016/s1365-1609(00)00052-6
3. Fardin N, Stephansson O, Jing L (2001) The scale dependence of rock joint surface roughness. International Journal of Rock Mechanics and Mining Sciences 38(5):659–669. pmid:11411401 doi: 10.1016/s1365-1609(01)00028-4
4. Jing L, Hudson JA (2004) Fundamentals of hydro-mechanical behaviour of rock fractures: Roughness characterization and experimental aspects. International Journal of Rock Mechanics and Mining Sciences 41(3):383. pmid:11411401 doi: 10.1016/j.ijrmms.2003.12.044
5. Qin S, Jiao JJ, Wang S, Long H (2001) A non-linear catastrophe model of instability of planar-slip slope and chaotic dynamical mechanisms of its evolution process. International Journal of Solids and Structures 38(44–45):8093–8109. doi: 10.1016/s0020-7683(01)00060-9
6. Sitharama TG, Sridevib J, Shimizuc N (2001) Practical equivalent continuum characterization of jointed rock masses. Int. J. Rock Mech. Min. Sci 38(3): 437–448.
7. Yan SL, Huang YY, Chen CY (2001) An equivalent model for jointed rock mass with planar non penetrative joint and its elastic parameters. J. of Huazhong Univ.of Sci. & Tech 29(6):64–67.
8. Yan SL, Huang YY, Chen CY (2001) An equivalent model for jointed rock mass with persistent joint and its elastic parameters. J. of Huazhong Univ.of Sci. & Tech 29(6):60–63
9. Rice JR, Lapusta N, Ranjith K (2002) Rate and state dependent friction and the stability of sliding between elastically deformable solids. Journal of the Mechanics and Physics of Solids 49(9):1865–1898. doi: 10.1016/s0022-5096(01)00042-4
10. Maghous S, Buhan DP, Dormieux L (2002) Non-linear global elastic behaviour of a periodically jointed material. Mechanics Research Communications 29(1): 45–51. doi: 10.1016/s0093-6413(02)00225-2
11. Grasselli G, Wirth J, Egger P (2002) Quantitative three-dimensional description of a rough surface and parameter evolution with shearing. International Journal of Rock Mechanics and Mining Sciences 39(6):789–800. doi: 10.1016/s1365-1609(02)00070-9
12. Grasselli G, Egger P (2003) Constitutive law for the shear strength of rock joints based on three-dimensional surface parameters. International Journal of Rock Mechanics and Mining Sciences 40(1):25–40. doi: 10.1016/s1365-1609(02)00101-6

13. Caia M, Kaisera PK, Unob H, Tasakab Y, Minamic M (2004) Estimation of rock mass deformation modulus and strength of jointed hard rock masses using the GSI system. Int. J. of Rock Mechanics and Mining Sciences 41(1):3–19. doi: 10.1016/s1365-1609(03)00025-x
14. Zhu DJ, Yang LD, Cai YC (2010) Mixed multi-weakness plane softening model for jointed rock mass. Chinese J. of Geotechnical Engineering 32(2):185–191.
15. Wang XG, Chen ZY, Liu WS (1992) Determination of joint persistence and shear strength parameters of rock masses by Monte-Carlo method. Chinese J. of Rock Mechanics and Engineering 11(4):345–355.
16. Einstein HH, Baecher GB (1983) Probabilistic and statistical methods in engineering geology specific methods and examples part I: exploration. Rock Mechanics And Rock Engineering 16:39–72. doi: 10.1007/bf01030217
17. Einstein HH, Venieziano D, Baecher BG, Oreilly JK (1983) The effect of discontinuity persistence on rock slope stability. Int. J. Rock Mech. Min. Sci. & Geomech. Abstr 20:227–236.
18. Liu YM, XIA CC (2010) Research on rock mass containing discontinuous joints by direct shear test based on weakening mechanism of rock bridge mechanical properties. Chinese Journal of Rock Mechanics and Engineering 29(7):1467–1472.
19. Zienkiewicz OC, Pande GN (1977) Time dependent multilaminate model of rocks-a numerical study of deformation and failure of rock masses. International Journal for Numerical and Analytical Methods in Geomechanics 1:219–247. doi: 10.1002/nag.1610010302

Chapter 4

STUDY ON CALCULATION OF ROCK PRESSURE FOR ULTRA-SHALLOW TUNNEL IN POOR SURROUNDING ROCK AND ITS TUNNELING PROCEDURE

Xiaojun Zhou[1], Jinghe Wang[1], Bentao Lin[2]

[1] Key Laboratory of Transportation Tunnel Engineering of Ministry of Education, School of Civil Engineering, Southwest Jiaotong University, Chengdu 610031, China

[2] The 2nd Institute of Civil and Architecture Engineering, China Railway Eryuan Engineering Corporation Ltd., Chengdu 610031, China

ABSTRACT

A computational method of rock pressure applied to an ultra-shallow tunnel is presented by key block theory, and its mathematical formula is proposed according to a mechanical tunnel model with super-shallow depth. Theoretical analysis shows that the tunnel is subject to asymmetric rock pressure due to oblique topography. The rock pressure applied to the tunnel crown and sidewall is closely related to the surrounding rock bulk density, tunnel size, depth and angle of oblique ground slope. The rock pressure applied to the tunnel crown is much greater than that to the sidewalls, and the load applied to the left sidewall is also greater than that to the right sidewall. Meanwhile, the safety of the lining for an ultra-shallow tunnel in strata with inclined surface is affected by rock pressure and tunnel support parameters. Steel pipe grouting from ground surface is used to consolidate the unfavorable surrounding rock before tunnel excavation, and the reinforcing scope is proposed according to the analysis of the asymmetric load induced by tunnel excavation in weak rock with inclined ground surface. The tunneling procedure of bench cut method with pipe roof protection is still discussed and carried out in this paper according to the special geological condition. The method and tunneling procedure have been successfully utilized to design and drive a real expressway tunnel. The practice in building the super-shallow tunnel has proved the feasibility of the calculation method and tunneling procedure presented in this paper.

INTRODUCTION

The fast development of economy requires much more rapid and convenient transportation systems. Therefore, expressways and high speed railways have been built in China. Since the most land of western and northern China belongs to mountainous area, it is extremely difficult to build railways and expressways in these regions. If a transportation line such as highway or railway will be built in mountainous areas, tunnels are frequently adopted to overcome height barriers in the line. When the conditions of the area where transportation lines pass through is complicated in geology and topography, special geological problems might be encountered during the design and construction of transportation tunnels. If the ground has inclined topography, namely the ground surface appears in oblique form, then tunnel is easily subject to asymmetric rock pressure and its support structure must be excogitated according to asymmetric rock pressure.

In order to realize rational design and safe construction of mountainous tunnels, many studies and experiments have been carried out by scholars across the world. Goodman et al. [1] investigated the modeling techniques of tunnels in jointed rock, and presented an experimental and a numerical method to analyze the tunnel excavation in jointed rockmass. Shen and Barton [2] analyzed the disturbed zone around tunnels in jointed rock mass; they classified the disturbed zone into failure, open, and shear ones around a tunnel in jointed rock mass. Zhou et al. [3] made an insight into the rock pressure on tunnel with shallow depth in geologically inclined bedding strata, and set up formulas to calculate the asymmetric rock pressure applied to a tunnel with shallow depth in stratified rock. Later, Zhou and Yang [4] discussed the asymmetric rock pressure applied to the shallow tunnel in strata with inclined ground surface, and proposed a method to calculate the asymmetric rock pressure applied to the tunnel in strata with inclined ground surface by key block theory. Yang et al. [5] analyzed the calculation of rock pressure applied to three tunnels with large transection and small neighborhood in shallow surrounding rock; they suggested that the conventional method to calculate the rock pressure applied to shallow tunnel with small neighborhood should be improved. He et al. [6] studied the asymmetrical load effect on tunnels in geologically inclined bedding strata. They found that the rock pressure at the left wall of a tunnel is greater than that at the right wall. As the dip angle of bedding strata increases, the asymmetrical load gradually tends to be symmetric. In Zhou's [7] recent research, he studied the method to calculate the rock pressure for shallow asymmetric tunnel, and presented a method to determine a proper depth for shallow asymmetric tunnel in strata with inclined ground surface.

Most of studies available focused on the stability and deformation of tunnels and surrounding rock in jointed rockmass. Only a few concerned the calculation method of rock pressure applied to shallow tunnel [8, 9]. When designing tunnel structures, the rock pressure applied to tunnel lining is often obtained by numerical simulation or field monitoring. However, this is often hard for designers to use. To date, the calculation method of asymmetric rock pressure applied to super-shallow tunnel in strata with inclined ground surface has not been addressed [10]. As we know, the most difficult problem during design and construction of a tunnel in strata with inclined topography is to deal with the extremely thin overburden depth. This means that an ultra-shallow-tunnel will be excogitated and built in strata with inclined geomorphology; consequently, the influence of rock excavation will extend to ground surface and possibly cause casualties during tunneling [11, 12].

This paper aims to find out a simple way to calculate the asymmetric rock pressure for design of tunnel lining in super-shallow surrounding rock. The driving procedure for the bored tunnel is determined, taking into account the poor geological condition. Finally, a typical ultra-shallow tunnel for an expressway in Sichuan, China was taken as an example to make its structural design and safe construction procedure, which verifies the effectiveness of the proposed method.

ANALYSIS OF ASYMMETRIC ROCK PRESSURE APPLIED TO SUPER-SHALLOW TUNNEL

In order to analyze the asymmetric load on the tunnel support with an ultra-shallow depth in strata with inclined ground surface, a mechanical model is set up according to the geological and topographical condition of a vehicular tunnel named Zagunao with two lanes in an expressway in Sichuan, the southwest province of China, as shown in Figure. 1.

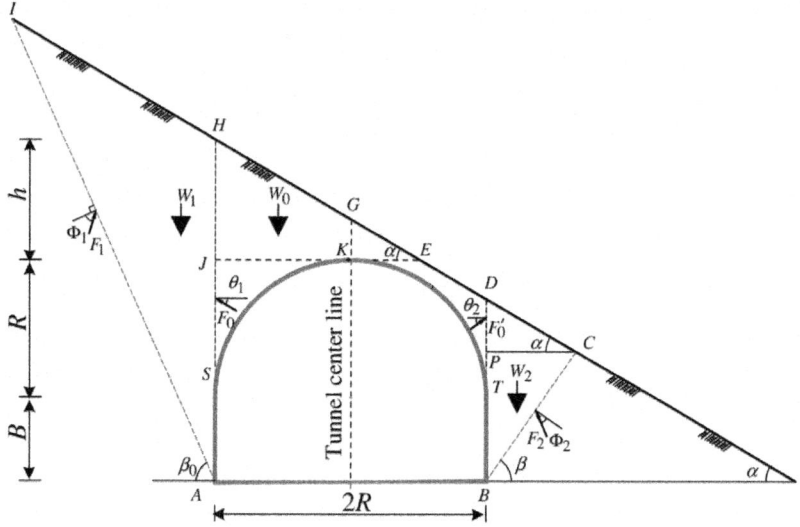

Figure. 1: Mechanical model of ultra-shallow tunnel.

According to the normal structural design of expressway tunnel with two lanes, the upper part of its reinforced concrete lining from its spring line appears in semi-circular arch with a radius of R and a width of 2R. Since the tunnel has an ultra-shallow overburden depth, excavation of surrounding rock must cause ground subsidence and its influence will extend to the surface. If it is supposed that there exist two breaking planes in the surrounding rock on each side of the tunnel, namely plane BC and plane AI, then the surrounding rock in ΔBCD and ΔAHI may tend to descend downward due to its rock gravity; synchronously, the surrounding rock above the tunnel crown namely in ΔJEH may also be caused to descend along breaking planes BD and AH by gravity as the tunnel will be driven. However, friction resistances must exist on each fracture plane; frictions will impede the slide of the surrounding rock in each triangular block. Let the surrounding rock in blocks exist in a critical equilibrium state; then their mechanical relations can be derived from Figure. 1.

In ΔAHI, the gravity W_1 of the surrounding rock circumscribed by block AHI is derived from the geometric relation as shown in Figure. 2, i.e.,

$$W_1 = \frac{1}{2}\gamma_1(h+R+B)^2, \qquad (1)$$

where α denotes the slope angle and β_0 the breaking angle between planes AI and MA, in degrees; γ_1 stands for the bulk density of the surrounding rock in block AHI, in kN/m³; and other symbols are delineated in Figures. 1 and 2.

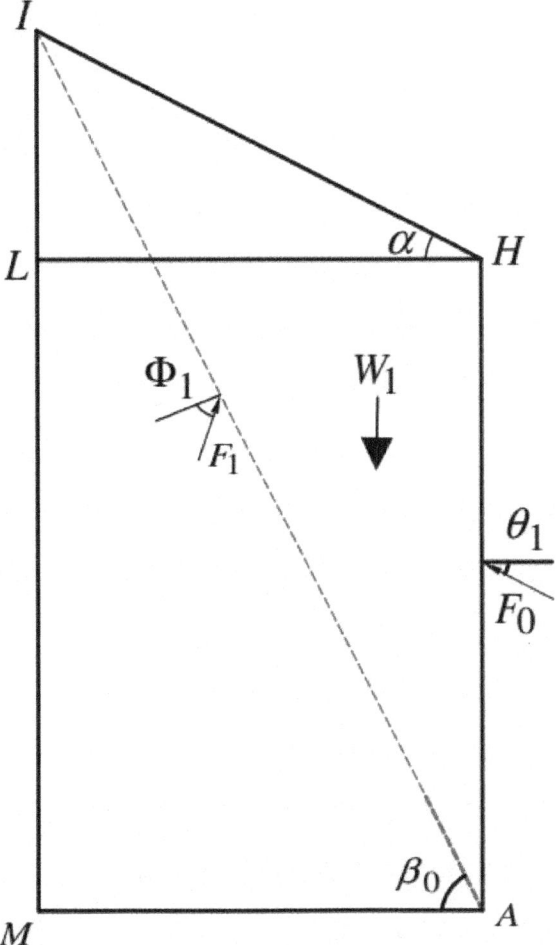

Figure. 2: Geometric relation of angle α and β₀.

In addition, three forces F_1, W_1, and F_0 constitute a vector triangle as shown in Figure. 3, where F_1 denotes the friction resistance applied to plane AI, F_0 the friction resistance to plane AH, and W_1 the gravity of the surrounding rock enclosed by block AHI. We can deduce from Figure. 3 by sine theorem that

$$\frac{F_0}{\sin(\beta_0 - \phi_1)} = \frac{W_1}{\sin\left[\frac{\pi}{2} - (\beta_0 - \theta_1) + \phi_1\right]} \qquad (2)$$

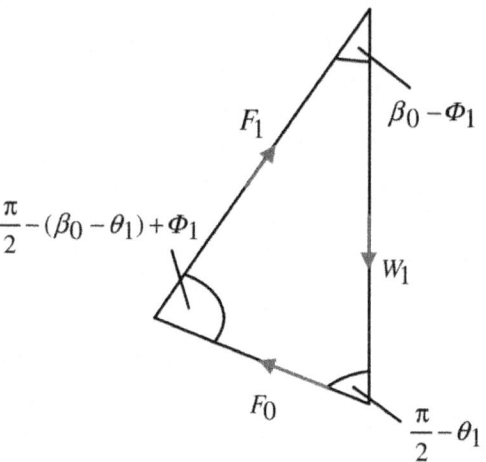

Figure. 3: Vector triangle of forces F_1, W_1, and F_0.

Substituting Eq. (1) into Eq. (2), we obtain

$$F_0 = \frac{\gamma_1(h+R+B)^2}{2\cos\theta_1} \cdot \frac{1}{\tan\beta_0 - \tan\alpha}$$
$$\times \frac{\tan\beta_0 - \tan\phi_1}{1 + \tan\beta_0 \tan\phi_1 + \tan\theta_1(\tan\beta_0 - \tan\phi_1)}. \tag{3}$$

Equation (3) shows that the friction resistance applied to breaking plane AH varies with the breaking angle β_0, so F_0 may get its maximum or minimum value as β_0 varies within 90°. Thus, if let

$$\lambda = \frac{1}{\tan\beta_0 - \tan\alpha}$$
$$\times \frac{\tan\beta_0 - \tan\phi_1}{1 + \tan\beta_0 \tan\phi_1 + \tan\theta_1(\tan\beta_0 - \tan\phi_1)}, \tag{4}$$

then Eq. (3) is converted into

$$F_0 = \frac{\gamma_1(h+R+B)^2}{2} \cdot \frac{\lambda}{\cos\theta_1}. \tag{5}$$

In order to get the extreme value of F_0, let

$$\frac{\partial F_0}{\partial \beta_0} = 0. \tag{6}$$

Then the following expression is derived:

$$\tan^2 \beta_0 - 2\tan\phi_1 \tan\beta_0$$
$$-\frac{\tan\phi_1 - \tan\alpha - (\tan\theta_1 + \tan\alpha)\tan^2\phi_1}{\tan\phi_1 + \tan\theta_1} = 0. \tag{7}$$

Equation (7) represents a quadratic equation with one variable β_0, and its maximum root is derived as follows:

$$\tan\beta_0 = \tan\phi_1 + \sqrt{\frac{(\tan\phi_1 - \tan\alpha)(1 + \tan^2\phi_1)}{\tan\phi_1 + \tan\theta_1}}. \tag{8}$$

It is known from Eq. (8) that the maximum β_0 for breaking plane AI is controlled by angle ϕ_1, α, and θ_1.

As for the triangular block BCD shown in Figure. 1, the surrounding rock may also slide along the fracture planes BC and BD due to its gravity W_2. In order to anatomize the friction resistance applied to breaking plane BD, the gravity W_2 must be obtained. Therefore, if extend straight line KE to point N, and line TD to N, then point N is the intersection between line EN and DN, as shown in Figure. 4.

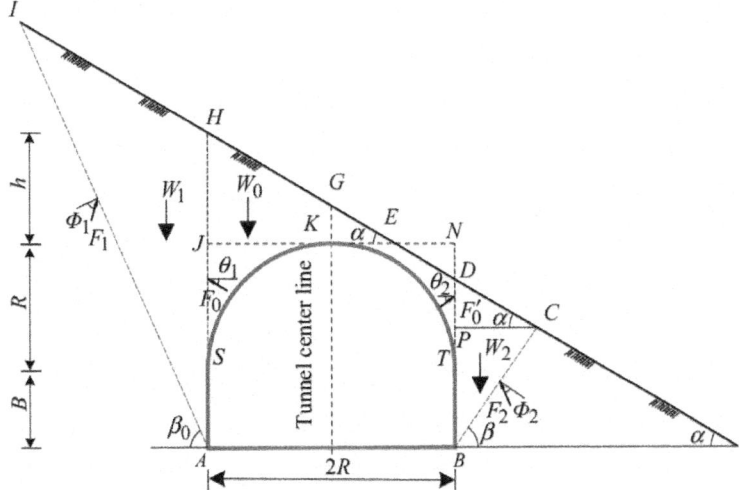

Figure. 4: Geometric relation of $\triangle$BDC and $\triangle$DPC.

In $\triangle$HJE, since HJ = h and JE = h·c tan α, we have
$$EN = 2R - h \cdot c \tan\alpha. \tag{9}$$

Furthermore, in $\triangle$EDN, we have ND = 2R tanα − h, and BD = B+R − (2R tan α − h). Thus, there is

$$BD = B + h + R(1 - 2\tan\alpha). \tag{10}$$

In addition, both $\triangle DPC$ and $\triangle BCP$ are in right triangles; therefore, following relations may exist, i.e.,

$$\tan\alpha = \frac{DP}{PC}, \tag{11}$$

$$\tan\beta = \frac{PB}{PC}, \tag{12}$$

$$PB = BD - DP. \tag{13}$$

Substituting Eqs. (11) and (12) into Eq. (13), we get

$$PC = \frac{B + h + R(1 - 2\tan\alpha)}{\tan\alpha + \tan\beta} \tag{14}$$

Therefore, the gravity W_2 of the triangular block BCD is derived as

$$W_2 = \frac{\gamma_2}{2}\frac{[B + h + R(1 - 2\tan\alpha)]^2}{\tan\alpha + \tan\beta}, \tag{15}$$

where γ_2 denotes the bulk density of the surrounding rock within the block BCD.

Similarly, the three forces F_2, W_2, and F_0' also constitute a vector triangle as shown in Figure. 5 and their relation is as follows:

$$\frac{F_0'}{\sin(\beta - \phi_2)} = \frac{W_2}{\sin\left[\frac{\pi}{2} - (\beta - \theta_2) + \phi_2\right]} \tag{16}$$

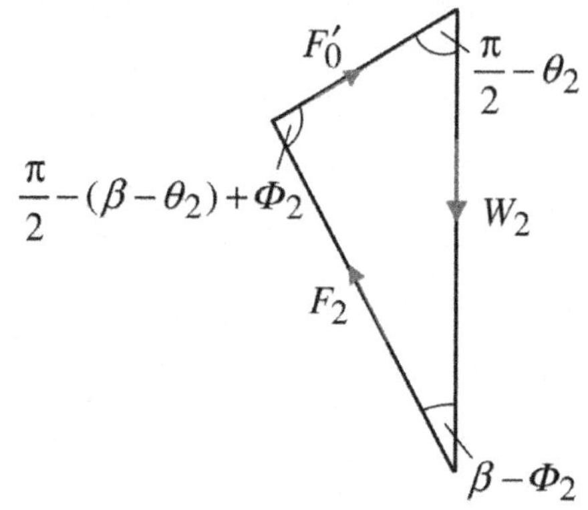

Figure. 5: Vector triangle of forces F_2, W_2, and F_0'.

Substituting Eq. (15) into Eq. (16), we get

$$F_0' = \frac{\gamma_2[B+h+R(1-2\tan\alpha)]^2}{2\cos\theta_2} \cdot \frac{1}{\tan\beta+\tan\alpha}$$
$$\times \frac{\tan\beta-\tan\phi_2}{1+\tan\beta\tan\phi_2+\tan\theta_2(\tan\beta-\tan\phi_2)}. \quad (17)$$

It is known from Eq. (17) that the friction resistance F_0' applied to the breaking plane BD varies with the breaking angle β, so it may get an extreme value as β varies within 90°. Let

$$\lambda' = \frac{1}{\tan\beta+\tan\alpha}$$
$$\times \frac{\tan\beta-\tan\phi_2}{1+\tan\beta\tan\phi_2+\tan\theta_2(\tan\beta-\tan\phi_2)}. \quad (18)$$

Then Eq. (17) is converted into

$$F_0' = \frac{\gamma_2[B+h+R(1-2\tan\alpha)]^2}{2} \cdot \frac{\lambda'}{\cos\theta_2}. \quad (19)$$

To derive the extreme value of F_0', let

$$\frac{\partial F_0'}{\partial \beta_0} = 0. \quad (20)$$

Then we obtain

$$\tan^2\beta - 2\tan\phi_2\tan\beta - \frac{\tan\phi_2+\tan\alpha-(\tan\theta_2+\tan\alpha)\tan^2\phi_2}{\tan\phi_2+\tan\theta_2} = 0. \quad (21)$$

Equation (21) is also a quadratic equation with one variable β, so its maximum root is derived as follows:

$$\tan\beta = \tan\phi_2 + \sqrt{\frac{(\tan\alpha+\tan\phi_2)(1+\tan^2\phi_2)}{\tan\phi_2+\tan\theta_2}}. \quad (22)$$

Equation (22) shows that the failure angle β is closely related to angles ϕ_2, θ_2, and α. As for the vertical load applied on the crown of the tunnel, W_0 contributes the great proportion. If the surrounding rock above its crown tends to slide along the assumed planes AH and BD, it will be resisted by the vertical components of friction resistance F_0 and F_0'. Their vertical components are derived as follows:

$$F_{0v} = \frac{\gamma_1}{2}(B+h+R)^2 \lambda \tan\theta_1, \tag{23}$$

$$F'_{0v} = \frac{\gamma_2}{2}[B+h+R(1-2\tan\alpha)]^2 \lambda' \tan\theta_2. \tag{24}$$

Then the downward load applied to the crown of the tunnel is

$$Q = W_0 - (F_{0v} + F'_{0v}), \tag{25}$$

where W_0 stands for the gravity of the surrounding rock above the tunnel crown, in kN; its mathematical expression can be derived from the geometrical relation as shown in Figure. 1. The gravity of the surrounding rock above the tunnel crown is obtained by deducting the area of tunnel transection from the area of trapezoid ABDH, namely the area above the tunnel crown is as follows

$$A = \frac{(AH+BD)}{2} \cdot AB - AB \cdot AS - \frac{\pi R^2}{2}. \tag{26}$$

Then gravity W_0 is obtained as follows:

$$W_0 = 2\gamma_0 Rh + 2\gamma_0 R^2(1-\tan\alpha) - \frac{\gamma_0}{2} \cdot \pi R^2, \tag{27}$$

where γ_0 stands for the bulk density of the surrounding rock above the tunnel crown, in kN/m³.

Substitution of Eqs. (23), (24), and (27) into (25) yields

$$Q = 2\gamma_0 Rh + 2\gamma_0 R^2(1-\tan\alpha) - \frac{\gamma_0}{2} \cdot \pi R^2$$
$$- \frac{\gamma_1}{2}(B+h+R)^2 \lambda \tan\theta_1$$
$$- \frac{\gamma_2}{2}[B+h+R(1-2\tan\alpha)]^2 \lambda' \tan\theta_2. \tag{28}$$

Then the vertical downward pressure applied to the crown is

$$q = \frac{Q}{2R}. \tag{29}$$

Substituting Eq. (28) into Eq. (29), we obtain the rock pressure applied on the tunnel crown as follows:

$$q = \gamma_0 h + \gamma_0 R(1-\tan\alpha) - \frac{\gamma_0}{4}\pi R$$
$$- \frac{\gamma_1}{4R}(B+h+R)^2 \lambda \tan\theta_1$$
$$- \frac{\gamma_2}{4R}[B+h+R(1-2\tan\alpha)]^2 \lambda' \tan\theta_2. \tag{30}$$

Equation (30) shows that the vertical pressure applied to the tunnel crown caused by gravity W_0 is related to parameters such as γ_0, h, α, β, R, θ_1, and θ_2; and its direction is always downward.

According to Figure. 1, the surrounding rock enclosed by $\triangle AHI$ and $\triangle BCD$ exert a lateral rock thrust on two side walls of the tunnel. As for $\triangle AHI$, the lateral load applied to the side wall is derived as follows.

In $\triangle AHI$ as shown in Figure. 6, according to the division of forces, the vertical component of friction resistance can be obtained. According to Figure. 3, there exists sine theorem, namely

$$\frac{F_1}{\sin\left(\frac{\pi}{2} - \theta_1\right)} = \frac{F_0}{\sin(\beta_0 - \phi_1)}. \tag{31}$$

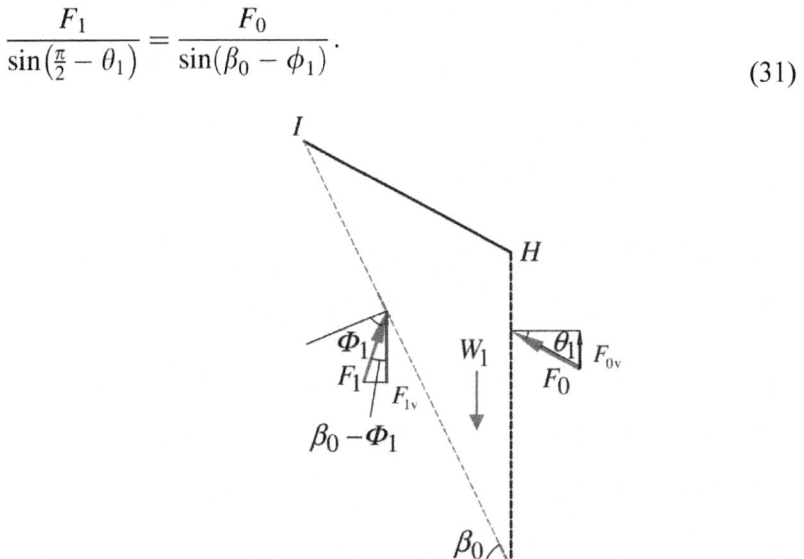

Figure. 6: Vertical components of F_1 and F_0.

Then friction resistance F_1 is

$$F_1 = \frac{\gamma_1}{2}(B+h+R)^2 \frac{\lambda}{\sin(\beta_0 - \phi_1)}. \tag{32}$$

In addition, the sine theorem also holds in Figure. 5, and there is a relation as follows:

$$\frac{F_2}{\sin\left(\frac{\pi}{2} - \theta_2\right)} = \frac{F_0'}{\sin(\beta - \phi_2)}. \tag{33}$$

And the friction resistance F_2 is

$$F_2 = \frac{\gamma_2}{2}[B + h + R(1 - 2\tan\alpha)]^2 \frac{\lambda'}{\sin(\beta - \phi_2)} \qquad (34)$$

The vertical components of friction resistance F_1 and F_2 tends to prevent the surrounding rock both in block AHI and BCD from sliding along the assumed fracture plane. The vertical components of F_1 and F_2 are as follows:

$$F_{1v} = F_1 \cos(\beta_0 - \phi_1), \qquad (35)$$

$$F_{2v} = F_2 \sin\left[\frac{\pi}{2} - (\beta - \phi_2)\right]. \qquad (36)$$

Substitution of Eqs. (32) and (34) into Eqs. (35) and (36) yields the following expressions:

$$F_{1v} = \frac{\gamma_1}{2}(B + h + R)^2 \frac{\lambda}{\tan(\beta_0 - \phi_1)}, \qquad (37)$$

$$F_{2v} = \frac{\gamma_2}{2}[B + h + R(1 - 2\tan\alpha)]^2 \frac{\lambda'}{\tan(\beta - \phi_2)}. \qquad (38)$$

As for block AHI, the downward load Q_1 is derived as

$$Q_1 = W_1 - (F_{1v} + F_{0v}). \qquad (39)$$

Thus, the downward load Q_1 caused by gravity W_1 is derived as

$$Q_1 = \frac{\gamma_1}{2}(B + h + R)^2$$
$$\times \left[\frac{1}{\tan\beta_0 - \tan\alpha} - \frac{\lambda}{\tan(\beta_0 - \phi_1)} - \lambda\tan\theta_1\right]. \qquad (40)$$

If Q_1 is totally applied to the left side wall, then the lateral pressure (e_1) applied to the left side wall AS is derived as

$$e_1 = \frac{\gamma_1}{2B}(B + h + R)^2 \eta, \qquad (41)$$

where η is the lateral pressure coefficient, and its specific expression is as follows:

$$\eta = \frac{1}{\tan\beta_0 - \tan\alpha} - \frac{\lambda}{\tan(\beta_0 - \phi_1)} - \lambda\tan\theta_1. \qquad (42)$$

For the lateral load applied to the right side wall of the tunnel, the forces and their relation are shown in Figure. 7.

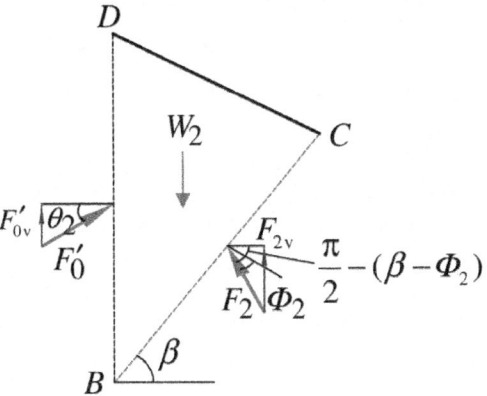

Figure. 7: Vertical components of $F_0{'}$ and F_2

According to the relation between gravity and friction resistances in Figure. 7, the downward load Q_2 is derived as

$$Q_2 = W_2 - (F_{2v} + F'_{0v}). \tag{43}$$

If Eqs. (24) and (36) are substituted into Eq. (43), then the downward load Q_2 caused by gravity W_2 is also obtained as

$$Q_2 = \frac{\gamma_2}{2}[B + h + R(1 - 2\tan\alpha)]^2$$
$$\times \left[\frac{1}{\tan\beta + \tan\alpha} - \frac{\lambda'}{\tan(\beta - \phi_2)} - \lambda'\tan\theta_2\right] \tag{44}$$

Then, the lateral pressure (e_r) applied to the right side wall BD is

$$e_r = \frac{\gamma_2}{2B}[B + h + R(1 - 2\tan\alpha)]^2 \eta', \tag{45}$$

where η' denotes lateral pressure coefficient, and its mathematical expression is as follows:

$$\eta' = \frac{1}{\tan\beta + \tan\alpha} - \frac{\lambda'}{\tan(\beta - \phi_2)} - \lambda'\tan\theta_2. \tag{46}$$

To sum up, the method to calculate the rock pressure applied to an ultra-shallow tunnel in strata with inclined ground has been derived from above analysis. It is apparent that the lateral pressure applied to two side walls of the tunnel is not identical. Therefore, a tunnel with ultra-shallow depth undergoes asymmetric rock pressure.

CALCULATION OF ASYMMETRIC ROCK PRESSURE FOR AN EXPRESSWAY TUNNEL

In order to analyze the rock pressure applied to an typical ultra-shallow tunnel, Zagunao tunnel in an expressway of Sichuan province in China is analyzed; its typical geological transection is shown in Figure. 8. There are two tunnels in the expressway; both are separated from each other. The net distance from outside wall of the left tunnel to that of the right tunnel is only 20 m. The main feature of this expressway tunnel is that its entry portal section lies in strata with a shallow and ultra-shallow overburden due to the requirement of line development.

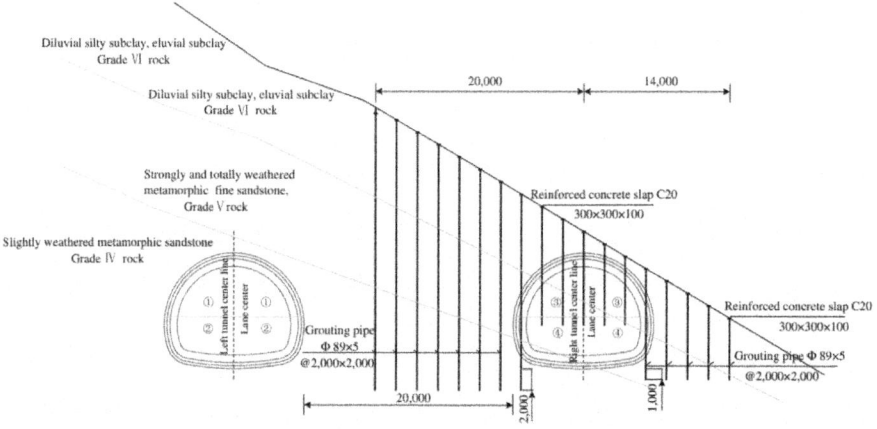

Figure. 8: Geological transection of an ultra-shallow tunnel in entry portal section (unit: mm).

Each tunnel has two vehicular lanes. The total length of the entry portal section with shallow depth is 25 m long in its longitudinal direction. In this shallow section, the left belongs to deep tunnel with an overburden from 13 to 36 m, but the right one has a very thin overburden that varies from 2 to 18 m. The surrounding rock mainly falls into diluvial silty subclay, eluvial subclay, and strongly weathered metamorphic fine sandstone. According to the Rock Quality Designation (RQD) and Basic Quality (BQ) system for rock mass classification [13, 14], and the Code for Design of Highway Tunnel, the surrounding rock belongs to grades IV, V, and VI [15, 16].

In order to make a concise analysis and simplify the computation process of the surrounding rock pressure, let all surrounding rock belong to grade VI. The physical and mechanical parameters of the surrounding rock and the basic size of the tunnel are shown in Table 1. Although this hypothesis seems to be simple, the result is still representative.

Table 1: Basic parameters of the surrounding rock and tunnel

γ_0(kN/m³)	γ_1(kN/m³)	γ_2(kN/m³)	θ_1(°)	θ_2(°)	ϕ_1(°)	ϕ_1(°)	a(°)	B(m)	R(m)	h(m)
15	16	14	20	18	40	35	28	4.11	6.77	6.0

Substituting all the parameters in Table 1 into the above-stated equations, we derived the rock pressure applied to the ultra-shallow tunnel, as shown in Figure. 9.

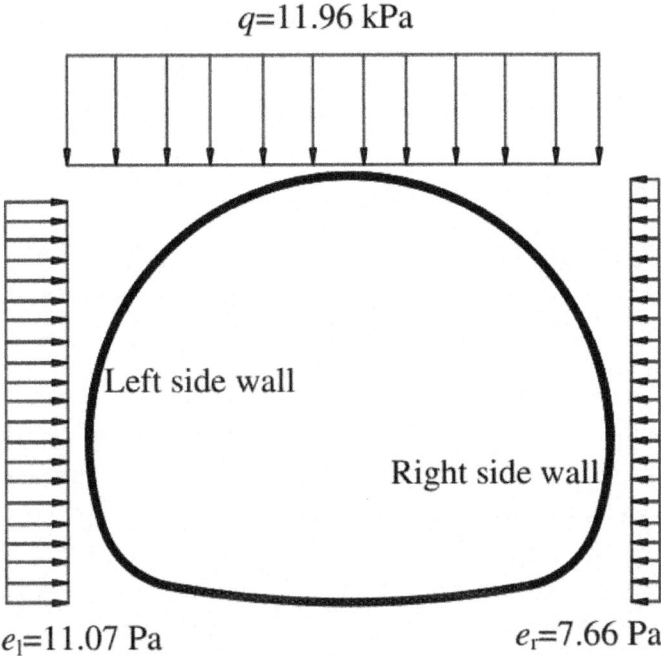

Figure. 9: Rock pressure applied to tunnel support.

It is clear from Figure. 9 that the right tunnel is subject to asymmetric pressure; and its crown is subject to the maximum vertical pressure with downward direction. The rock pressure applied to its left sidewall is greater than that to the right one in magnitude, but smaller than that on its crown. In addition, if these parameters are substituted into Eqs. (8) and (22), then the breaking plane angle are obtained as $\beta_0 = 56.28°$ and $\beta = 63.86°$. This indicates that excavation of the surrounding rock inside the tunnel contour may possibly cause overburden rock to slide or subside along the assumed breaking plane as shown in Figure. 10.

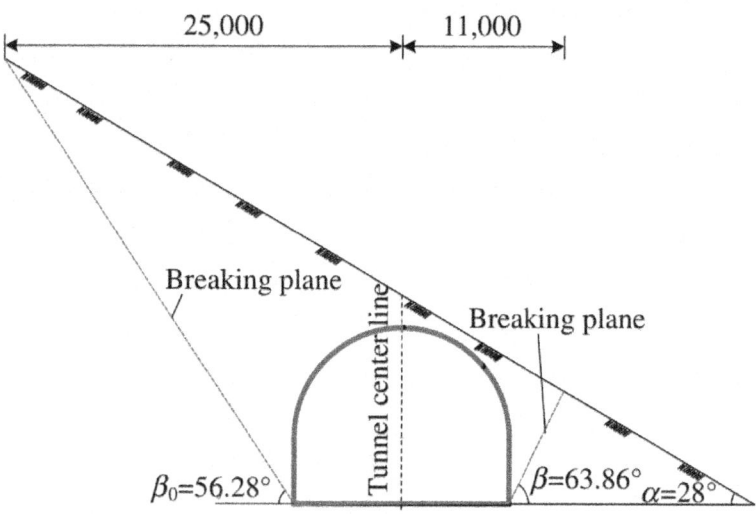

Figure. 10: Slide plane of the surrounding rock (unit: mm).

In order to illustrate the assumed plane of the surrounding rock, the arc invert of the tunnel is simplified to a straight floor but this does not affect the analysis of rock pressure.

According to the calculated results of the rock pressure applied to tunnel and the assumed fracture plane, the structural design of the ultra-shallow tunnel and its ground treatment for the right tunnel have been carried out. The tunnel support and its designed parameters are shown in Figure. 11 in detail.

The tunnel is supported with composite lining which consists of primary support and inner lining. According to the load and structure model in structure mechanics [15, 16], the internal forces in the primary and secondary support are calculated. By assuming the rock pressure calculated by the method presented in this paper is totally applied to the primary support, the safety factors in the typical cross section of the primary support is obtained by allowable stress design and ultimate strength design method [15, 16]. The obtained results for primary support are shown in Figure. 12. Since the minimum safety factor reaches 6.56, the designed primary support is safe when it is subject to asymmetric rock pressure. In addition, if all the pressure is applied to the inner lining, then safety factors of the secondary lining are also calculated by using the same model as shown in Figure. 13. It is apparent that the secondary lining is also secure when it is subject to asymmetric rock pressure.

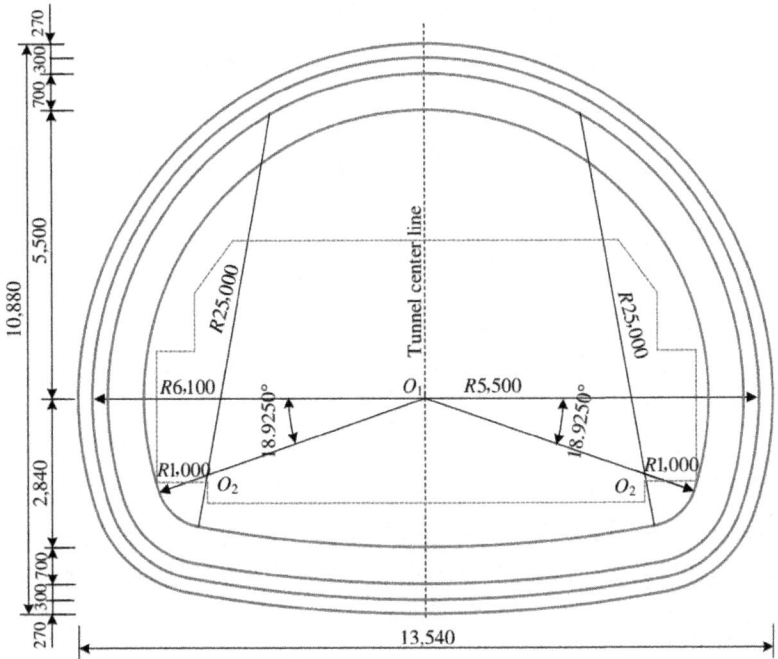

Figure. 11: Tunnel support and its transection (unit: mm). Grouting pipe ϕ42 mm × 5 mm, L = 4,500 mm @400 mm; Hollow grouting bolt ϕ25, L = 6,000 mm @300; I-shaped steel arch, I20, longitudinal @600 mm. Shotcrete, C25, δ = 270, ϕ8 steel mesh @150 × 150. Allowable camber 300 mm. Geotextile 350 g/m^2, δ = 1.2 mm PVC-P. Reinforced concrete, C25, δ = 700 mm.

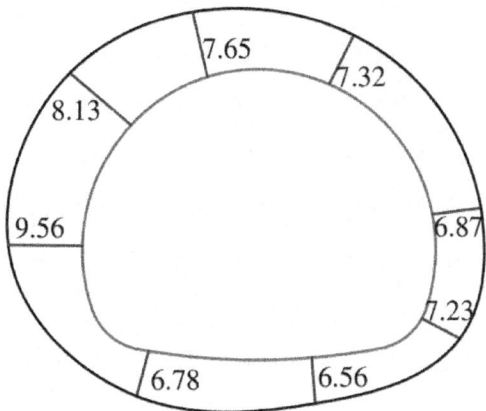

Figure. 12: Safety factor of primary support.

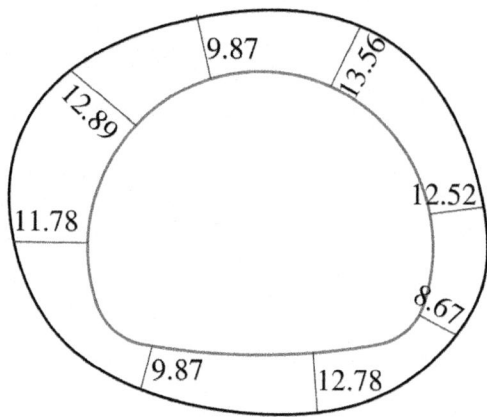

Figure. 13: Safety factors of inner lining.

According to the rock pressure calculation of the ultra-shallow tunnel, the surrounding rock above the tunnel vault tends to slide along the two assumed fracture planes, and it will result in ground subsidence; therefore, ground treatment should be conducted in light of the above calculated results so as to lessen the rock settlement and construction risk during tunneling [17, 18]. According to the calculated maximum fracture angle of plane shown in Figure. 10, the ground treatment range must be greater than the subsidence scope; in addition, the supposed slide planes must also be consolidated. In consideration of the geology and geomorphology of the tunnel, cement grouting method is adopted to reinforce the surrounding rock with steel pipes from the ground surface. Since sandstone possesses large porosity, grouting can effectively enhance its shear strength; the detail of the grouting pipe is shown in Figure. 14. Since the portal section of the right tunnel is 25 m long in its longitudinal direction, cement grouting was carried out only in this section.

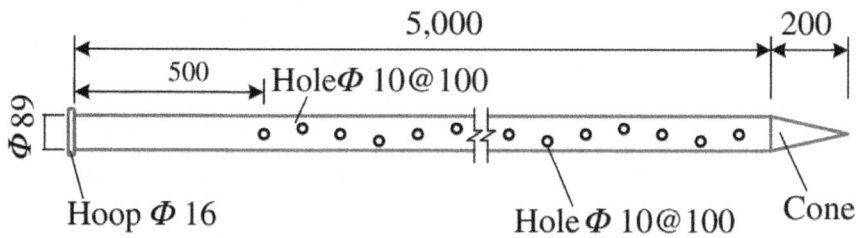

Figure. 14: Detail of grouting pipe (unit: mm).

In order to analyze the effect of cement grouting in strata, the friction angle of the slide plane will be raised after cement grouting in the surrounding rock with steel pipes from ground surface. This means that their value will be enhanced. If the computational friction angle of the surrounding rock rises to $\phi1=55°$ and $\phi2=50°\phi1=55°$ and $\phi2=50°$, then $\theta_1 = 38.5°$ and $\theta_2 = 35°$. If these parameters are substituted into Eqs. (8) and (22), then the maximum fracture angles are obtained as $\beta_0' = 68.47°$ and $\beta' = 69.51°$. By comparing this result with the angle of slide plane before grouting, we obtain their relations as follows

$\beta_0'=1.22\beta_0$, (47)

$\beta'=1.09\beta$. (48)

This apparently shows that the shear strength of the surrounding rock gets enhanced after the ground has been reinforced with cement grouting. A comparison of ground treatment effect is shown in Figure. 15. From Figure. 15 we can infer that the vertical and lateral rock pressure applied to the tunnel support decrease largely as well after the surrounding rock is treated with steel pipe grouting from ground surface.

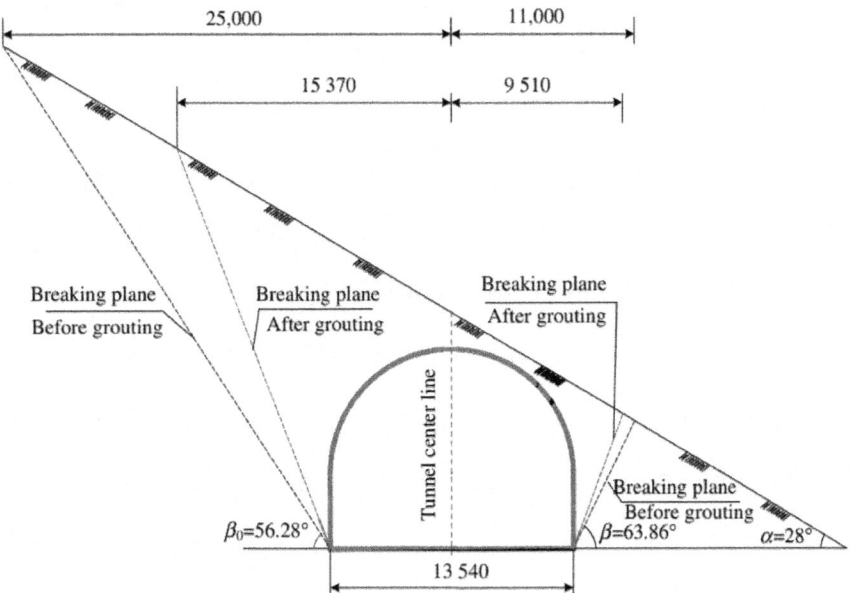

Figure. 15: Effect of ground treatment by cement grouting (unit: mm).

During the operation of in situ grouting, the grouting pressure must be kept within 1.0–1.5 MPa. In view of the in situ surface condition and ground geology at the tunnel site, the actual ground treatment and the details of designed parameters and grouting scope are shown in Figure. 8. Since the right tunnel more approaches to the ultra-shallow depth in the strata in comparison with the right one, ground treatment was only performed in its surrounding rock. As the left tunnel is deeply buried and the net distance between the two tunnels reaches 20 m, there is no need to treat the surrounding rock around the right tunnel.

CONSTRUCTION PROCEDURE OF THE ULTRA-SHALLOW TUNNEL

The surrounding rock around the two expressway tunnels falls into three grades in terms of its uniaxial compressive strength and lithological integrity, namely, grades IV, V, and VI according to its RQD [12, 13]. The main lithology comprises sub-clay and metamorphic sandstone. The upper strata belongs to diluvial silty subclay and eluvial subclay, the middle strata belongs to totally and strongly weathered metamorphic fine sandstone, and the lower strata exists in slightly weathered metamorphic sandstone. Furthermore, the maximum area of tunnel transection amounts to 123 m^2, and the maximum tunnel span reaches 13.54 m. Since the left tunnel has higher overburden depth than the right one, in order to reduce tunneling risk such as rock collapse and cave-in, the two tunnels are driven with the conventional drill and blast method. The excavation sequence of the surrounding rock inside tunnel contour is divided into four main steps. First, the left tunnel is excavated, and then the right one. The detailed construction process and its support system are both shown in Figure. 8. The whole working face of each tunnel is divided into two parts, first the upper part is headed, and then the lower part: The circled digits in Figure. 8 represent the tunneling sequence of the surrounding rock within the tunnel contour by conventional tunneling method [18–20]. The longitudinal profile of the tunnel construction process is shown in Figure. 16.

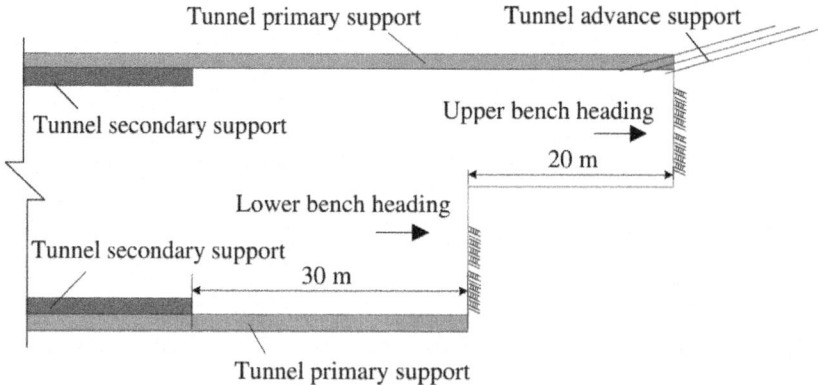

Figure. 16: Conventional tunneling process.

During the construction period, in situ ground settlement was measured in order to analyze the ground subsidence. The monitored final settlement of ground is shown in Figure. 17.

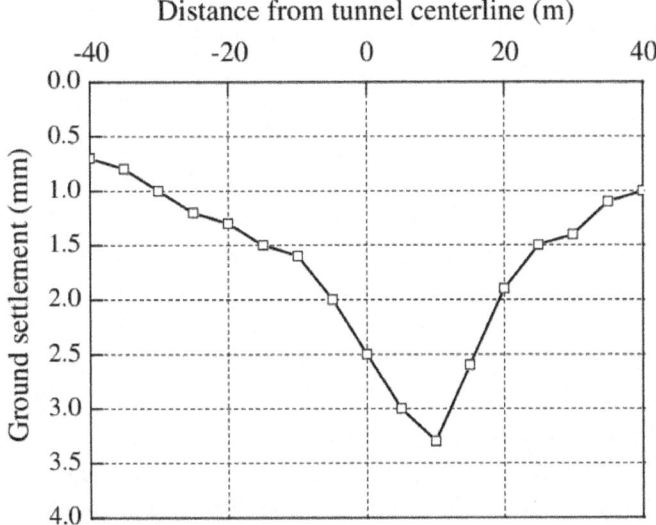

Figure. 17: Ground settlement due to tunnel excavation.

It is known from Figure. 17 that the maximum value of ground settlement reaches 3.7 mm; it does not occur at the center point above the tunnel crown, and just takes place at the point 5 m away from centerline. This phenomenon is principally caused by uneven pressure in the surrounding rock. It must be mentioned that the ground settlement was measured after the ground treatment had been completed. The measurement reflects that performing ground treatment and tunnel primary support has achieved a desired effect. Therefore, the ground treatment, tunnel support, and tunneling procedure are practicable.

Through the above technical scenario, the construction of Zagunao tunnel, a typical ultra-shallow expressway tunnel in Sichuan Province, has been successfully completed. During its whole construction process, no disaster or safety accident occurred. This, in turn, verified that the design and construction program for the ultra-shallow expressway tunnel was feasible; furthermore, the goal of safe and economical construction for the ultra-shallow tunnel was fulfilled.

CONCLUSIONS

This paper mainly deals with the calculation of rock pressure for ultra-shallow tunnel and its ground treatment, including its construction procedure. For this kind of tunnel, the rock pressure applied to the tunnel support is related to parameters such as ground slope angel α, rock fracture angle β, tunnel size B, R, H, and its depth h. After theoretical analysis on the calculation method of asymmetric rock pressure, the ground treatment of ultra-shallow tunnel and its tunneling procedure, some useful conclusions can be drawn as follows.

- The vertically downward load caused by gravity of the surrounding rock above the tunnel crown is much greater than the lateral load applied to tunnel side walls. When calculating the lateral pressure applied to super-shallow tunnel sidewalls, its value is mainly derived from the lateral surrounding rock, not from the surrounding rock above the tunnel crown. This is an innovative outcome in comparison with conventional methods.
- The rock pressure applied to the left sidewall is greater than that applied to the right one, and the lateral forces are not identical; therefore, the tunnel is subject to asymmetric pressure under ultra-shallow depth in topographically inclined strata.
- As for expressway tunnels, if there are two tunnels in a trunk line, and the external tunnel lies in an ultra-shallow condition, then its support should be designed according to the ultra-shallow condition, and the internal tunnel can be designed in terms of in situ geological condition. The construction method is determined in accordance with surrounding rock conditions. In order to control ground settlement, and keep safe tunneling, the priority of excavation should be given first to internal tunnel, and then to the external one.
- If surrounding rock is poor and unfavorable, it is necessary to consolidate the strata using techniques such as ground rockbolt, cement and mortar grouting, steel pipe grouting, jet grouting with high pressure, pile, ground freezing. Meanwhile, ground surface and water should also be treated with a proper method as well so as to keep tunneling safe.
- The structural design and construction technique for the ultra-shallow tunnel in an expressway in Sichuan province turns out to be a great success, and can be used to guide the design and construction of other super-shallow tunnels under similar situations.

ACKNOWLEDGMENTS

The work is financially supported by the National Natural Science Foundation of China (No. 51378436) and the Fundamental Research Funds for the Central Universities (SWJTU11ZT33).

REFERENCES

1. Goodman RE, Heuze HE, Bureau GJ (1997) On modeling techniques for the study of tunnels in jointed rock. 14th symposium on rock mechanics, pp 441–479
2. Shen B, Barton N (1997) The disturbed zone around tunnels in jointed rock masses. Int J Rock Mech Min Sci 34(1):117–125
3. Zhou XJ, Li ZL, Yang CY, Gao Y (2006) Rock pressure on tunnel with shallow depth in geologically inclined bedding strata. J Southwest Jiaotong Univ (English edition) 14(1):52–62
4. Zhou XJ, Yang CY (2007) Asymmetric rock pressure on shallow tunnel in strata with inclined ground surface. J Southwest Jiaotong Univ (English edition) 15(3):203–207
5. Yang XL, Jin QY, Ma J (2012) Pressure from surrounding rock of three tunnels with large section and small spacing. J Cent South Univ (English Edition) 19:2380–2385
6. He BG, Zhang ZQ, Chen Y (2012) Unsymmetrical load effect of geologically inclined bedding strata on tunnels of passenger dedicated lines. J Mod Transp 21(1):24–30
7. Zhou XJ (2011) Study on calculation of rock pressure and determination of depth for shallow asymmetric tunnel. Adv Mater Res 261–263:1034–1038
8. Zhou XJ, Gao B, Gao Y (2005) Safety study on tunnel with shallow depth in geologically oblique bedding strata. Prog Saf Sci Technol 5:869–875
9. Zhou XJ, Gao Y, Li ZL, Yang CY (2006) Experimental study on the uneven rock pressure and its distribution applied on a tunnel embedded in geologically bedding strata. Mod Tunn Technol 43(1):12–21 (in Chinese)
10. Bhawani Singh, Goel RK (2011) Engineering rockmass classification. Butterworth-Heinemann
11. Wu AQ, Liu FZ (2012) Advancement and application of the standard of engineering classification of rock masses. Chin J Rock Mech Eng 31(8):1513–1523 (in Chinese)
12. Hack R (1997) Rock mass strength by rock mass classification. South African rock engineering congress. Johannesburg, pp 346–356
13. Jaeger JC, Cook NGW, Zimmerman RW (2007) Fundamentals of rock mechanics, 4th edn. Blackwell Publishing Ltd, Oxford
14. Sheng MR (2000) Rock mass mechanics. Tongji University Press, Shanghai (in Chinese)

15. Vocational standard of the P.R.C. Code for design of highway tunnel (JTGD70-2004). China communication press, Beijing, 2004 (in Chinese)
16. Vocational standard of the P.R.C. Code for design of railway tunnel (TB10003-2005). China railway press, Beijing 2005 (in Chinese)
17. Goel RK, Bhawani Singh, Zhao J (2011) Underground infrastructures: planning, design and construction. Butterworth-Heinemann
18. Gioda G, Locatelli L (1999) Back Analysis of the measurements performed during the excavation of a shallow tunnel in sand. Int J Numer Anal Meth Geomech 23:1407–1425
19. Chehade FH, Shahrour I (2008) Numerical analysis of the interaction between twin tunnels: influence of the relative position and construction procedure. Tunn Undergr Space Technol 23:210–214
20. Miura K, Yagi H, Shiroma H, Takekuni K (2003) Study on design and construction method for the New Tomei-Meishin expressway tunnels. Tunn Undergr Space Technol 18:271–281

Chapter 5

A GENERALIZED PLASTICITY-BASED MODEL FOR SANDSTONE CONSIDERING TIME-DEPENDENT BEHAVIOR AND WETTING DETERIORATION

Meng-Chia Weng

Department of Civil and Environmental Engineering, National University of Kaohsiung, 700, Kaohsiung University Rd, Kaohsiung 81148, Taiwan, ROC

ABSTRACT

Based on the concept of generalized plasticity, this study proposes a constitutive model to describe the time-dependent behavior and wetting deterioration of sandstone. The proposed model (1) exhibits nonlinear elasticity under hydrostatic and shear loading, (2) follows the associated flow rule for viscoplastic deformation, (3) adopts a creep modulus that varies with the stress ratio, (4) considers the primary and secondary creep behaviors of rock, and (5) considers the effect of wetting deterioration. This model requires 13 material parameters, comprising 3 for elasticity, 7 for plasticity, and 3 for creep. All parameters can be determined easily by following the suggested procedures. The proposed model is first validated by comparison with triaxial tests of sandstone under different hydrostatic stress and cyclic loading conditions. In addition, the model is versatile in simulating time-dependent behavior through a series of multistage creep tests. Finally, to consider the effects of wetting deterioration, triaxial and creep tests under dry and water-saturated conditions are simulated. Comparison of the simulated and experimental data shows that the proposed model can predict the behavior of sandstone in dry and saturated conditions.

INTRODUCTION

The theory of generalized plasticity was first introduced by Zienkiewicz and Mroz (1984) to simulate soil behavior and was later elaborated by Pastor and

Zienkiewicz (1986) and Pastor et al. (1990). In contrast to other plastic models, this theory does not explicitly define the yield and plastic potential surfaces. Instead, it adopts the gradients of these functions so that simple models within this framework can consider material behavior responses under loading. Generalized plasticity considers plastic deformation at any stress level for stress increments in both loading and unloading conditions. These features enable the generalized plasticity model to predict the stress–strain behavior of numerous soil types with good accuracy under various types of loading. Researchers have recently developed various constitutive relationships based on this framework to describe sophisticated features encompassing soil behavior, including anisotropy (Pastor1991; Pastor et al. 1992), unsaturated conditions (Bolzon et al. 1996; Manzanal et al. 2011b), degradation phenomena (Fernandez Merodo et al. 2004), and the effects of stress levels and densification on sand (Ling and Liu 2003; Ling and Yang 2006; Manzanal et al. 2006, 2011a). Notably, the transformation of this theory from the defining space to general Cartesian stress space is one of the key steps in extending it to computational implementations (Chan et al. 1988).

Other than soil, Weng and Ling (2012) adopted the generalized plasticity concept to investigate nonlinear elasticity behavior in rock. The proposed model produces reasonable predictions of the elastoplastic deformation of sandstone under varying stress paths, cyclic loadings, and postpeak behavior. In addition to simulating immediate rock deformation, the time-dependent deformation (i.e., creep deformation) of rock is a major concern in engineering practice (Cristescu 1989; Hoxha et al. 2005; Tomanovic 2006; Xie and Shao2006; Sterpi and Gioda 2009; Weng et al. 2010a, b). According to previous studies on creep deformation of sandstone (Tsai et al. 2008; Weng et al. 2010a), viscoplastic flows indicate that the viscoplastic potential surface has a similar shape to the plastic potential surface, but the size of the viscoplastic potential surface changes with time. The plastic potential surface has a time-independent size. Meanwhile, through calculation of the irreversible work, direct evidence of orthogonality between the yield surface and the plastic flow, as well as the viscoplastic flow, has been observed. Thus, it is reasonable to state that the yield surface, plastic potential, and viscoplastic potential all have the same geometry. Consequently, the associated flow rules are applicable for modeling the time-dependent deformational behavior of sandstone.

Based on these characteristics of sandstone, this study extends the work of Weng and Ling (2012) to develop an elastic–viscoplastic model that incorporates the generalized plasticity concept. This study also presents an assessment of the validity of the proposed model by comparing simulated and actual deformations in various multistage creep tests. Moreover, the strength and stiffness of sandstone are significantly reduced because of the wetting process (Dyke and Dobereiner 1991; Hawkins and McConnell 1992; Jeng et al. 2004). This phenomenon commonly occurs in sandstone of medium to moderate strength. To evaluate the performance of the proposed model for the wetting deterioration of sandstone, this study employs the proposed model to simulate triaxial and creep tests of deformational behaviors under dry and water-saturated conditions.

MODEL CONCEPT

Based on the concept of generalized plasticity, the total strain increment can be divided into elastic and plastic components as follows:

$$d\varepsilon = d\varepsilon^e + d\varepsilon^p, \tag{1}$$

where $d\varepsilon$, $d\varepsilon^e$, and $d\varepsilon^p$ are the increments of the total, elastic, and plastic strain tensors, respectively.

The elastic and plastic strain increments can be obtained from

$$d\varepsilon^e = \boldsymbol{C}^e : d\sigma \tag{2}$$

and

$$d\varepsilon^p = d\lambda \boldsymbol{n}_g = \frac{1}{H_{L/U}} \left(\boldsymbol{n}_{gL/U} \otimes \boldsymbol{n}\right) : d\sigma, \tag{3}$$

where C^e is the elastic constitutive tensor, $d\sigma$ is the increment of the stress tensor, n_g is the unit vector defining the plastic flow direction, n represents the loading-direction vector, $d\lambda$ is a plastic scalar, and HL/UHL/U is the plastic modulus, which can be assumed directly without introducing a hardening rule. Subscripts "L" and "U" indicate loading and unloading, respectively.

To consider time-dependent deformation, the plastic strain increment $d\varepsilon^p$ can be substituted by the viscoplastic strain increment $d\varepsilon^{vp}$. Equation (3) can then be modified to

$$d\varepsilon^{vp} = \frac{1}{H_{L/U}} \left(\boldsymbol{n}_{gL/U} \otimes \boldsymbol{n}\right) : d\sigma + \frac{G(t)}{H_c} \left(\boldsymbol{n}_c \otimes \boldsymbol{n}\right) : d\sigma, \tag{4}$$

where H_c is the creep modulus, $G(t)$ is a time-dependent function, and n_c is the viscoplastic flow vector. The concept employed in Eq. 4 is similar to the viscoelastic model in rheology. The first term $\frac{1}{H_{L/U}}\left(\boldsymbol{n}_{gL/U} \otimes \boldsymbol{n}\right) : d\sigma$ represents instantaneous deformation, whereas the second term $\frac{G(t)}{H_c}\left(\boldsymbol{n}_c \otimes \boldsymbol{n}\right) : d\sigma$ corresponds to long-term deformation, including the primary and secondary creep behavior of the rock.

Based on this concept of generalized plasticity, the yield and viscoplastic potential surfaces are not directly specified, but the scalar functions for plastic modulus $H_{L/U}$, creep modulus H_c, and direction tensors ***n***, ***n**$_g$*, and ***n**$_c$* are required. To incorporate the deformation characteristics of sandstone into the generalized plasticity, this study proposes (and subsequently defines) the major constituents of the model, including nonlinear elasticity, dilatancy, plastic modulus, and a time-dependent function.

Nonlinear Elastic Behavior

According to hyperelasticity theory, the strain tensor is related to the derivatives of the energy density function as follows:

$$\varepsilon^e = \frac{\partial \Omega}{\partial \sigma}, \tag{5}$$

where Ω is the energy density function. Based on experimental sandstone results, this study adopts the following energy density function for Ω, which has been proposed by previous studies (Weng et al. 2010a; Weng and Ling 2012):

$$\Omega = b_1 I_1^{3/2} + b_2 I_1^{-1} J_2 + b_3 J_2, \tag{6}$$

where b_1, b_2, and b_3 are material parameters, I_1 is the first stress invariant ($I_1 = \sigma_{kk} = 3p'$), and J_2 is the second deviatoric stress invariant ($J_2 = \frac{1}{2} s_{ij} s_{ji}$, where s_{ij} is the deviatoric stress tensor). After substituting Eq. (6) into Eq. (5), the elastic strain tensor ε^e_{ij} takes the following form:

$$\varepsilon^e_{ij} = \frac{\partial \Omega}{\partial \sigma_{ij}}$$
$$= \left({}^3\!/\!{}_2 b_1 I_1^{1/2} - b_2 I_1^{-2} J_2 + J_2 \right) \delta_{ij} + \left(b_2 I_1^{-1} + b_3 \right) s_{ij}, \tag{7}$$

where δ_{ij} is the Kronecker delta tensor.

Equation (7) shows that the increment of the elastic strain tensor $d\varepsilon e_{ij}$ is

$$d\varepsilon^e_{ij} = \frac{\partial^2 \Omega}{\partial \sigma_{ij} \partial \sigma_{kl}} d\sigma_{kl} = C^e_{ijkl} d\sigma_{kl}, \tag{8}$$

$$d\varepsilon^e_{ij} = \left[\Phi_1 \delta_{ij} \delta_{kl} + \Phi_2 \delta_{ij} s_{kl} + \Phi_3 \delta_{ik} \delta_{jl} \right.$$
$$\left. - \frac{1}{3} \Phi_3 \delta_{ij} \delta_{kl} + \Phi_2 \delta_{kl} s_{ij} \right] d\sigma_{kl},$$

where $\Phi_1 = {}^3\!/\!{}_4 b_1 I_1^{-1/2} + 2 b_2 I_1^{-3} J_2$, $\Phi_2 = -b_2 I_1^{-2}$ and $\Phi_3 = b_2 I_1^{-1} + b_3$. Equations (7) and (8) are derived from rigorous elastic theory, and satisfy the principle of thermodynamics, which indicates that energy is conserved during any type of loading. A similar relationship based on hyperelasticity was also proposed by Houlsby et al. (2005); they adopted a power function of the stress to describe the nonlinear elastic stiffness of soil. Mira et al. (2009) combined the work of Houlsby et al. (2005) and generalized plasticity to successfully predict soil behavior under cyclic loading.

Based on Eq. (7), the elastic strain induced in the shear-loading stage can be calculated as shown in Eqs. (9) and (10):

$$\varepsilon_v^e = -3b_2\left(\sqrt{J_2}/I_1\right)^2, \qquad (9)$$

$$\gamma^e = \frac{2}{\sqrt{3}}\left(b_2 I_1^{-1} + b_3\right)\sqrt{J_2}. \qquad (10)$$

Equations (9) and (10) show two features of the proposed model: (1) shear loading induces elastic dilative deformation, and (2) the elastic shear stiffness increases with the application of increasing hydrostatic pressure. In addition, greater values for the parameters b_1, b_2, and b_3 indicate that increasing elastic strain is generated by the model.

Dilatancy and Viscoplastic Flow

For stress–dilatancy relationships, this study adopts a function similar to that of Pastor et al. (1990), relating the dilatancy dg and stress ratio η as follows:

$$d_g = \frac{d\varepsilon_v^p}{d\gamma^p} = (1+\alpha)(M_g - \eta), \qquad (11)$$

where $d\varepsilon_v^p$ and $d\gamma^p$ are the incremental plastic volumetric and shear strain, respectively. The term Mg is the threshold of shear dilation in the triaxial plane. When η=Mg, dg equals zero and volumetric strain does not occur. The sandstone converts from compression to dilation when η>Mg. α is a model parameter.

Based on the definition by Weng and Ling (2012), the stress ratio η here is defined as

$$\eta = q/q_f, \qquad (12)$$

where $q = \sqrt{3J_2}$ and qf is the shear strength. The linear strength criterion, known as the Drucker–Prager criterion, is adopted as follows:

$$q_f = \sqrt{3J_{2f}} = \sqrt{3}(\alpha_d I_1 + k_d) \qquad (13)$$

where the parameters α_d and k_d are the slope and cohesive intercept of the failure envelope, respectively. If the shear strength exhibits a nonlinear failure envelope, use of the Hoek–Brown criterion (Hoek and Brown1980) for rock is recommended.

To further investigate the variation of dilatancy with the stress ratio, the actual behavior of sandstone was compared with the proposed model. The plastic flow angle β1β1 is defined by

$$\tan \beta_1 = \frac{d\gamma^p}{d\varepsilon_v^p} = \frac{1}{d_g}. \qquad (14)$$

When β_1 ranges from 0° to 90°, $d\varepsilon^p_v$ indicates compression. Conversely, when β_1 is >90°, $d\varepsilon^p_v$ is dilative. Figure 1 shows the typical variation of the plastic angle β_1 with the stress ratio. At low stress ratio, the plastic angle β_1 is smaller than 90°, and gradually increases with the shear stress ratio. When the stress ratio η equals Mg, β_1 becomes 90° and the volumetric deformation begins to dilate. In addition to Mg, the parameter α affects the slope of the proposed model; as α increases, β_1 becomes flatter.

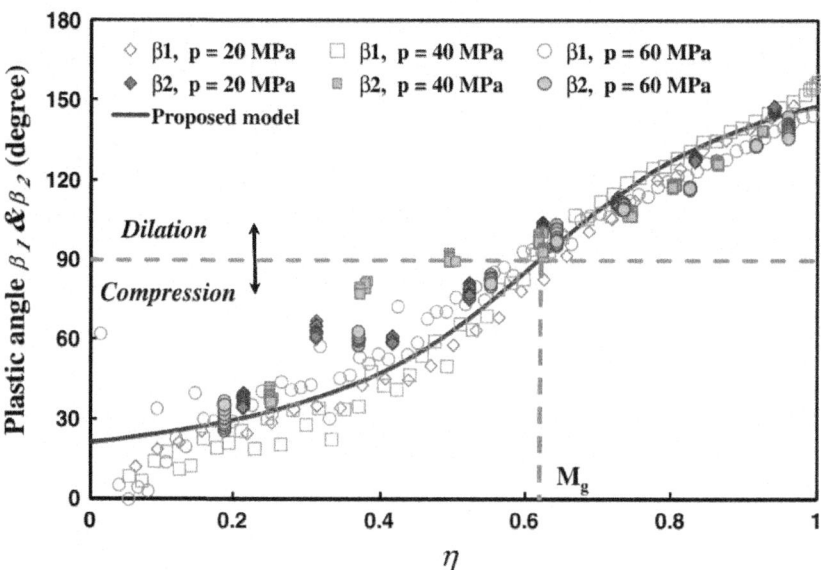

Figure. 1: Variations of the plastic flow angle ₁ and viscoplastic flow angle ₂ under shear loading with different confining pressures. is defined as q/q_f and serves as an index of the degree of shear loading. Dilation occurs when is >90°.

Furthermore, the viscoplastic flow angle β_2 is defined as

$$\tan \beta_2 = \frac{d\gamma^{vp}_{t_0 \to t}}{d\varepsilon^{vp}_{v(t_0 \to t)}} \qquad (15)$$

where $d\gamma^{vp}_{t_0 \to t}$ is the creep shear-strain increment from time t0 to t, $d\varepsilon^{vp}_{v(t_0 \to t)}$ is the creep volumetric-strain increment from time t_0 to t, and t_0 is the initiation time of one particular loading step in a sequence of loading steps during testing under either increasing or constant loading (creep test condition). The time t is an arbitrary time after creep begins. Figure 1 shows the variation in the viscoplastic flow angle β_2 with the stress ratio η. The tendency of the viscoplastic flow angle β_2 is relatively consistent with that of the plastic flow angle β_1 (Figure. 1), indicating that the viscoplastic flow vector is likely the same as the plastic flow (i.e., $n_c = n_g$). In addition, the proposed model reasonably simulates these two variations. Based on this assumption, Eq. (4) can be modified to

$$d\varepsilon^{vp} = \frac{1}{H_{L/U}} \left(n_{gL/U} \otimes n \right) : d\sigma + \frac{G(t)}{H_c} \left(n_{gL/U} \otimes n \right) :$$

$$d\sigma = \frac{1}{H_{L/U}} \left(1 + \frac{H_{L/U}}{H_c} G(t) \right) \left(n_{gL/U} \otimes n \right) :$$

$$d\sigma = \frac{1}{H_{vp}} \left(n_{gL/U} \otimes n \right) : d\sigma, \tag{16}$$

where H_{vp} is the viscoplastic modulus. To express the stress increments as a function of strain increments, Eq. (16) is inverted and expressed as

$$d\sigma = D^{evp} : d\varepsilon,$$

$$D^{evp} = D^e - \frac{D^e : n_{gL/U} \otimes n : D^e}{H_{vp} + n : D^e : n_{gL/U}}, \tag{17}$$

where D^{evp} and D^e are the elasto–viscoplastic and elastic tensors, respectively.

According to Pastor et al. (1990), the viscoplastic flow direction under loading and unloading $n_{gL/U}$ in triaxial space is

$$n_{gL/U} = \left(\frac{d_g}{\sqrt{1 + d_g^2}}, \frac{1}{\sqrt{1 + d_g^2}} \right)^T. \tag{18}$$

Similarly, the loading-direction vector can be expressed as

$$n = \left(\frac{d_f}{\sqrt{1 + d_f^2}}, \frac{1}{\sqrt{1 + d_f^2}} \right)^T, \tag{19}$$

where $d_f = ((1 + \alpha))(M_f - \eta)$ and Mf is a material parameter.

According to Jeng et al. (2002), Weng et al. (2005), and Tsai et al. (2008), triaxial results show that the plastic potential surface of sandstone coincides with the yield surface in the prepeak stage. Therefore, the associated flow rule, n=$n_{gL/U}$ and M_f=M_g, can be used when formulating the constitutive model for sandstone. However, n should be specified differently from $n_{gL/U}$, and the nonassociated flow rule should be followed if the softening (postpeak) behavior is considered (Weng and Ling 2012).

Plastic Modulus for Loading and Unloading

Figure 2 shows the variation of the plastic modulus of sandstone at various stages of shear loading. This Figure indicates that the plastic modulus decreases as the stress ratio rises. In particular, the modulus decreases by approximately three orders of magnitude when approaching the failure state (η=0.8−1). Based on this tendency, the function of the plastic modulus for sandstone under loading can be expressed as

$$H_L = H_0 \sqrt{p'/p_{atm}} H_f H_s, \qquad (20)$$

$$H_f = (1 - \eta^2), \qquad (21)$$

$$H_s = \exp(-\beta_0 \xi_s), \qquad (22)$$

where H0H0 is a multiplication factor related to the initial plastic modulus, H_f and H_s are plastic coefficients, p_{atm} is the atmospheric pressure, β_0 is a material parameter, and $\xi_s = \int |d\gamma^p| = \int d\xi_s$ is the accumulated plastic shear strain.

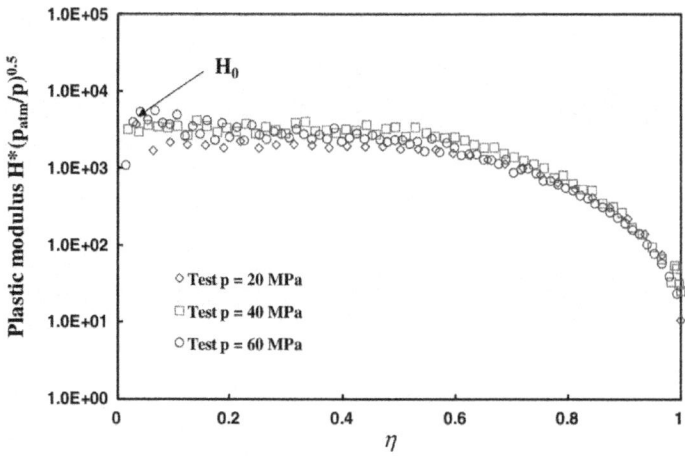

Figure. 2: Variation of the normalized plastic modulus $H_L \sqrt{p_{atm}/p'}$ under shear loading with different confining pressures.

The original model suggests that plastic strain also occurs during the unloading process; the unloading plastic modulus H_U can be expressed as

$$H_U = H_{U0}, \qquad (23)$$

where H_{U0} is a material parameter.

Creep Modulus for Time-Dependent Behavior

For time-dependent deformation, Figure. 3 shows the variation of the creep modulus of sandstone for various stress ratios. This Figure shows variations similar to that of the plastic modulus (Figure. 2). Thus, the function of the creep modulus for sandstone can be written as

$$H_c = H_{c0} \sqrt{p'/p_{atm}} H_f' H_s, \qquad (24)$$

$$H_f' = (1 - \eta)^2, \qquad (25)$$

where Hc0 is a factor related to the initial creep modulus. Considering the primary and secondary creep behaviors of rock (Goodman 1989), this study proposes the following time-dependent function:

$$G(t) = 1 - \exp[c_1(t - t_0)] + c_2 \eta (t - t_0), \qquad (26)$$

where c_1 and c_2 are material parameters. The term $1-\exp[-c_1(t-t_0)]$ is adopted to describe the primary creep behavior, while the term $c_2\eta(t-t_0)$ corresponds to secondary creep deformation. As η increases, the secondary creep deformation develops more rapidly.

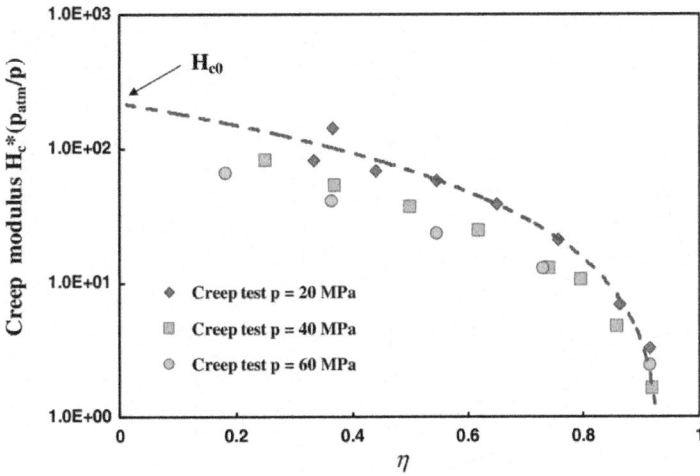

Figure. 3: Variation of the normalized creep modulus $H_c(p_{atm}/p')$ under shear loading with different confining pressures.

PARAMETER DETERMINATION

There are a total of 13 material parameters (b_1, b_2, b_3, α_d, k_d, M_g, α, H_0, β_0, H_{U0}, H_{c0}, c_1, c_2) to be determined from experimental results. The influence of these parameters on the deformation behavior is relatively straightforward. The parameters b_1, b_2, and b_3 are elastic parameters; $\alpha_d \alpha d$ and kd are strength parameters; M_g and α are parameters related to the stress–dilatancy relationship; H_0 and β_0 are parameters representing the variation of the loading plastic modulus; H_{U0} is a parameter related to the unloading plastic modulus; and H_{c0}, c_1, and c_2 are time-dependent parameters. To obtain the values of these parameters, it is recommended to conduct three triaxial tests with various hydrostatic pressures and one multistage creep test. The difference between the hydrostatic pressures should be sufficiently great to encompass the range of stress levels of interest. Furthermore, the test with the medium hydrostatic pressure should be conducted with multiple unloading–reloading procedures to distinguish and separate elastic deformation from total deformation. The creep test should involve at least three shear-loading stages.

The following section demonstrates how these parameters can be determined from laboratory experiments using a sample of Mushan sandstone (MS), a weak rock commonly found in mountainous areas of northern Taiwan. The porosity of the sampled specimen, denoted as MS-A, is ~14.1 %, and the dry density is ~2.28 g/cm³. The average uniaxial compressive strength is 37.1 MPa in dry conditions and 28.9 MPa in saturated conditions. Petrographic analysis shows that the percentages of grains, matrices, and voids are 59.9, 26.0, and 14.1 %, respectively. The average grain diameter is ~0.24 mm. Mineralogically, MS-A sandstone consists of 90.7 % quartz and 9.0 % rock fragments, and is classified as lithic greywacke (Weng et al. 2008). The following subsections describe the determination of each parameter.

- Elastic Parameters b_1, b_2, and b_3

The parameter b_1 controls the interaction between hydrostatic stress and elastic volumetric strain. This parameter can be determined by fitting the elastic volumetric unloading–reloading regression curve. The parameter b_2 is related to the coupling between shear stress and elastic dilation, and can be obtained by fitting the unloading–reloading regression curve with normalized stress ($\sqrt{J_2}/I_1$) using Eq. (9). After obtaining the parameter b_2, the parameter b_3 can be obtained by fitting the shear stress and elastic shear strain curve using Eq. (10).

- Strength Parameters α_d and k_d

The parameters α_d and k_d can be determined by fitting the failure envelope of sandstone using Eq. (13).

- Stress–Dilatancy Parameters Mg and α

The term Mg is determined by the threshold of shear dilation in a diagram of plastic flow angle versus stress ratio (Figure. 1). Alternatively, this parameter can be obtained directly from the plastic volumetric strain curve under shear loading. The parameter α is determined by fitting the curve of plastic flow angle versus stress ratio (Figure. 1). Another method for obtaining α is from the slope of the graph between dilatancy d_g and (Mg/η).

- Loading Plastic Modulus Parameters H_0 and β_0

The initial plastic modulus parameter H_0 can be determined based on the initial stress ratio in a diagram of the normalized plastic modulus $H_L\sqrt{p_{atm}/p'}$ versus the stress ratio (Figure. 2) or by fitting the initial slope of both the plastic shear and volumetric strain curves under shear loading. In this study, the plastic strain is the unrecoverable deformation under short-term loading (a few minutes). The parameter β_0 controls the degree of plastic modulus degradation; a higher β_0 value induces further significant modulus degradation. This parameter can be determined by matching the shear or volumetric strain curve as the stress ratio η approaches 1.

- Unloading Plastic Modulus Parameter H_{U0}

The parameter H_{U0} can be determined by matching the slope of the unloading curves.

- Time-Dependent Parameters H_{c0}, c_1, and c_2

The initial creep modulus parameter H_{c0} can be determined at the initial stress ratio in a diagram of the normalized creep modulus $H_c\sqrt{p_{atm}/p'}$ versus the stress ratio (Figure. 3). The parameter c_1 controls the retardation time during the primary creep behavior; as c_1 increases, the primary creep deformation develops more rapidly. Moreover, the parameter c_2 influences the secondary creep behavior; as c_2 becomes greater, the secondary creep deformation increases more rapidly. The influence of c_1 and c_2 on the creep strain is shown in Figure. 4. The parameter c_1 can be determined by matching the initial slope of the creep strain versus time curve, and the parameter c_2 can be determined by matching the final slope of the creep strain versus time curve.

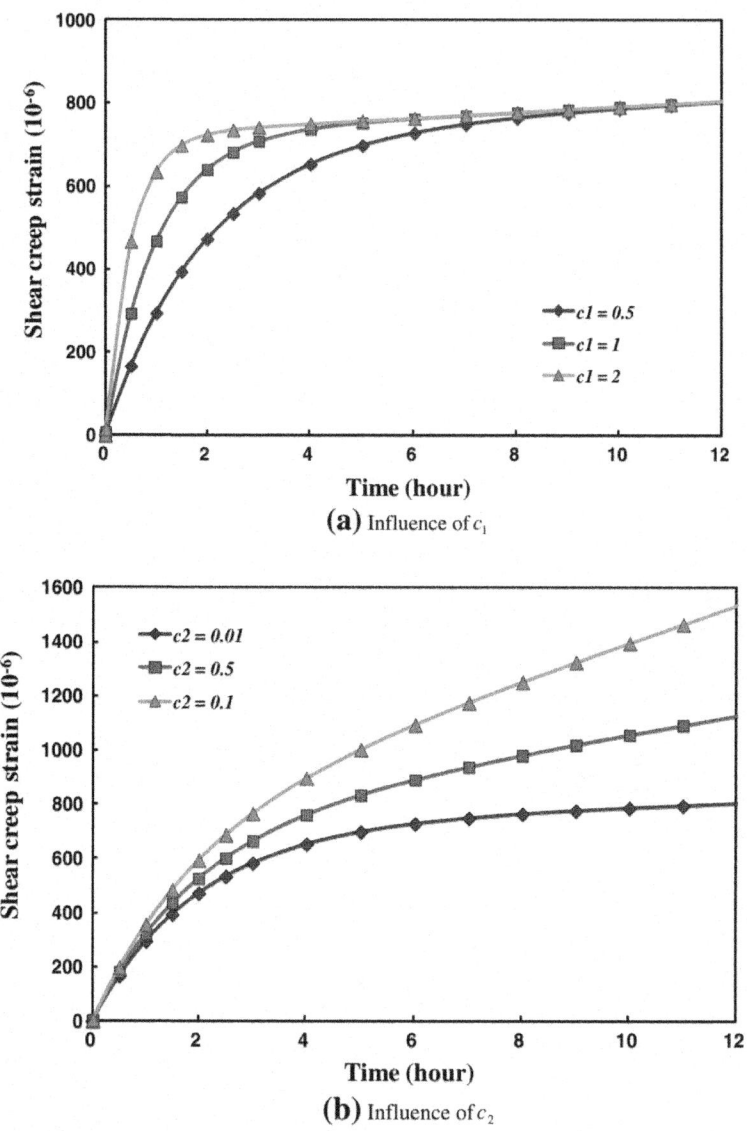

Figure. 4: Influence of the material parameters c_1 and c_2 on the simulated creep deformation based on the proposed model.

Using these procedures, the corresponding material parameters of MS-A sandstone in dry conditions can be obtained from one triaxial test with a pure shear stress path (PS test) and one creep test, both under hydrostatic pressure of 40 MPa. Table 1 presents a summary of these parameter values.

Table 1: Material parameters for different types of sandstone used in the proposed model

Model property	Parameter	MS-A sandstone	MS-B sandstone (dry)	MS-B sandstone (saturated)
Elastic component	$b_1(MPa)^{-1/2}$	130×10^{-6}	173×10^{-6}	215×10^{-6}
	b_2	$1{,}463 \times 10^{-6}$	$2{,}200 \times 10^{-6}$	$3{,}300 \times 10^{-6}$
	$b_3(MPa)^{-1/2}$	29×10^{-6}	35×10^{-6}	50×10^{-6}
Failure criterion	α_d	0.39	0.35	0.32
	$k_d(MPa)$	8.2	9.71	9.06
Plastic component	M_g	0.62	0.61	0.65
	α	2.9	2.0	4.0
	$H_0(MPa)$	4,590	3,067	2,530
	β_0	120	400	410
	$H_{U0}(MPa)$	180,000		
Time-dependent component	$H_c(MPa)$	540	750	125
	$c_1(h)^{-1}$	0.97	2.5	2.5
	$c_2(h)^{-1/2}$	0.012	0.02	0.018

MODEL VALIDATION ON IMMEDIATE DEFORMATION

To assess the validity of the proposed model, this section shows how the proposed model can simulate the immediate deformation behavior under various hydrostatic stress and cyclic loadings.

Five PS tests of MS-A sandstone with hydrostatic stress ranging from 20 to 60 MPa were simulated. Figure 5a shows the measured and simulated shear-induced shear strains under various hydrostatic stress conditions, exhibiting satisfactory agreement. Figure 5b shows the measured and simulated shear-induced volumetric strains under varying hydrostatic stress conditions. These simulated results are consistent with the measured results.

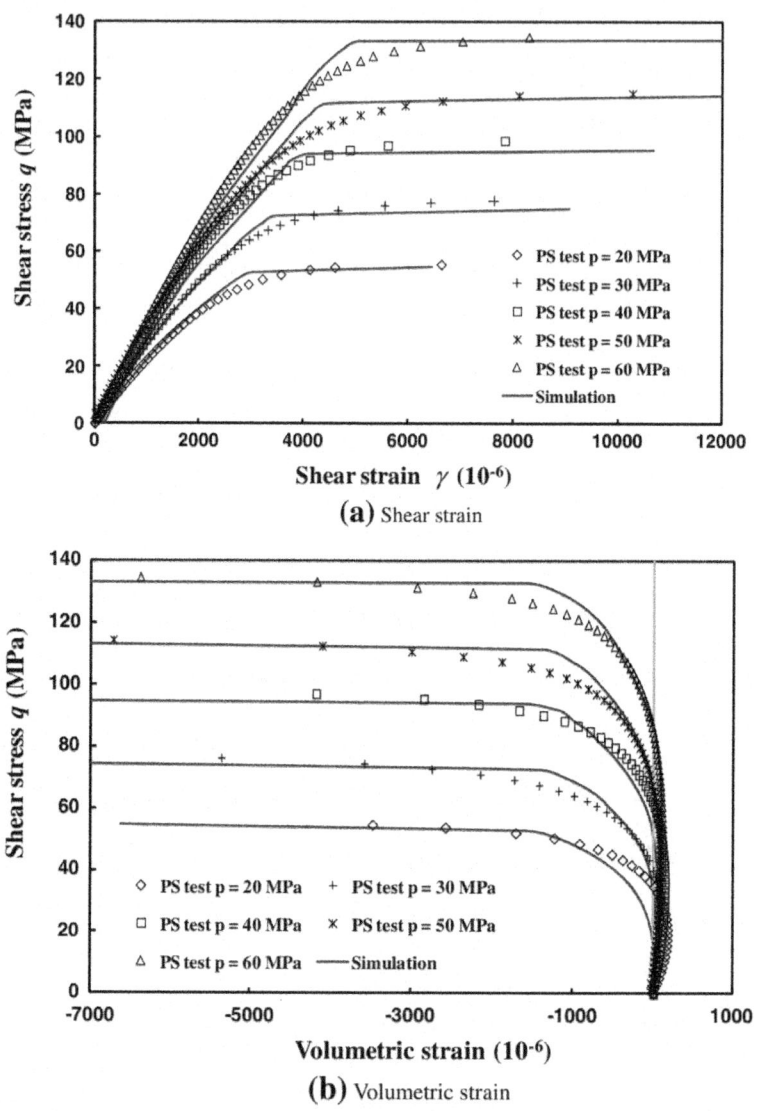

Figure. 5: Simulation of stress–strain relationships on dry MS-A sandstone under different hydrostatic pressures.

This study also used the proposed model to simulate three stress-controlled unloading–reloading cycles in a PS test. Figure 6 shows a comparison of the simulated and experimental results under constant hydrostatic stress of 60 MPa. Although the unloading–reloading-induced deformations are not large in either the shear or volumetric strain, the proposed model can simulate the cyclic behavior satisfactorily.

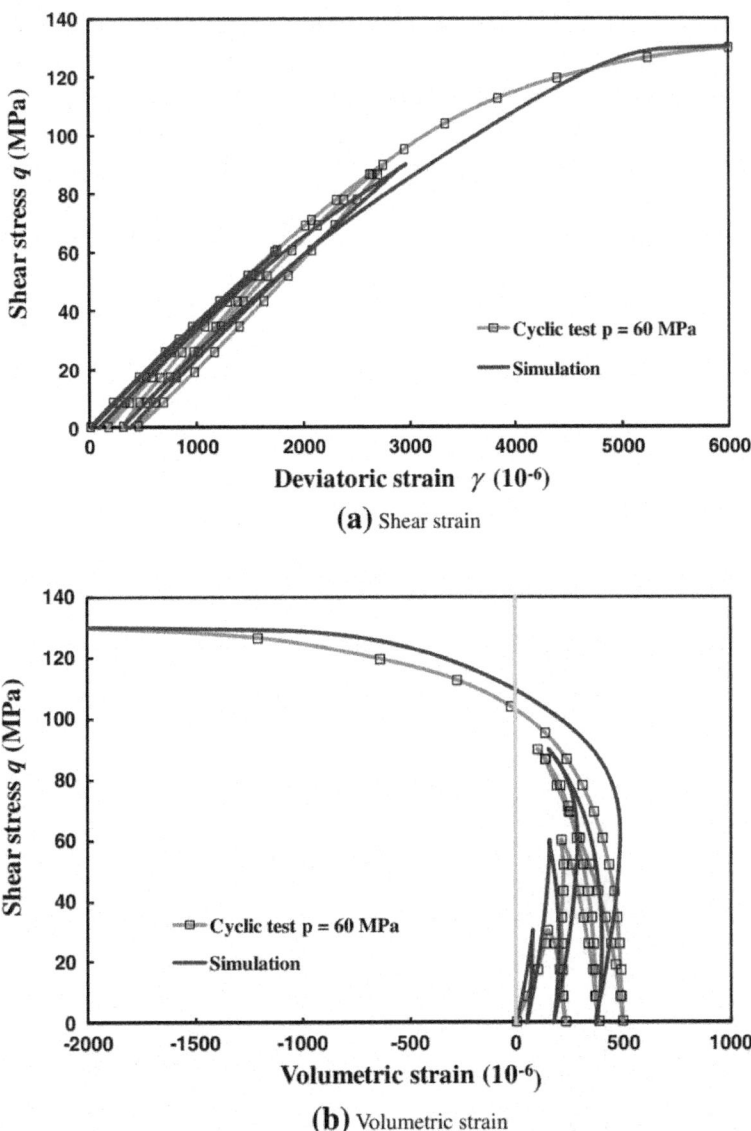

Figure. 6: Simulation of loading–unloading–reloading behavior under hydrostatic stress of 60 MPa.

This study also validated the proposed model by comparing with triaxial test results for different stress paths using two other sandstone samples. Weng and Ling (2012) provided additional details regarding these simulations. The comparisons showed that the proposed model satisfactorily captures the instantaneous deformation of sandstone.

VALIDATION OF TIME-DEPENDENT BEHAVIOR

This section describes simulations of a series of multistage creep tests to further assess the validity of the proposed model regarding time-dependent behavior. First, this section presents a multistage, long-term creep experiment under hydrostatic stress of 40 MPa (Figure. 7). This experiment has five stages of sustained loading with stress ratio η ranging from 0.37 to 0.92. The comparison of the simulated results with the actual creep behavior of the studied sandstone is as follows:

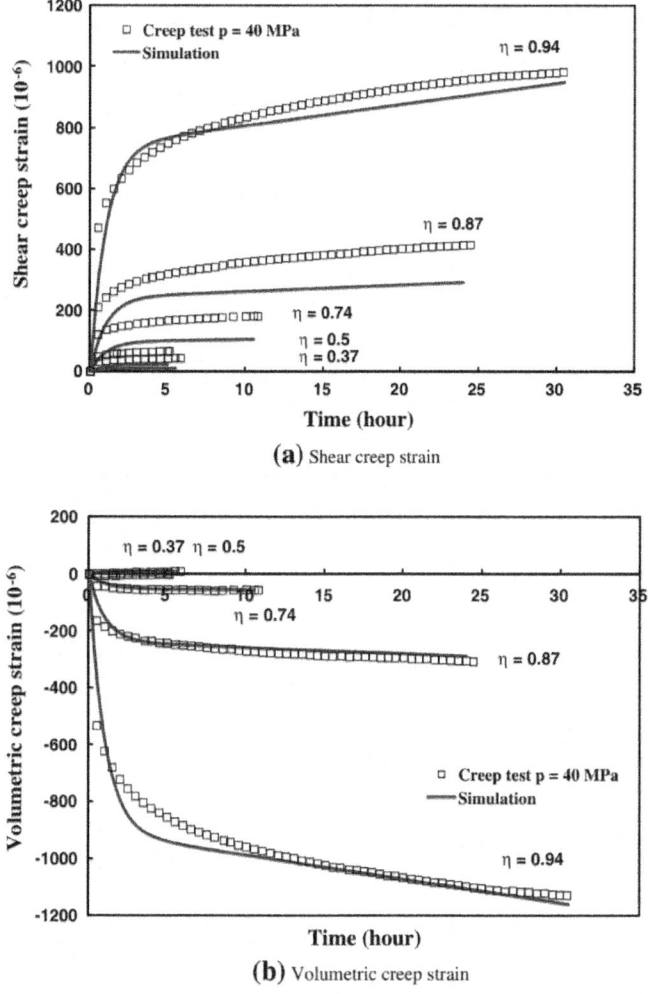

(a) Shear creep strain

(b) Volumetric creep strain

Figure. 7: Comparison of volumetric creep strain and shear creep strain predicted by the proposed model and data obtained from multistage creep tests under hydrostatic stress of 40 MPa.

Figure 7 shows the creep strain predicted by the proposed model and data obtained from the multistage creep test. Table 1 presents the parameter values required for the proposed model.

Regarding the material behavior during shearing, Figure. 7a and b show the shear and volumetric strain versus time during the creep stage, respectively. These Figures show that a higher stress ratio increases the magnitude of the creep-induced shear strain (Figure. 7a). Figure 7b shows that the simulated volume contracts under lower shear stress and converts to dilative behavior with an increasing shear-stress ratio. The behaviors of the material can be simulated by the proposed constitutive model. Although a minor discrepancy between the simulated and actual behavior occurs in the primary creep deformation, which is attributable to the creep modulus function in Eq. (24) underestimating the creep strain for low stress ratios, the simulated secondary creep deformation is consistent with the experimental results. Figure 8 shows an additional comparison of the simulated and actual behavior of the total deformation induced by a multistage creep experiment. This Figure shows the strain induced by shearing, including the total strain, elastic component, and viscoplastic component. Regarding the elastic component of deformation, the simulations are consistent with the actual results. In addition, viscoplastic volumetric strain is induced by increasing either the shearing or the creep under constant shear loading (Figure. 8b). To increase the shearing, the material first undergoes shear contraction and gradually transitions to shear dilation. Similarly, for creep deformation under constant shear stress, the material transitions from initially contracting to dilating. The proposed constitutive model captures all these material behaviors well.

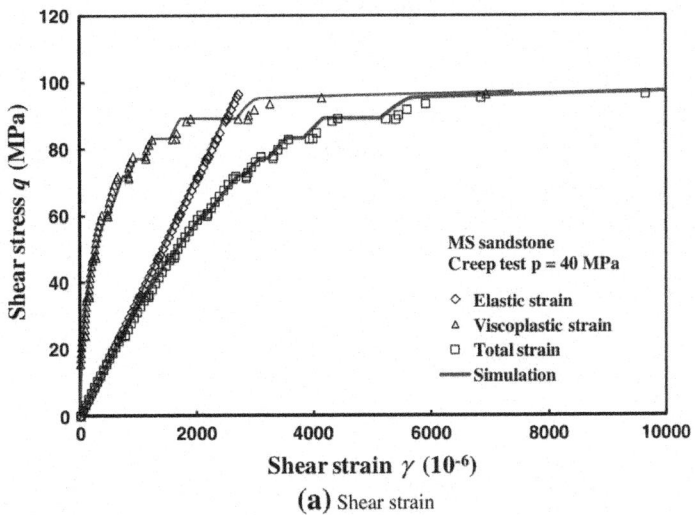

(a) Shear strain

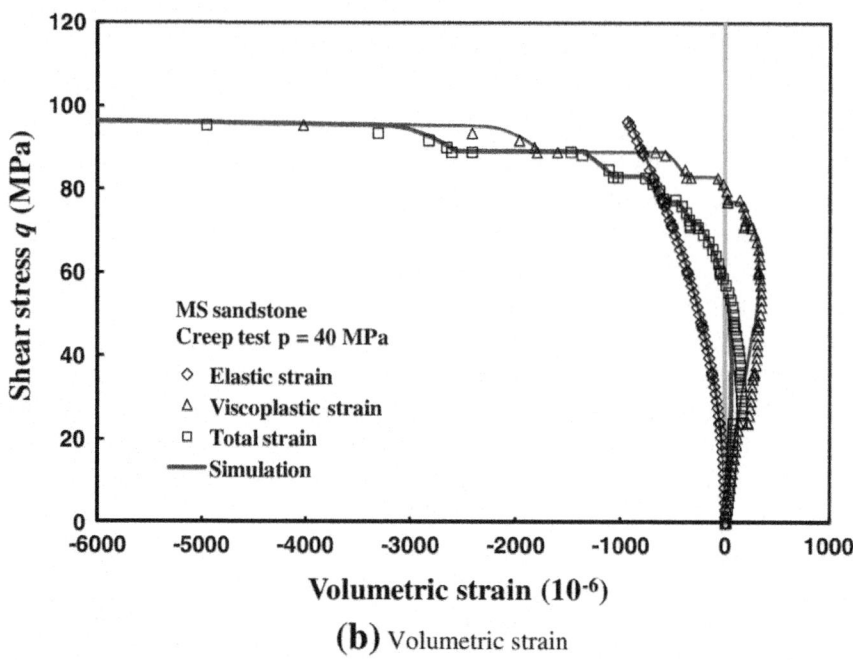

(b) Volumetric strain

Figure. 8: Simulation of the total strain in a multistage creep test, including the decomposed elastic and viscoplastic components inuced by the shear stress.

To further evaluate the predictive capability of the proposed model under different hydrostatic stresses, two additional creep tests at hydrostatic stresses of 20 and 60 MPa were simulated based on the same set of parameters presented in Table 1. Figures 9 and 10 show the simulated creep strains under different stress ratios η from 0.18 to 0.92. Comparison of the simulated results in Figures. 7, 9, and 10 indicates that the creep strain increases as the shear stress increases, but decreases under higher hydrostatic stress; such comparison also shows that the influence of the shear stress on the creep behavior is more significant than the influence of the hydrostatic stress on the creep behavior. The total deformation in the three multistage creep tests was simulated and is presented in Figure. 11. This Figure shows that the proposed model is capable of providing reasonable simulations under various situations.

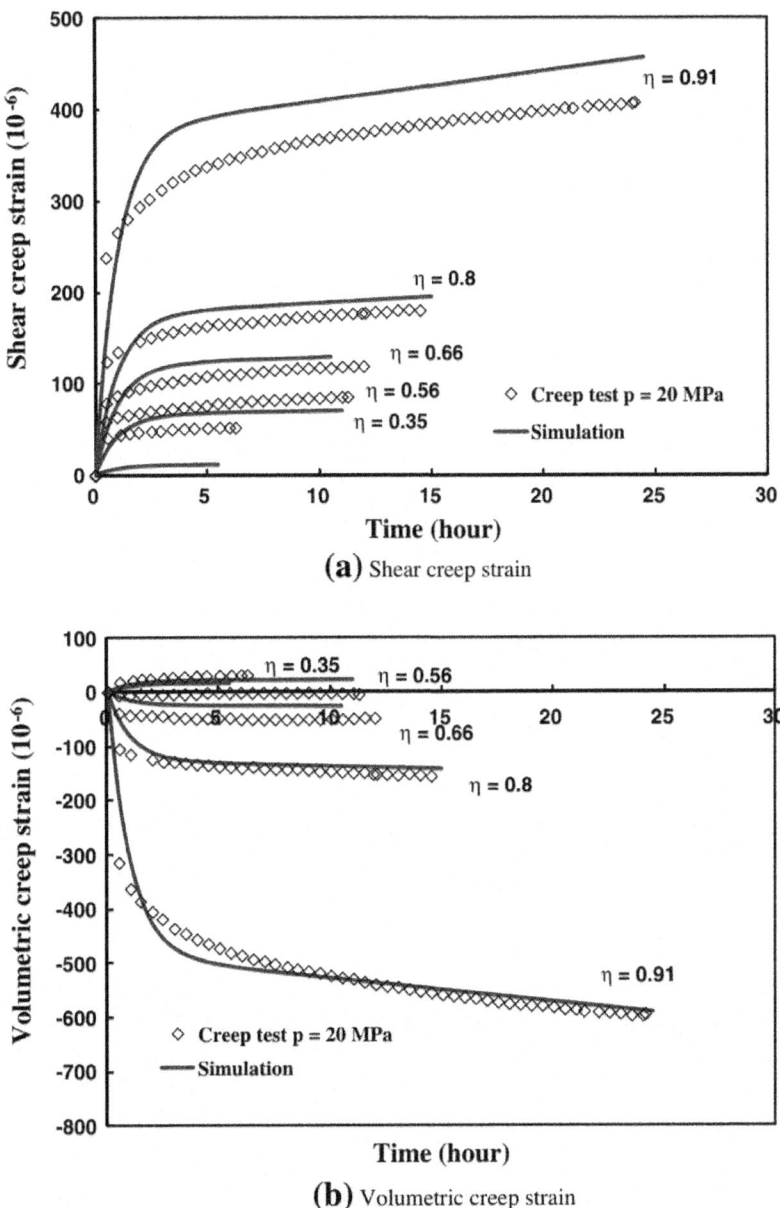

Figure. 9: Comparison of multistage creep strain predicted by the proposed model and test data under hydrostatic stress of 20 MPa.

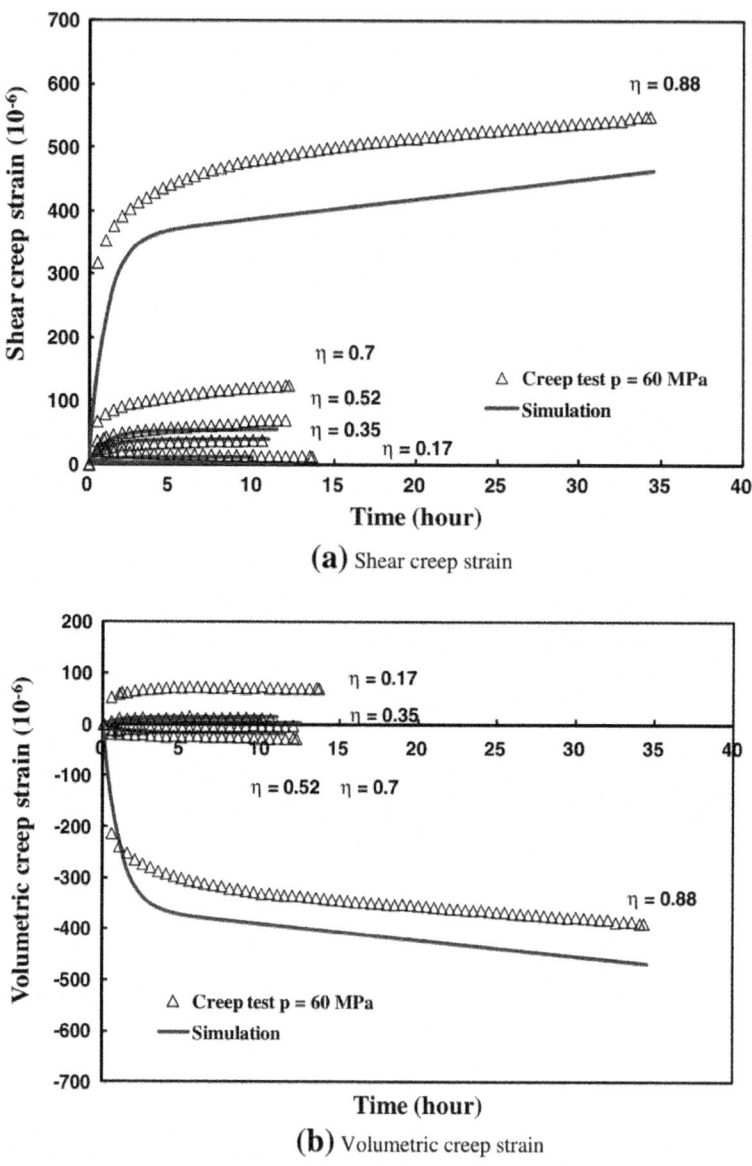

Figure. 10: Comparison of multistage creep strain predicted by the proposed model and test data under hydrostatic stress of 60 MPa.

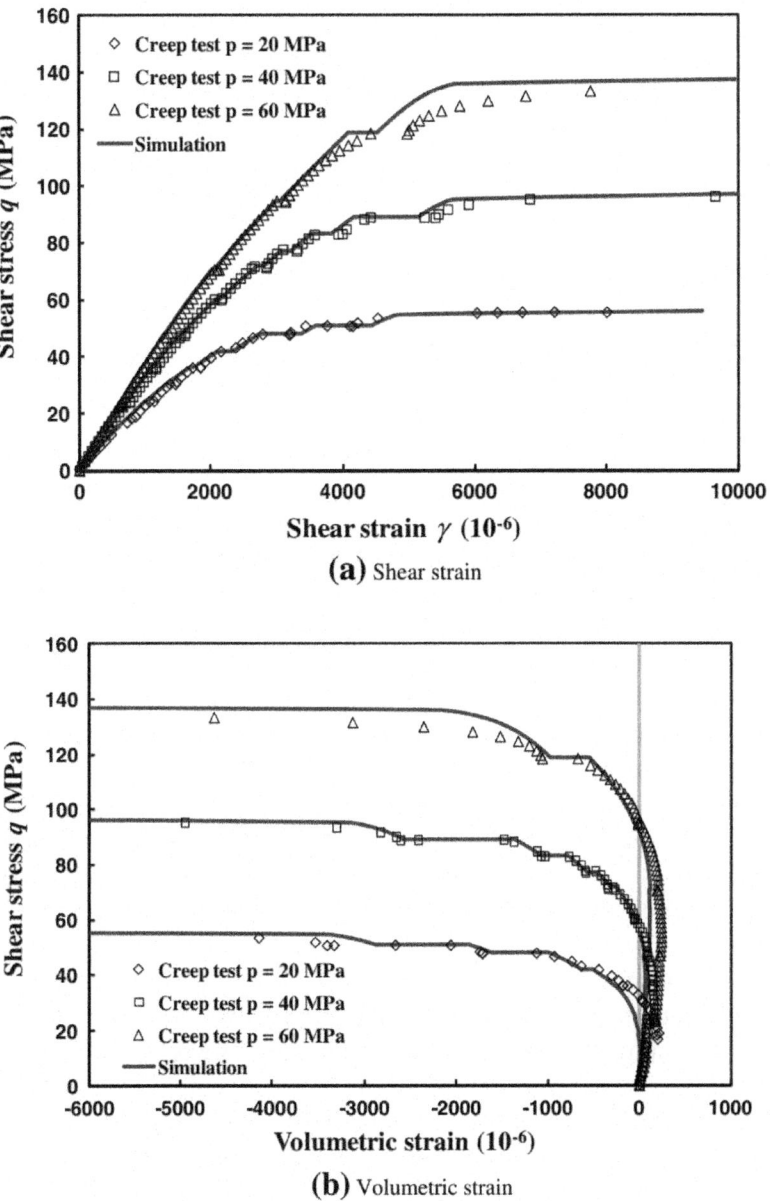

Figure. 11: Simulation of strains for multistage creep tests under different hydrostatic pressures.

WETTING DETERIORATION

Wetting deterioration is a decrease in material strength and stiffness caused by water penetration. To consider the wetting deterioration of sandstone, this study used other MS sandstone specimens, which are denoted as MS-B, for triaxial tests and multistage creep tests under dry and water-saturated conditions. The MS-B specimens have the following mean physical properties: porosity of 14.0 % and dry density of 2.28 g/cm^3. The average uniaxial compressive strength is 27.2 MPa in dry conditions and 12.9 MPa in saturated conditions. Based on petrographic analyses, the percentages of grains, matrices, and voids are 60.0, 26.0, and 14.0 %, respectively. The average grain diameter is ~0.34 mm. Mineralogically, MS-B sandstone consists of 88.5 % quartz and 7.2 % rock fragments, and is classified as lithic greywacke.

Figure 12 shows the failure envelopes of sandstone under dry and saturated conditions. The two failure envelopes remain linear under different hydrostatic pressures, and the saturated sandstone has lower shear strength than dry sandstone does. Table 1 presents the corresponding parameters, α_d and k_d, both of which are reduced because of wetting deterioration. In addition to the strength, Figure. 13 shows the variations of plastic deformation under dry and saturated conditions. The Figure shows that the values of the plastic angle β_1 under dry and saturated conditions exhibit similar tendencies, but the saturated condition has a lower value than the dry condition under the same stress ratio (Figure. 13a), thereby indicating that the dilation threshold in a saturated condition occurs later than that under a dry condition. Figure 13b shows the variations of the plastic modulus under both conditions. The plastic modulus decreases as the stress ratio increases. When approaching a failure state, the saturated modulus decreases by approximately three orders of magnitude and the dry modulus also exhibits a similar tendency. Figure 14 presents the creep deformation under dry and saturated conditions. Greater creep strains can be induced under saturated conditions, especially when the loading approaches the shear strength. The creep volumetric strain (Figure. 14b) is initially compressive and then becomes dilative at later stages of loading. The amount of creep volumetric strain when approaching the shear strength is considerably greater than that under lower levels of shear stress.

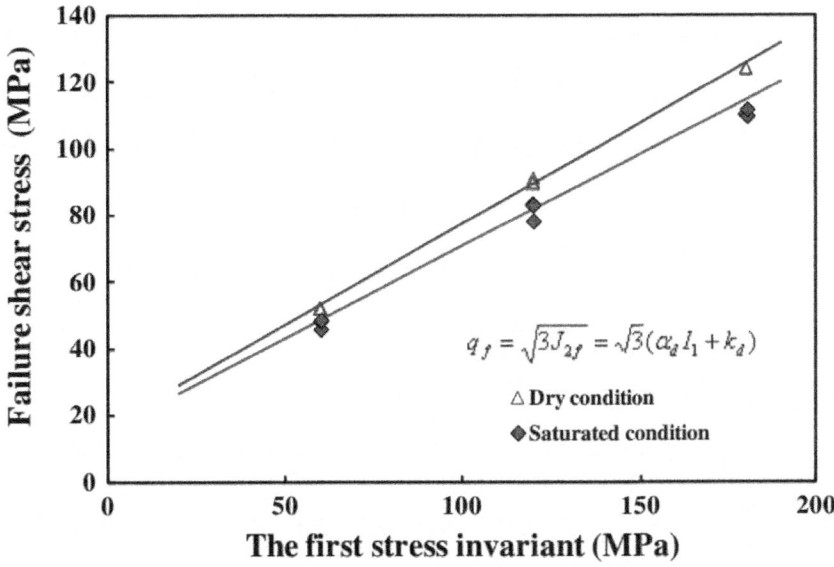

Figure. 12: Failure envelopes of MS-B sandstone under dry and saturated conditions.

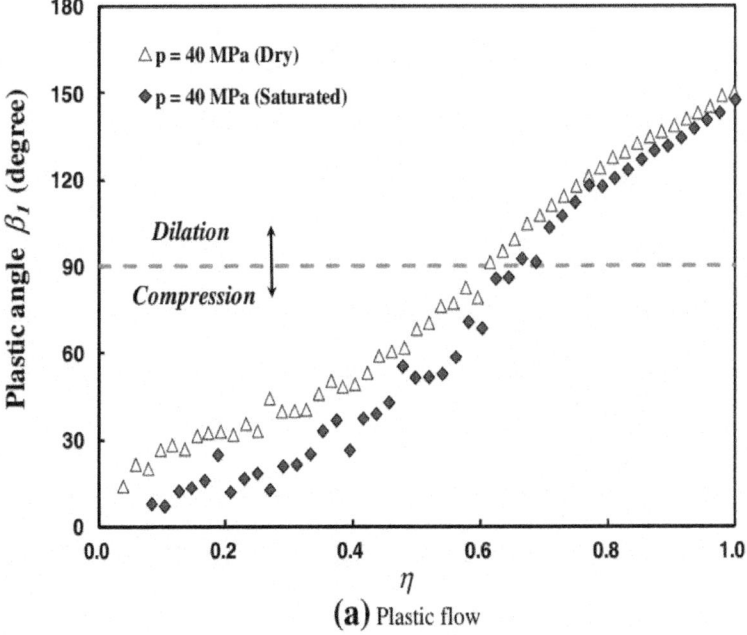

(a) Plastic flow

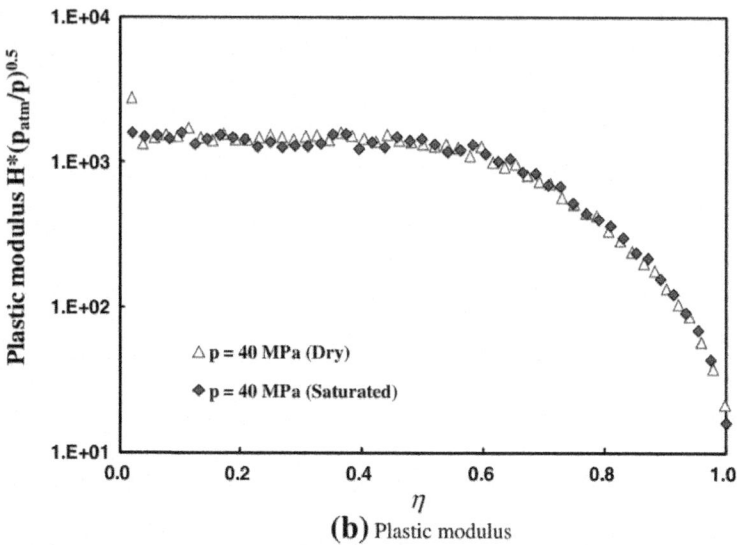

(b) Plastic modulus

Figure. 13: Variations of plastic flow angle β_1 and normalized plastic modulus $H_L\sqrt{p_{atm}/p'}$ under dry and saturated conditions.

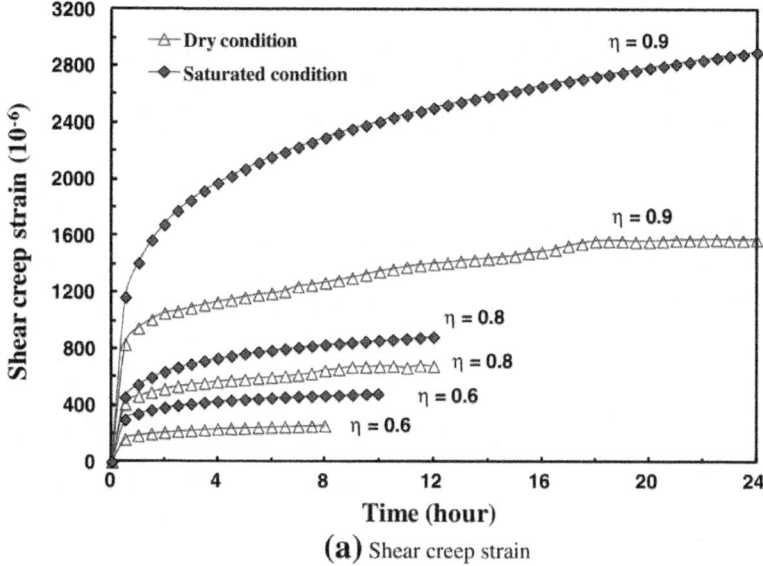

(a) Shear creep strain

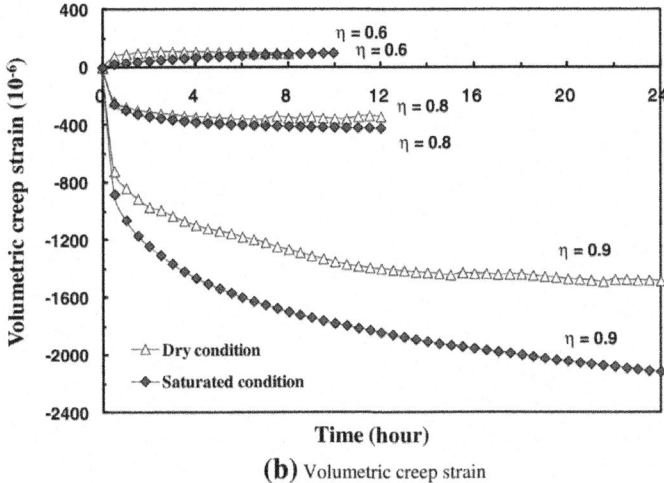

(b) Volumetric creep strain

Figure. 14: Comparison of multistage creep strain under dry and saturated conditions.

Figures 15 and 16 present simulated results for PS tests under dry and saturated conditions, respectively, to enable an evaluation of the validity of the model. The hydrostatic pressure ranges from 20 to 60 MPa. The corresponding parameters of MS-B sandstone are shown in Table 1. The Figures show that the proposed model can reasonably predict the deformation behavior of sandstone caused by wetting deterioration. In addition, Figure. 17 shows the simulated stress–strain curves of the multistage creep tests. This simulation, which is also shown in Figure. 17, is congruent with the experimental data. In summary, the proposed model can predict the behavior of sandstone in dry to saturated conditions.

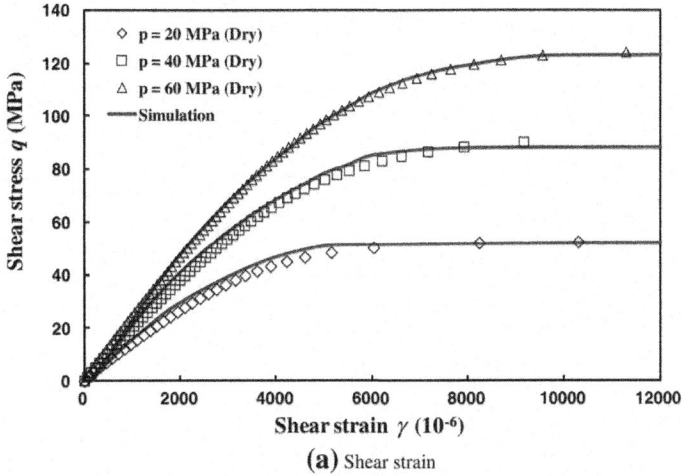

(a) Shear strain

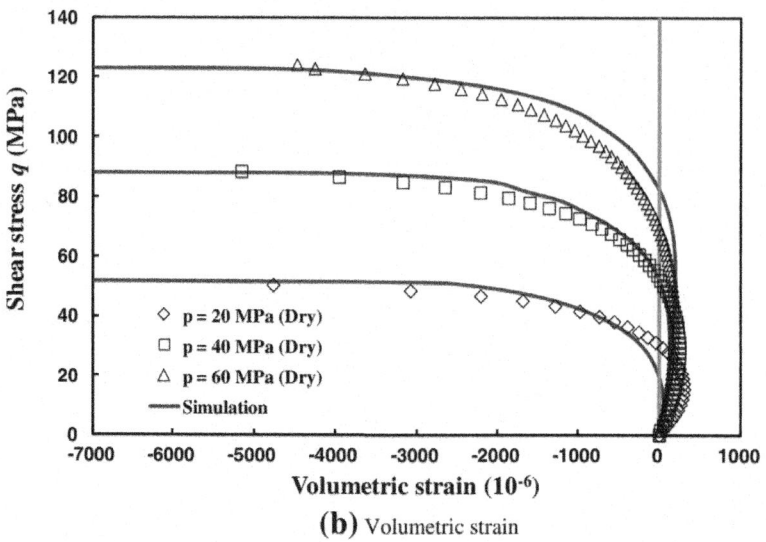

(**b**) Volumetric strain

Figure. 15: Simulation of stress–strain curves for dry MS-B sandstone under different hydrostatic pressures.

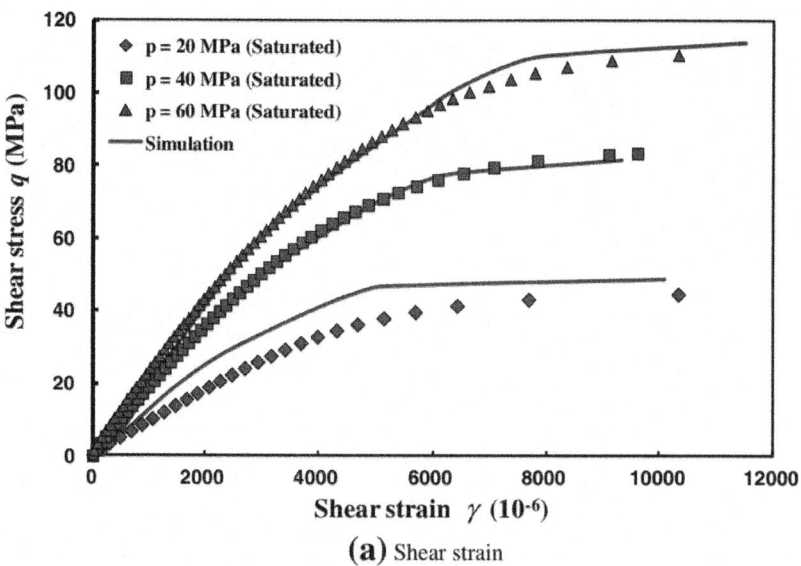

(**a**) Shear strain

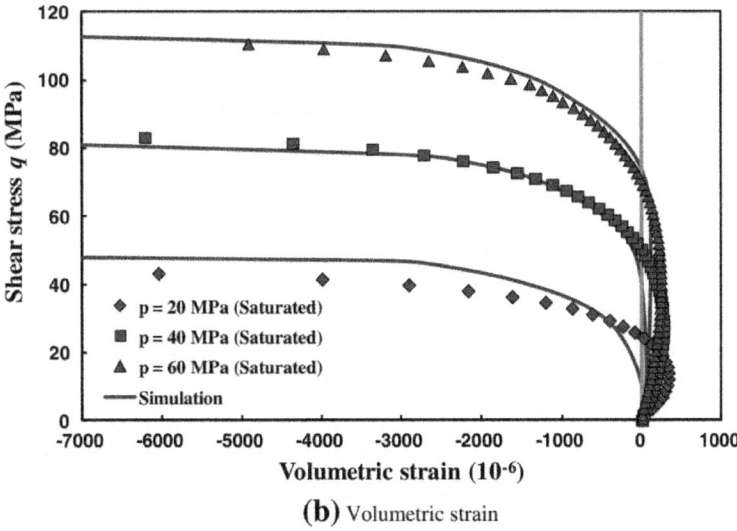

(b) Volumetric strain

Figure. 16: Simulation of stress–strain curves for saturated MS-B sandstone under different hydrostatic pressures.

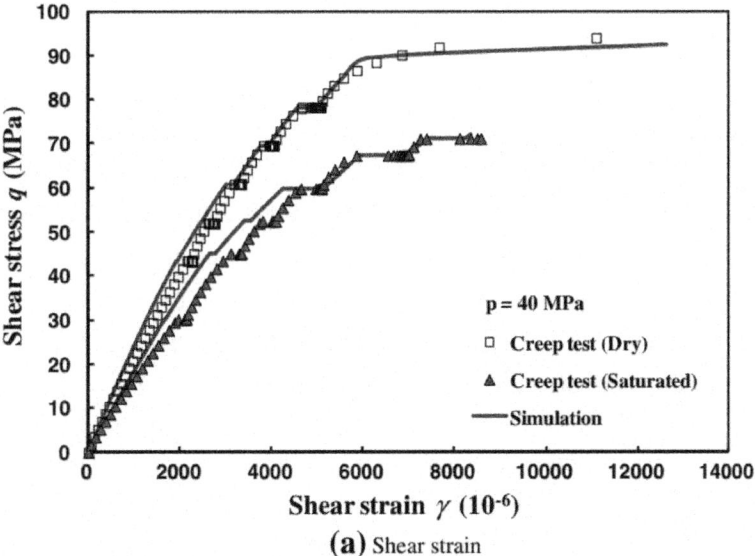

(a) Shear strain

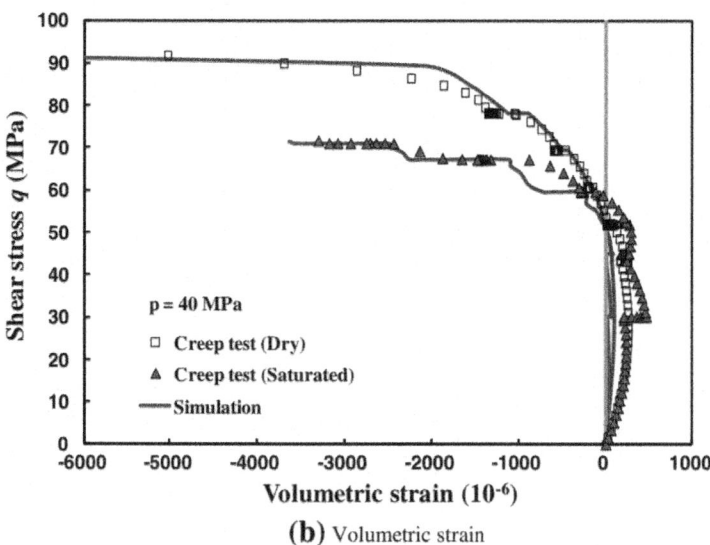

(b) Volumetric strain

Figure. 17: Simulation of strains in the multistage creep test under dry and saturated conditions.

CONCLUSIONS

This study extends previous research on predicting the time-dependent behavior and wetting deterioration of sandstone and presents a constitutive model based on nonlinear elasticity and generalized plasticity. The proposed model

- exhibits nonlinear elasticity under hydrostatic and shear loading,
- follows the associated flow rule for viscoplastic deformation,
- adopts a creep modulus that varies according to the stress ratio,
- considers both the primary and secondary creep behavior of rock, and
- considers the effect of wetting deterioration.

This model involves 13 material parameters, comprising 3 for elasticity, 7 for plasticity, and 3 for creep. All parameters can be determined straightforwardly by following the recommended procedures.

For prediction of immediate deformation, this study validates the proposed model by comparison with triaxial test results of MS-A sandstone under various hydrostatic stress and cyclic loading conditions. The proposed model is versatile in simulating the time-dependent behavior of sandstone through a series of multistage creep tests. Furthermore, to consider the effect of wetting deterioration, this study uses MS-B sandstone in triaxial and creep tests under dry and water-saturated conditions. Comparison of the simulated and experimental data shows that the proposed model can predict the behavior of sandstone in dry to saturated conditions. Future studies should extend the presented q-p' formulation to the multiaxial stress space and incorporate this constitutive model into finite-element software for analytical use in relevant rock engineering applications.

REFERENCES

1. Bolzon G, Schrefler BA, Zienkiewicz OC (1996) Elasto–plastic constitutive laws generalised to partially saturated states. Géotechnique 46(2):279–289
2. Chan AHC, Zienkiewicz OC, Pastor M (1988) Transformation of incremental plasticity relation from defining space to general Cartesian stress space. Commun Appl Numer Meth 4:577–580
3. Cristescu ND (1989) Rock rheology. Kluwer Academic, Dordrecht
4. Dyke CG, Dobereiner L (1991) Evaluating the strength and deformability of sandstones. Q J Eng Geol 24:123–134
5. Fernandez Merodo JA, Pastor M, Mira P, Tonni L, Herreros MI, Gonzalez E, Tamagnini R (2004) Modelling of diffuse failure mechanisms of catastrophic landslides. Comput Meth Appl Mech Eng 193:2911–2939
6. Goodman RE (1989) Introduction to rock mechanics. Wiley, New York
7. Hawkins AB, McConnell BJ (1992) Sensitivity of sandstone strength and deformability to changes in moisture content. Q J Eng Geol 25:115–130
8. Hoek E, Brown ET (1980) Empirical strength criterion for rock masses. J Geotech Eng Div (ASCE) 106(GT9):1013–1035
9. Houlsby GT, Amorosi A, Rojas E (2005) Elastic moduli of soils dependent on pressure: a hyperelastic formulation. Geotechnique 55(5):383–392
10. Hoxha D, Giraud A, Homand F (2005) Modelling long-term behaviour of a natural gypsum rock. Mech Mater 37:1223–1241
11. Jeng FS, Weng WC, Huang TH, Lin ML (2002) Deformational characteristics of weak sandstone and impact to tunnel deformation. Tunn Undergr Space Technol 17:263–264
12. Jeng FS, Weng WC, Lin ML, Huang TH (2004) Influence of petrographic parameters on geotechnical properties of Tertiary sandstones from Taiwan. Eng Geol 73:71–91
13. Ling HI, Liu H (2003) Pressure-level dependency and densification behavior of sand through a generalized plasticity model. J Eng Mech 129(8):851–860

14. Ling HI, Yang S (2006) Unified sand model based on the critical state and generalized plasticity. J Eng Mech 132(12):1380–1391
15. Manzanal D, Fernández Merodo JA, Pastor M (2006) Generalized plasticity theory revisited: new advances and applications. In: Proceedings of 17th European young geotechnical engineer's conference Zagreb, Zagreb, Croatia, pp 238–246
16. Manzanal D, Fernández Merodo JA, Pastor M (2011a) Generalized plasticity state parameter-based model for saturated and unsaturated soils. Part 1: saturated state. Int J Numer Anal Meth Geomech 35:1347–1362
17. Manzanal D, Pastor M, Fernández Merodo JA (2011b) Generalized plasticity state parameter-based model for saturated and unsaturated soils. Part II: unsaturated soil modeling. Int J Numer Anal Meth Geomech 35:1899–1917
18. Mira P, Tonni L, Pastor M, Fernandez Merodo JA (2009) A generalized midpoint algorithm for the integration of a generalized plasticity model for sands. Int J Numer Meth Eng 77:1201–1223
19. Pastor M (1991) Modelling of anisotropic sand behaviour. Comput Geotech 11(3):173–208
20. Pastor M, Zienkiewicz OC (1986) A generalized plasticity, hierarchical model for sand under monotonic and cyclic loading. 2nd International symposium numerical models in geomechanics, Ghent, pp 131–149
21. Pastor M, Zienkiewicz OC, Chan AHC (1990) Generalized plasticity and the modelling of soil behaviour. Int J Numer Anal Meth Geomech 14:151–190
22. Pastor M, Zienkiewicz OC, Guang-Duo X, Peraire J (1992) Modelling of sand behaviour: cyclic loading, anisotropy and localization. In: Kolymbas Gudehus (ed) Modern approaches to plasticity. Springer, Berlin
23. Sterpi D, Gioda G (2009) Visco-plastic behaviour around advancing tunnels in squeezing rock. Rock Mech Rock Eng 42:319–339
24. Tomanovic Z (2006) Rheological model of soft rock creep based on the tests on marl. Mech Time-Depend Mater 10:135–154
25. Tsai LS, Hsieh YM, Weng MC, Huang TH, Jeng FS (2008) Time-dependent deformation behaviors of weak sandstones. Int J Rock Mech Min Sci 45:144–154
26. Weng MC, Ling HI (2012) Modeling the behavior of sandstone based on generalized plasticity concept. Int J Numer Anal Meth Geomech. doi:10.1002/nag.2127
27. Weng MC, Jeng FS, Huang TH, Lin ML (2005) Characterizing the

deformation behavior of Tertiary sandstones. Int J Rock Mech Min Sci 42:388–401
28. Weng MC, Jeng FS, Hsieh YM, Huang TH (2008) A simple model for stress-induced anisotropic softening of weak sandstones. Int J Rock Mech Min Sci 45:155–166
29. Weng MC, Tsai LS, Hsieh YM, Jeng FS (2010a) An associated elastic–viscoplastic constitutive model for sandstone involving shear-induced volumetric deformation. Int J Rock Mech Min Sci 47:1263–1273
30. Weng MC, Tsai LS, Liao CY, Jeng FS (2010b) Numerical modeling of tunnel excavation in weak sandstone using a time-dependent anisotropic degradation model. Tunn Undergr Space Technol 25:397–406
31. Xie SY, Shao JF (2006) Elastoplastic deformation of a porous rock and water interaction. Int J Plast 22:2195–2225
32. Zienkiewicz OC, Mroz Z (1984) Generalized plasticity formulation and applications to geomechanics. Mechanics of engineering materials. Wiley, New York, pp 655–679

Chapter 6

APPLICATION OF GEOSTATISTICAL MODELS FOR ESTIMATING SPATIAL VARIABILITY OF ROCK DEPTH

Pijush Samui[1], Thallak G. Sitharam[2]

[1]Centre for Disaster Mitigation and Management, VIT University, Vellore, India
[2]Department of Civil Engineering, Indian Institute of Science, Bangalore, India

ABSTRACT

Rock depth information of a site is a significant factor for geotechnical engineering and earthquake ground response analysis. In this paper, reduced level of rock at Bangalore is arrived from the 652 boreholes in the area covering 220 km^2. Geostatistical modeling based on kriging (simple and ordinary) techniques has been applied for estimating reduced level of hard rock in Bangalore. The models are used to compute variance of estimated reduced level of the rock. A new type of cross-validation analysis proves the robustness of the developed models. The comparison between the simple and ordinary kriging model demonstrates that the ordinary kriging model is superior to simple kriging model in predicting reduced level of rock in the subsurface of Bangalore.

INTRODUCTION

Rock depth for a site is a very useful parameter to the geotechnical and earthquake engineers to find their basic requirement of hard strata and ground motion at rock level. In most geotechnical investigations, knowledge of the hard strata or hard rock depth is essential to design a suitable foundation. In ground response analysis, Peak Ground Acceleration (PGA) at rock level and response spectrum for the particular site is evaluated. This information is also essential to evaluate liquefaction hazards of a site and to estimate earthquake induced forces on structures.

With an objective of evaluating hard rock depth in Bangalore, an attempt has been made to develop a two dimensional map of reduced level of rock for Bangalore based on kriging (simple and ordinary) technique. Rock is identified by borelogs data available in the area and identified by visual observation of the cores taken at these locations. Hard disintegrated rock is also identified as rock and depth from the ground level has been used to evaluate reduced level of rock at any location.

The kriging method was developed during the 1960s and 1970s and has been acknowledged as a good spatial interpolator [1-3]. The most important features of this method are that the interpolator 1) is linear and unbiased 2) gives minimum estimation error, and 3) is exact and gives an evaluation of uncertainty for interpolated values. This technique is widely used in the field of earth sciences, including mining, geochemistry, remote sensing, etc. A major advantage of kriging is that it is more flexible than other interpolation methods such as inverse distance weighting, deterministic splines and Thiessen polygons. The weights are not selected arbitrarily, but depend on how the variable of interest (in this case hard rock elevation) varies in space. In kriging, the variable weights have been used based on the scale of variability whereas in Thiessen polygon, one has to apply same weights, whether the function exhibits small-or largescale variability. The main goal of kriging is to predict (in the interpolation sense) a variable where no measurements were made using the semivariogram as a model characterizing spatial variability. Semivariogram is the analytical tool used to evaluate and quantify the degree of spatial autocorrelation. It contains three elements of information. First of all the semivariogram is an appreciation of the dispersion (or scatter) of the parameters, which equates to the half variance. Secondly, it gives an autocorrelation distance that represents the radius of influence of a measurement made at a given point. Thirdly, it provides the type of variability that indicates how values fluctuate in space.

A valid estimation of the variance of estimated reduced level of rock depths is important for developed ordinary as well as simple kriging model. So, the developed models have been used to compute variance of estimated reduced level of the rock depths. In general, the variance depends both on the semivariogram and the location of the measurements. A number of publications can be found in the literature which presents the theory and application of kriging [4-12].

In this paper, a semivariogram model has been developed along with the kriging model for the reduced level of the rock in the subsurface of Bangalore. Semivariogram analysis is used to detect trends in structure or internal properties of deposits and estimated values can be obtained at points where no data is available. A new method for cross-validation analysis of developed models has been proposed. The cross-validation of the model has been done based on the examination of residuals. A comparative study has been also done between developed simple kriging and ordinary kriging.

SUBSURFACE OF BANGALORE AND GIS MODEL DEVELOPMENT

Bangalore covers an area of over 220 square kilometres and ground Reduced levels (GRL) also vary a lot in the city. It varies from 810 m in north-east part to 940 m in south-western part of Bangalore. Ground reduced levels do not vary much in the central and north-western parts of the city. The population of greater Bangalore exceeds 6 million and it is the fifth biggest city in India. It is situated on latitude of 1208' North and longitude of 77037' East. From geology, subsurface of Bangalore region is a Gneiss complex, which was formed by several tectonic-thermal events with large influx of sialic material, occurring between 3 to 3.4 billion years ago giving rise to an extensive group of gray gneisses designated as the "older gneiss complex". These gneisses act as the basement for a widespread belt of schist's. The younger group of gneissic rocks mostly of granodiomitic and granitia composition is found in the eastern part of the state, representing remobilized parts of an older crust with abundant additions of newer granite material, for which the name "younger gneiss complex" has been given [13]. The soil is mostly a residual soil from granite gneiss due to weathering action. In the most recent past, there were more than 400 lakes, and more than 340 lakes were dried up and have been encroached for residential and industrial development. In the old tank beds, silty sand/clay is also found as overburden.

A Geographic Information System (GIS) model (see **Figure 1**) of Bangalore with several layers on a scale of 1:20,000 has been developed with a purpose of carrying out microzonation of Bangalore. The Bangalore map forms the base layer for GIS. The map entities have been developed for locating boreholes to the utmost accuracy and at each location borelogs have been attached along with geotechnical data of each layer up to the hard rock. The digitized map has several layers of information. Some of the important layers considered are the boundaries (outer and Administrative), Highways, Major roads, Minor roads, Streets, Rail roads, Water bodies, Drains, Ground Contours and Borehole locations. The locations of boreholes are shown in **Figure 1** along with ground reduced level with an interval of 10 m (see **Figure 2**). Distribution of collected boreholes in Bangalore is shown in **Figure 3**, indicating a very good distribution of the boreholes in each quadrant of Bangalore from the city center. **Figure 1** also depicts grids of 1kmX1km along with the corporate boundary of Bangalore and outer boundary circumscribing the ring road. **Figure 1** gives a clear view of the spatial distribution of boreholes in Bangalore region. An average of about three boreholes data is available within the grid of 1 km × 1 km.

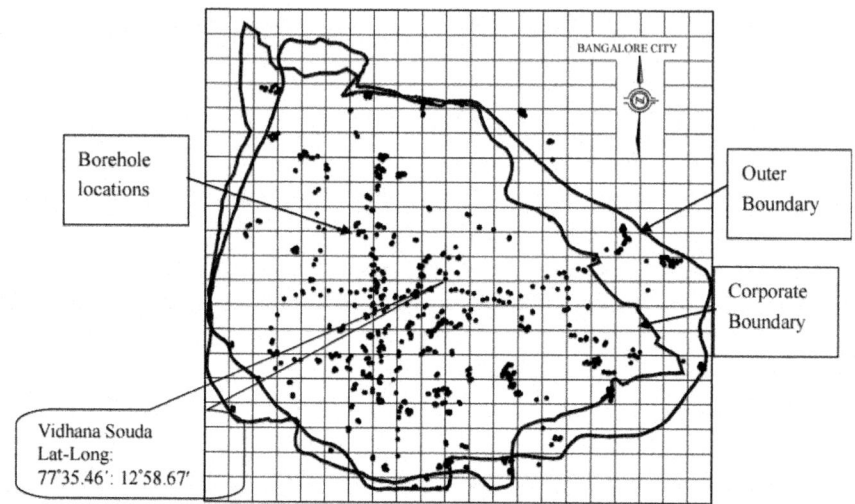

Figure 1. Borehole location in Bangalore Map (scale: 1:20000).

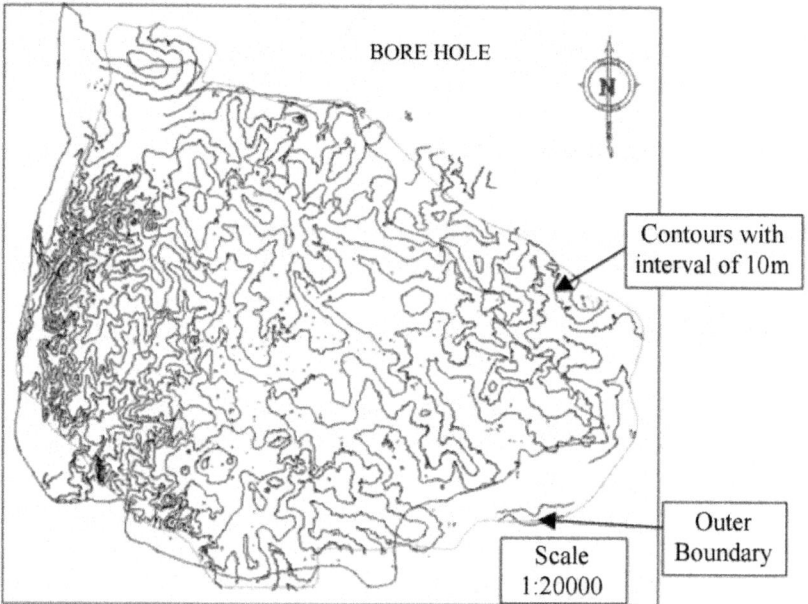

Figure 2. GIS model of borehole locations with respect to contours.

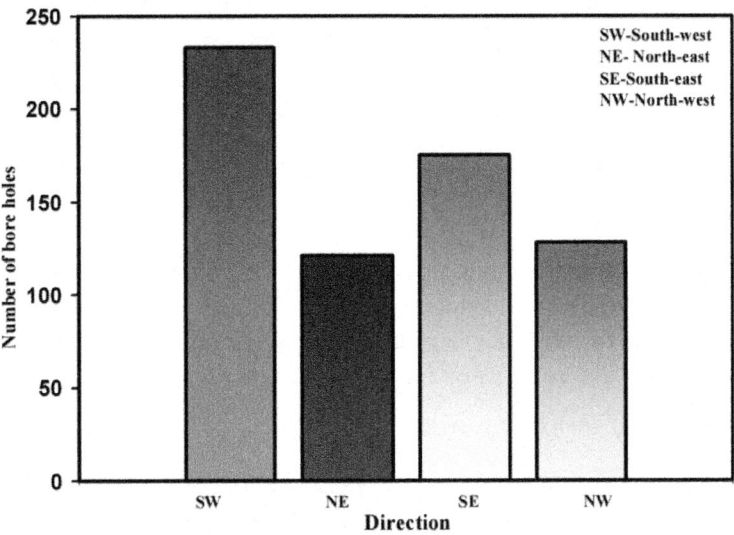

Figure 3. Distribution of boreholes in quadrants of Bangalore.

Geotechnical data for 652 boreholes was collated from archives of only two organizations; Torsteel Research Foundation in India and Indian Institute of Science. This data was generated for geotechnical investigations carried out for several major projects in Bangalore including Bangalore metro project. The data collected is of very high quality and collected during the years 1995-2003. The data in the model is on average to a depth of 30m below the ground level. Each borelog contains information about depth, density of the soil, total stress, effective stress, fines content and N values, depth of ground water table and rock depth. In GIS model, the boreholes are represented as three dimensional object spanning below the map layer. These three dimensional objects are generated with several layers with a bore location in each layer overlapping one below the other and each layer representing 0.5m interval of the subsurface. Each layer of this model is attached with borelog data at that depth. The data consists of visual soil classification, borehole location, ground water level, date and time during which test has been carried out, other physical and engineering properties of soil and rock depth. As such when this model is viewed in three dimensions, subsurface information on any borehole at any depth can obtained by clicking at that level. The hard rock has been identified by visual observation of the cores taken at these locations. Rock depth from ground level is the difference between the ground reduced level at borehole location and reduced level of the hard rock at the same borehole location. The reduced level of the hard rock at borehole location is the difference between the ground reduced level at borehole location and depth of overburden thickness up to hard rock in the same borehole. The depth of overburden is estimated from the available borelogs. The term hard rock in this paper corresponds to engineering bed rock (shear wave velocity ≈700 m/sec) as against seismic bed rock (shear wave velocity ≈3000 m/sec).

METHODOLOGY

In this paper, ordinary kriging and simple kriging are adopted for evaluating reduced level of the rock in subsurface of the Bangalore. For both the methods, there is a need to introduce the terminology of the covariance function or semivariogram. The covariance function between two points is defined as:

$$C(h) = E\left[(d(x)-m)(d(x')-m)\right] \quad (1)$$

where m is the mean of d(x) and C(h) is the covariance function with lag h, with h being the distance between two samples x and x':

$$h = \|x-x'\| = \sqrt{(x-x')^2 + (y-y')^2} \quad (2)$$

The semivariogram for an intrinsic random function [14,15] is defined as:

$$\gamma(h) = 0.5 * E\left[(d(x)-d(x'))2\right] \quad (3)$$

Figure 4 illustrates the different components of the semivariogram.

If the variance exists (the random function is second order stationary), the relation between the covariance function and the semivariogram is as follows:

$$\gamma(h) = C(0) - C(h) \quad (4)$$

where, g(h) is the semivariogram and C(0) is the variance.

For semivariogram, the model used in this analysis is Gaussian model. The general equation for this model looks like:

$$\gamma(h) = c_0 + c\left[\frac{3h}{2a} - \frac{1h^3}{2a^3}\right] \text{ for } h \leq a \quad (5)$$

$$\gamma(h) = c_0 + c \text{ for } h \leq a \quad (6)$$

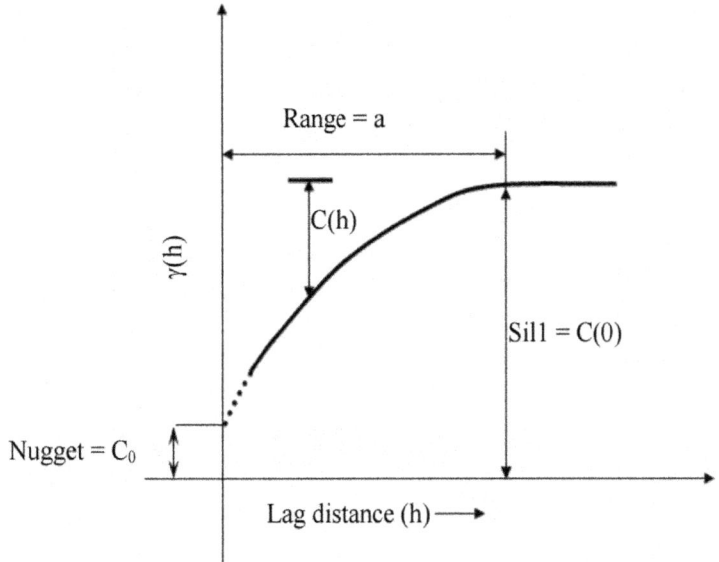

Figure 4. A typical semivariogram.

For semivariogram model, geoanisotropic model has been used to reduce the anisotropy into isotropy by a linear transformation of coordinates [16]. Once the model of semivariogram is constructed, the weights are computed for kriging. The method for ordinary kriging, simple kriging and cross-validation of the models are given as below:

Ordinary Kriging

Ordinary kriging is a linear geostatistical method. It gives local estimation by interpolation. The basic equation used in ordinary kriging is as follows:

$$d(x,y) = \sum_{i=1}^{n} w_i d_i \qquad (7)$$

where n is the number of scatter points in the subsurface of the Bangalore. d_i is reduced level of the rock at point i in the subsurface of the Bangalore and w_i is weights assigned for point i. This equation is essentially the same as the equation used for inverse distance weighted interpolation except that rather than using weights based on an arbitrary function of distance. The weights used in kriging are based on the model semivariogram. For example, to interpolate reduced level of the rock (d_p), at a point 'P' in the subsurface of the Bangalore based on the surrounding points P_1, P_2, and P_3, the weights w_1, w_2, and w_3 must be found. The weights are found through the solution of the simultaneous equations:

$$w_1\gamma(h_{11}) + w_2\gamma(h_{12}) + w_3\gamma(h_{13}) = \gamma(h_{1p}) \tag{8}$$

$$w_1\gamma(h_{12}) + w_2\gamma(h_{22}) + w_3\gamma(h_{23}) = \gamma(h_{2p}) \tag{9}$$

$$w_1\gamma(h_{13}) + w_2\gamma(h_{23}) + w_3\gamma(h_{33}) = \gamma(h_{3p}) \tag{10}$$

where (h_{ij}) is the model semivariogram evaluated at a distance equal to the distance between points i and j in the subsurface of the Bangalore. For example (h_{1p}) is the model semivariogram evaluated at a distance equal to the separation of points P_1 and P. Since it is necessary that the weights sum to unity, a fourth equation is added.

$$w_1 + w_2 + w_3 = 1.0 \tag{11}$$

Since there are now four equations and three unknowns, a slack variable, is added to the equation set. The final set of equations is as follows:

$$w_1\gamma(h_{11}) + w_2\gamma(h_{12}) + w_3\gamma(h_{13}) + \gamma = \gamma(h_{1p}) \tag{12}$$

$$w_1\gamma(h_{12}) + w_2\gamma(h_{22}) + w_3\gamma(h_{23}) + \gamma = \gamma(h_{2p}) \tag{13}$$

$$w_1\gamma(h_{13}) + w_2\gamma(h_{23}) + w_3\gamma(h_{33}) + \gamma = \gamma(h_{3p}) \tag{14}$$

$$w_1 + w_2 + w_3 = 1.0 \tag{15}$$

The equations are then solved for the weights w_1, w_2, and w_3. The d_p value of the point p is then calculated as:

$$d_p = w_1 d_1 + w_2 d_2 + w_3 d_3 \tag{16}$$

where, d_1, d_2 and d_3 are reduced level of the rock at point P_1, P_2 and P_3 respectively. The variance can be calculated at each interpolation point as:

$$S_z^2 = w_1\gamma(h_{1p}) + w_2\gamma(h_{2p}) + w_3\gamma(h_{3p}) + \gamma \tag{17}$$

when interpolating to an object using the kriging method, variance data set is always produced along with the interpolated data set. In some cases, specific spatial data distributions give rise to negative kriging weights, causing interpolated values to be negative or out of data limits and physically not compatible with data. For this reason negative and particular positive weights are set to 0 according to the rules proposed by Deutsh [17], ensuring the sum of weights is equal to one.

Simple Kriging

Simple kriging uses the average of the entire data set. The basic equation used in simple kriging is as follows:

$$d_{sk}(x,y) = m + \sum_{i=1}^{n} \lambda_i [d_i - m] \qquad (18)$$

where λ_i is weight assigned for point i and m is the average value.

The weights (λ_i) are found through the solution of the simultaneous equations:

$$\gamma_1 [C(0) - \gamma(h_{11})] + \gamma_2 [C(0) - \gamma(h_{12})]$$
$$+ \gamma_3 [C(0) - \gamma(h_{13})] = C(0) - \gamma(h_{1p}) \qquad (19)$$

$$\gamma_1 [C(0) - \gamma(h_{21})] + \gamma_2 [C(0) - \gamma(h_{22})]$$
$$+ \gamma_3 [C(0) - \gamma(h_{23})] = C(0) - \gamma(h_{2p}) \qquad (20)$$

$$\gamma_1 [C(0) - \gamma(h_{31})] + \gamma_2 [C(0) - \gamma(h_{32})]$$
$$+ \gamma_3 [C(0) - \gamma(h_{33})] = C(0) - \gamma(h_{3p}) \qquad (21)$$

The variance of estimation then become equal to

$$S_{sk}^2 = \left(1 - \sum_{i=1}^{n} \lambda_i\right) C(0) + \sum_{i=1}^{n} \lambda_i \gamma(h_{ip}) \qquad (22)$$

Cross-Validation of the Models

A new type of cross-validation analysis for kriging has been presented in this study. In practice, cross-validation is based on statistical tests involving the residuals. Residuals are differences between observation and model predictions. The detailed description of residuals in the case of kriging is given by Kitanidis [16]. It has been assumed that the n measurements are available at a time, in a given sequence. The kriging estimate of z at the second point, x, from the first measurement, x_1 is calculated. So, one can write $z_2 = z(x_1)$ and $\sigma_2 = 2\gamma(x_1 - x_2)$. Where, z_2 is the kriged value at the point x_2. The actual error

$(\delta_2) = z(x_2) - z_2$ is normalized by the standard error (σ_2) and this normalized value of the error is given by:

$$\varepsilon_2 = \frac{\delta_2}{\sigma_2} \qquad (23)$$

For the k-th measurement location, the actual error (δ_k) and normalized error (ε_2) can be written as, respectively:

$$\delta_k - z(x_k) - \hat{z}_k, \text{ for } k = 2, \cdots, n \qquad (24)$$

$$\varepsilon_k = \frac{\delta_k}{\sigma_k}, \text{ for } k = 2, \cdots, n \qquad (25)$$

A cross-validation Q1 and Q2 are used to check the statistical distribution of the residuals between the observed data and kriged values at the original observation location by using the same kriging parameters and semivariogram model parameters. To perform Q1 and Q2 cross validation, a normalized residual array (ε_k) is constructed as suggested by Kitanidis (1997). Q1 is the mean of the residual (ε_k) and it is written as:

$$Q1 = \frac{1}{n-1} \sum_{k=2}^{n} \varepsilon_k \qquad (26)$$

Under the null hypothesis, Q1 is normally distributed with mean 0 and variance $\frac{1}{n-1}$. The probability density function (pdf) of Q1 is:

$$f(Q1) = \frac{1}{\sqrt{\frac{2\pi}{(n-1)}}} \exp\left(-\frac{Q1^2}{\frac{2}{(n-1)}}\right) \qquad (27)$$

where, n is the number of data points. If the experimental value of Q1 turns out to be acceptable close to zero then this test gives no reason to question the validity of the model. The Q2 is the variance of ε_k and it is written as:

$$Q2 = \frac{1}{n-1} \sum_{k=2}^{n} \varepsilon_k^2 \qquad (28)$$

$(Q2)*(n-1)$ approximately follows the chi-square distribution with parameter(n-1). Where, n is the number of data points. The mean and variance of Q2 are 1 and $\frac{2}{n-1}$ respectively. The pdf of Q2 is given by the following equation:

$$f(Q2) = \frac{(n-1)^{\frac{n-1}{2}} Q2^{\frac{n-3}{2}} \exp\left(-\frac{(n-1)Q2}{2}\right)}{2^{\frac{n-1}{2}} \Gamma\left(\frac{n-1}{2}\right)}$$

(29)

where, G is the gamma function. For robust model, the experimental value of Q2 should be close to one. In this work, kriging model and cross-validation have been programmed using MATLAB software.

RESULTS AND DISCUSSION

In case of ordinary kriging, the semivariogram of reduced level of rock obtained from the experimental values is shown in **Figure 5**. The Gaussian model has been plotted in **Figure 5** and gives a reasonable fit to the values obtained. The range, sill and nugget of the semivariogram are 0.95, 1.202 and 0.097 respectively. In the semivariogram, on the x-axis "relative to the full length scale" means normalized lag distance. The estimation of reduced level of rock has been done by using developed model of semivariogram (shown in **Figure 5**). **Figure 6** shows the kriging surface of the reduced level of rock for ordinary kriging. For simple kriging analysis, the experimental semivariogram has been calculated using the reduced level of rock data. A Gaussian model has been fitted with parameters: 0.95 for range, 1.178 for sill and 0.079 for nugget. By using the model of the semivariogram shown in **Figure 7**, reduced level of rock has been estimated in the subsurface of the Bangalore. The result is shown in **Figure 8**.

One of the most important finding of this study is that the semivariogram for both models is free from white noise or a pure nugget effect. The pure nugget effect corresponds to the total absence of auto-correlation. A weighted nonlinear least squares method has been used to fit semivariogram model. The points closer to the origin are given higher weights than points further away, because they are inherently more accurate, as they are calculated using more data pairs. Gaussian model, which has been used for semivariogram, ensures well-conditioned kriging matrix. So this study did not exhibit any numerical stability problem.

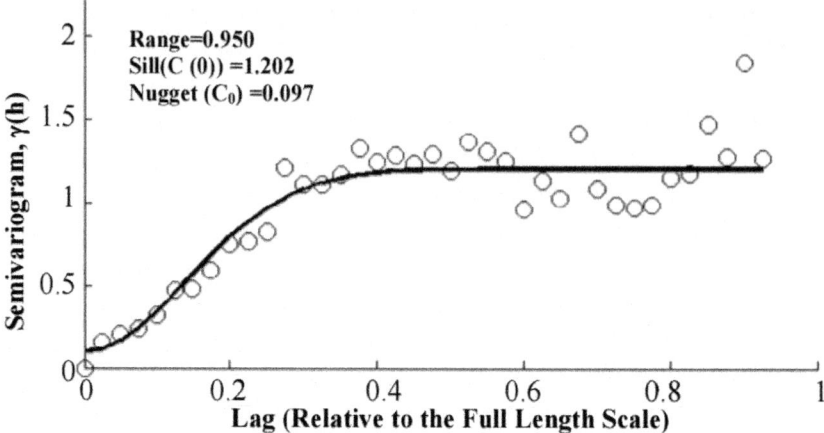

Figure 5. Semivariogram model using ordinary kriging.

For both models, the semivariogram stops increasing beyond a certain distance. This semivariogram is called "transition" models, and corresponds to a random function which is not only intrinsic but also stationary. A function is stationary if it consists of small scale fluctuations about some well-defined mean value. For a stationary function, the length scale at which the sill is obtained describes the scale at which two measurements of the variable become practically uncorrelated. The advantage of intrinsic model is that it has been used to summarize incomplete information and patterns in noisy data. It has been also used to interpolate unknown data from observation of data. The semivariogram has a sill, which indicates that having extreme values has a very low probability. The both kriging models gave a unique solution. The reason for unique solution is explained below.

Considering a case of two different measurements of reduced level of rock obtained at the same location, if the semivariogram for both models is continuous (with zero nugget) the function $d(x)$ will be continuous. This means that one of the two measurements is redundant. Thus, one must be discarded; otherwise, a unique solution cannot be obtained because the determinant of the matrix coefficients of the kriging system vanishes. This problem is solved by adding a nugget term to the semivariogram. As a result, the Gaussian model adopted shows the nugget effect. In practical sense, nugget effect gives the kriging equations a stability and robustness. Without a nugget effect, inverting the kriging matrices may lead to computational round-off errors. Nugget effect also confirms that the contour map of estimate has a discontinuity at each observation point. As the sampling distance decreases, it is possible to obtain a better estimate of nugget effect. But the cost of the exploration program increases enormously. The model for the experimental semivariogram has been chosen based on the examination of residuals (differences between observation and model predictions). The predicted values from simple and ordinary kriging are different because of their different properties. As a result, the residuals from simple and ordinary kriging are different. For this reason, the semivariograms are different for simple and ordinary kriging. In this study, the residuals are always uncorrelated. The lack of correlation in the residuals has been explained below:

If the residuals are correlated, one can use this correlation to predict the value of ε_k from the value of $\varepsilon_2, \ldots\ldots, \varepsilon_{k-1}$ using a linear estimator. So, one can reduce further the mean square error of estimation of $d(x_k)$. But this is impossible because $d(x_k)$ is already the minimum-variance estimate. Thus, the residuals must be uncorrelated.

Kriging maps (Figures 6 and 8) provide a qualitative difference between the ordinary kriging and simple kriging methods. The advantage of this study is that it provides the magnitude of the variance of the estimated reduced level of rock. **Figure 9** is the variance map of reduced level of rock generated by ordinary kriging. **Figure 10** is a variance map of reduced level of rock generated by simple kriging. Using kriging variance map (Figures 9 and 10), one can give an indication of the quality of estimate. From Figures 9 and 10, it has been seen that the variance increases with increasing distance between estimated points and the actual point. The overall pattern of Figures 9 and 10 give an indication of where in the field adequate or inadequate sampling occurred. In the Figures 9 and 10, it is clear that the variance of the estimated data from simple kriging analysis is always greater than the variance of the estimated data from ordinary kriging analysis. Prediction of variance also depends on the behavior of the semivariogram at the origin, and it is known that without a nugget effect the predicted variance is often underestimated.

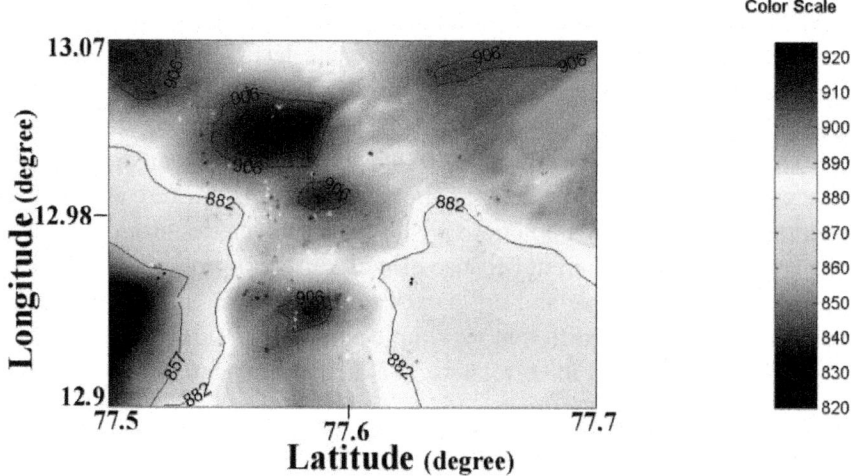

Figure 6. Map of the reduced level of rock for Bangalore using ordinary kriging.

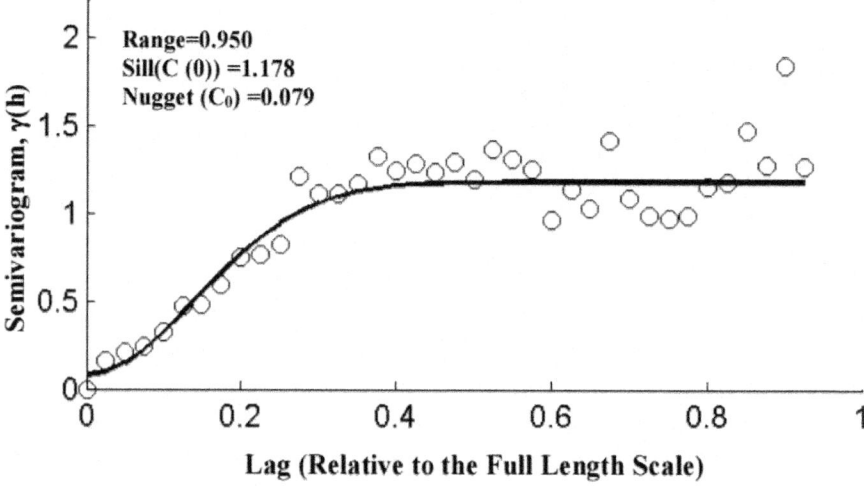

Figure 7. Semivariogram model using simple kriging.

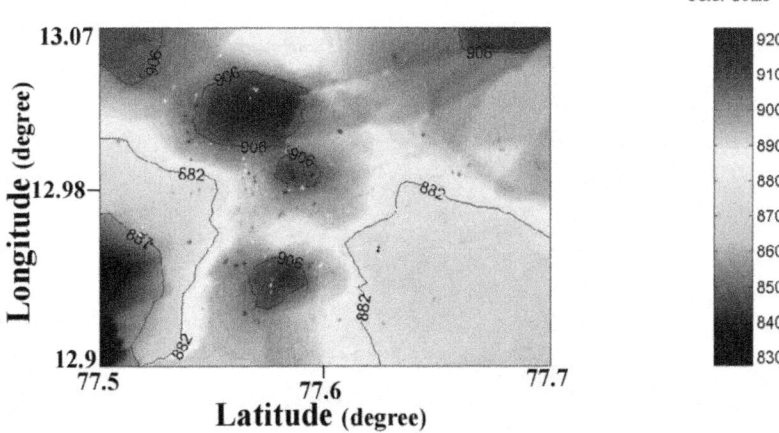

Figure 8. Map of the reduced level of rock for Bangalore using simple kriging.

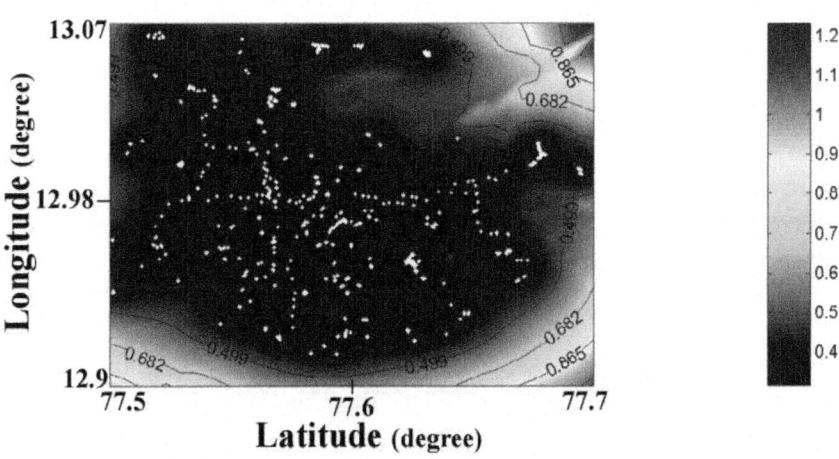

Figure 9. Map of the variance of the estimated reduced level of rock for Bangalore using ordinary kriging.

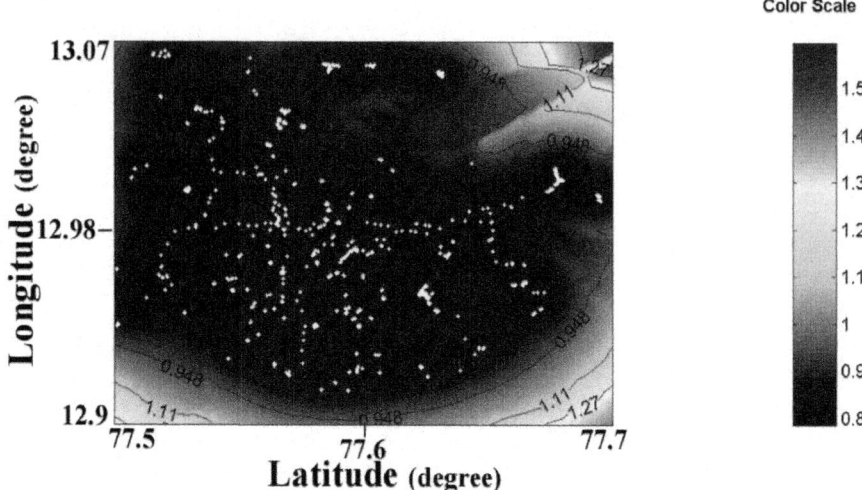

Figure 10. Map of the variance of the estimated reduced level of rock for Bangalore using simple kriging.

In case of cross-validation of kriging model, the acceptable region is defined in the Figures 11 to 14 (between the two vertical lines). For a good model, the Q1 as well as Q2 must fall in this acceptable regions as shown in Figures 11-14. In case of ordinary kriging, the value of Q1 and Q2 is 0.002, 1.069 respectively. The Q1 and Q2 values are well within the acceptable region (shown in Figures 11 and 12). In case of simple kriging, the value of Q1 and Q2 is 0.01, 0.911 respectively. The Q1 and Q2 fall in the acceptable region (shown in Figures 13 and 14). For both the models, the value of Q1 and Q2 are close to 0 and 1 respectively. The cross validation indicates that the developed ordinary kriging as well as simple kriging models are robust models for the estimation of the reduced level of rock in the subsurface of Bangalore. However it is clear from the results that the ordinary kriging seems to be predicting better than simple kriging.

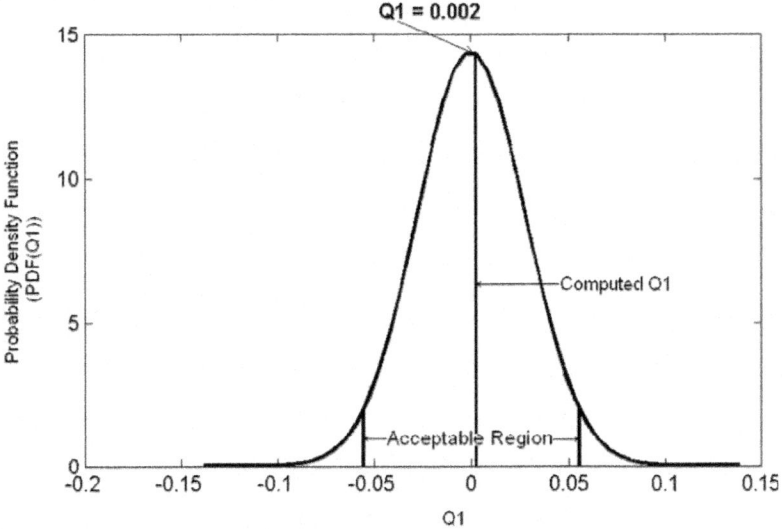

Figure 11. Distribution of Q1 for ordinary kriging.

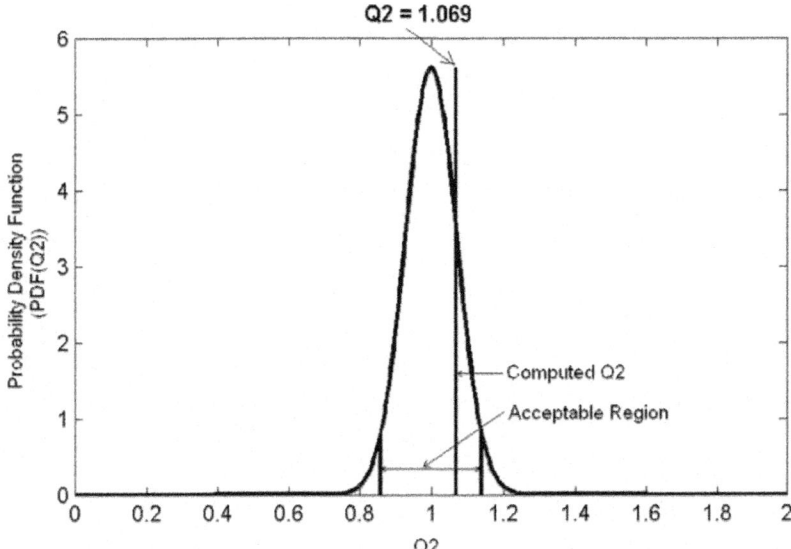

Figure 12. Distribution of Q2 for ordinary kriging.

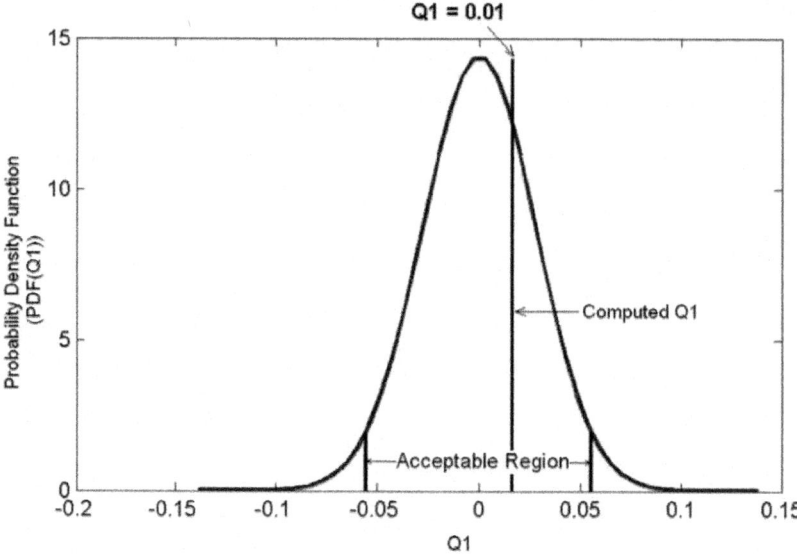

Figure 13. Distribution of Q1 for simple kriging.

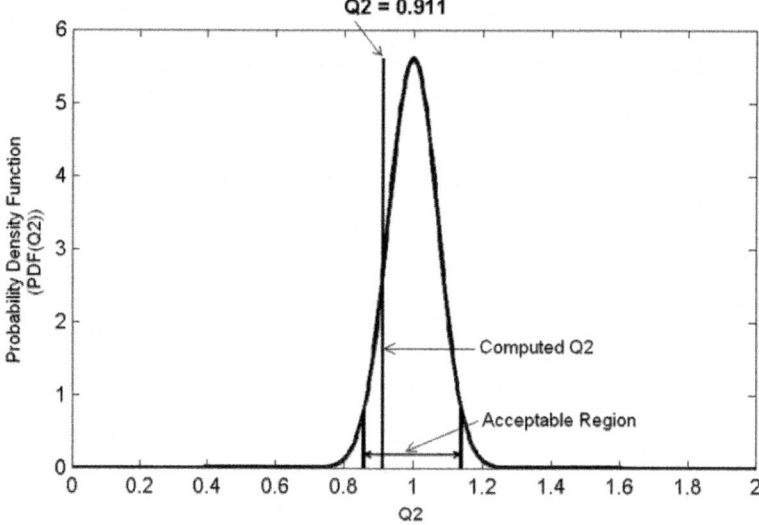

Figure 14. Distribution of Q2 for simple kriging.

Table 1. Comparison between ordinary kriging and simple kriging model

Bore hole No.	Longitude (degree)	Latitude (degree)	Actual reduced level of the rock (m)	Predicted value(m) by simple kriging	Predicted value(m) by ordinary kriging
275-1	77.5765	12.9448	885.2	896.2	890.4
965-3	77.6237	12.9447	884.46	875.46	883.46
15-1	77.6641	12.9924	893.5	888.5	896.5
104-1	77.5874	12.9331	896.6	908.6	903.6
344-6	77.5368	13.0293	900	893	896

In order to compare between the ordinary and simple kriging models, five points have been chosen randomly from known reduced level of the rock of 652 points in the subsurface model of Bangalore. The predicted values of these points are shown in **Table 1**. It can be seen from the table that the ordinary kriging model has given better prediction than simple kriging model. For the data sets used in this paper, ordinary kriging has shown to be a better estimator than simple kriging in terms of reduced kriging variance and the comparison between an estimated and actual value. This result is expected, since the simple kriging uses the average of the entire data set while ordinary kriging uses a local average (the average of the scatter points in the kriging subset for a particular interpolation point). The reduced level of rock at a half-space point could be more accurately estimated from the reduced level of rock at neighboring half-space points than that at distinct location. As a result, simple kriging is less accurate than ordinary kriging.

First, confirm that you have the correct template for your paper size. This template has been tailored for output on the A4 paper size. If you are using US letter-sized paper, please close this file and download the file for "MSW US ltr format".

Maintaining the Integrity of the Specifications

The template is used to format your paper and style the text. All margins, column widths, line spaces, and text fonts are prescribed; please do not alter them. You may note peculiarities. For example, the head margin in this

CONCLUSIONS

This study has demonstrated the usefulness of kriging as a tool to determine the reduced level of the rock in Bangalore considering a large data set (652 points) distributed over 220 sq·km area. Geostatistics has permitted the development of a semivariogram model for predicting reduced level of the rock in Bangalore.

The power of geostatistics has become even more apparent through the estimated reduced level of the rock in a way that is consistent with what we know from the field data. By the use of semivariogram, it is possible to make estimation of the reduced level of the rock at points of the site where reduced levels of the rock were not known. The models have been developed by using simple as well as ordinary kriging methods. Variance map for both kriging techniques have been generated and presented. A new type of cross-validation analysis (Q1 and Q2) which proves the robustness of the developed kriging models has been also presented in this study. Ordinary kriging yielded better results than simple kriging for estimation of reduced level of the rock in the subsurface of Bangalore.

REFERENCES

1. G. Matheron, "Principles of Geostatistics," Society of Economic Geologists, Vol. 58, No. 8, 1963, pp. 1246- 1266. doi:10.2113/gsecongeo.58.8.1246
2. E. H. Isaaks and R. M. Srivastava, "An Introduction to Applied Geostatics," Oxford University Press, New York, 1989.
3. J. C. Davis, "Statistics and Data Analysis in Geology," 3rd Edition, Wiley, New York, 2002.
4. J. P. Delhomme, "Spatial Variability and Uncertainty in Groundwater Flow Parameters: A Geostatistical Approach," Water Resources Research, Vol. 15, No. 2, 1979, pp. 269-280.doi:10.1029/WR015i002p00269
5. F. Gambolati and G. Volpi, "Groundwater Contour Mapping in Venic by Stochastic Interpolators 1. Theroy". Water Resources Research, Vol. 15, No. 2, 1979, pp. 281-290.doi:10.1029/WR015i002p00281
6. M. Soulie, "Geostatistical Applications in Goetechnics," Geostatistics for Naturai Resources Characterization: Part 2, NATO ASI Series," Reidel Publishing Company, Dordrecht, 1983, pp. 703-730.
7. P. H. S. W. Kulatilake and A. Ghosh, "An Investigation into Accuracy of Sparial Variation Estimation Using Static Cone Penetrometer Data," Proceeding of the First Internatioanl Symposium on Penetration Testing, Orlando, 1988, pp. 815-821.
8. P. H. S. W. Kulatilake, "Probabilistic Potentiometric Surface Mapping," Journal of Geotechnical & Geoenvironmental Engineering, Vol. 115, No. 11, 1989, pp. 1569-1587.doi:10.1061/(ASCE)0733-9410(1989)115:11(1569)
9. M. Soulie, P. Montes and V. Sivestri, "Modeling Spatial Variability of Soil Parameters," Canadian Geotechnical Journal, Vol. 27, No. 5, 1990, pp. 617-630. doi:10.1139/t90-076

10. P. Chiasson, J. Lafleur, M. Soulie and K. T. Law, "Characterizing Spatial Variability of Clay by Geostatistics," Canadian Geotechnical Journal, Vol. 32, No. 1, 1995, pp. 1-10.doi:10.1139/t95-001
11. M. W. O'Neill and L. M. Yoon, "Spatial Variability of CPT Parameters at University of Houston NGES," Probabilistic Site Characterization at the National Geotechnical Experimental Sites, Geotechnical Special Publication, Vol. 121, 2004, pp.1-12.
12. K. M. Dawson and L. G. Baise, "Three-Dimensional Liquefaction Potential Analysis Using Geostatistical Interpolation," Soil Dynamics and Earthquake Engineering, Vol. 52, No. 5, 2005, pp. 369-381. doi:10.1016/j.soildyn.2005.02.008
13. B. P. Radhakrishna and R. Vaidyanadhan. "Geology of Karnataka," Geological Society of India, Bangalore, 1997.
14. G. Matheron,"Théorie des Variables Régionalisées in Traité d'Informatique Géologique," Masson, Paris, 1972, pp. 306-378.
15. A. Guillaume, "Introduction a la Géologie Quantitative," Masson, Paris, 1977.
16. P. K. Kitanidis, "Introduction to Geostatistics: Applications in Hydrogeology," University Press, Cambridge, 1997, pp. 86-95.
17. C. V. Deutsh, "Correcting for Negative Weights in Ordinary Kriging," Computers & Geosciences, Vol. 22, No. 7, 1996, pp. 765-773. doi:10.1016/0098-3004(96)00005-2

Chapter 7

PROPOSAL OF A NEW PARAMETER FOR THE WEATHERING CHARACTERIZATION OF CARBONATE FLYSCH-LIKE ROCK MASSES: THE POTENTIAL DEGRADATION INDEX (PDI)

M. Cano, R. Tomás

Departamento de Ingenierı́a Civil, Escuela Polité́cnica Superior, Universidad de Alicante, 03080 Alicante, Spain

ABSTRACT

The susceptibility of clay bearing rocks to weathering (erosion and/or differential degradation) is known to influence the stability of heterogeneous slopes. However, not all of these rocks show the same behavior, as there are considerable differences in the speed and type of weathering observed. As such, it is very important to establish relationships between behaviors quantified in a laboratory environment with that observed in the field. The slake durability test is the laboratory test most commonly used to evaluate the relationship between slaking behavior and rock durability. However, it has a number of disadvantages; it does not account for changes in shape and size in fragments retained in the 2 mm sieve, nor does its most commonly used index (Id_2) accurately reflect weathering behavior observed in the field. The main aim of this paper is to propose a simple methodology for characterizing the weathering behavior of carbonate lithologies that outcrop in heterogeneous rock masses (such as Flysch slopes), for use by practitioners. To this end, the Potential Degradation Index (PDI) is proposed. This is calculated using the fragment size distribution curves taken from material retained in the drum after each cycle of the slake durability test. The number of slaking cycles has also been increased to five. Through laboratory testing of 117 samples of carbonate rocks, extracted from strata in selected slopes, 6 different rock types were established based on their slaking behavior, and corresponding to the different weathering behaviors observed in the field.

INTRODUCTION

In heterogeneous slopes, degradation mechanisms (and the choice of subsequent remedial works to be adopted) are closely linked to the weathering behavior of the different lithologies which make up the slope. In these slopes, instabilities resulting from differential erosion/degradation are common (Cano and Tomás2013a, b). This is of most importance in areas with an abrupt topography, where any linear works result in numerous cuttings and slopes. Throughout the service life of a transportation corridor, or any other asset whose construction results in the creation of slopes which are permanently exposed to the atmosphere, these instabilities result in significant maintenance and repair costs, and may pose a significant safety hazard. In the most extreme cases, these processes may trigger the failure of the entire slope, either suddenly or gradually. Additionally, these degradation processes may often be considered as failure mechanisms in their own right, rather than just triggering factors. Thus, the problem of slope stability over time should also be considered, where protection against weathering processes cannot be guaranteed (Mišccevic and Vlastelica 2014).

The phenomenon of slaking consists of the disintegration of clay-bearing rocks due to their interaction with water, which is common when they are exposed to the atmosphere. In areas whose climate is characterized by the absence of frosts and high temperature gradients, the weathering of the different lithologies is mainly caused by drying–wetting cycles due to rainfall and atmospheric moisture, meaning that in these areas the study of slaking is intrinsically linked to weathering behavior. The study area, located on the Mediterranean coast of Alicante, precipitation is scarce, irregular and random. The summer drought extends from three to 5 months, with few rainy days. At the end of this period, the autumnal heavy downpours cause numerous episodes of flooding. Cloudiness and fog is also scarce, so the number of clear days is very high, with nearly 2900 h of annual sunlight. The annual average temperature is 18.3 °C and there is practically no meteorological winter (AEMET 2012, Table 1). The potential evapotranspiration is high, with a Thornthwaite index of 896 mm, so there is a strong water deficit during most of the year (INGEMISA and Auernheimer1991).

Table 1: Normal climatic values of Alicante from 1981 to 2010, AEMET (2012)

Month	T(°C)	TM (°C)	Tm (°C)	R(mm)	RM (mm)	Rm (mm)	RMd (mm)	H(%)	DR	DF	I
January	11.7	17.0	6.3	22.8	82.0	0.0	54.8	67	3.6	0.4	181.2
February	12.3	17.6	7.1	22.1	95.1	0.0	40.5	66	3.0	0.3	180.3

March	14.2	19.6	8.9	23.0	79.7	0.2	32.8	65	3.4	0.0	226.9
April	16.1	21.3	10.9	28.7	91.7	1.2	40.8	63	4.1	0.0	247.0
May	19.1	24.1	14.1	27.8	88.7	0.0	42.1	64	4.0	0.0	277.4
June	22.9	27.8	18.1	11.9	56.8	0.0	45.1	63	1.8	0.0	302.3
Juliet	25.5	30.3	20.7	3.8	41.3	0.0	27.5	65	0.6	0.0	330.1
August	26.0	30.8	21.2	6.8	39.5	0.0	36.4	67	1.1	0.0	303.9
September	23.5	28.5	18.5	55.5	309.3	Imperceptible	270.2	69	3.3	0.0	249.9
October	19.7	24.9	14.5	47.4	271.1	0.1	220.2	70	4.5	0.0	216.7
November	15.4	20.5	10.3	35.9	117.0	0.0	68.3	69	4.2	0.0	173.4
December	12.6	17.7	7.4	25.4	170.9	0.6	119.8	68	3.8	0.1	163.8
Year	18.3	23.3	13.2	311.1	653.1	108.9	270.2	66	37.5	0.9	2850.9

T monthly/annual average temperature, *TM* monthly/annual average of daily maximum temperature, *Tm* monthly/annual average of daily minimum temperature, *R* = monthly/annual average rainfall, *RM* monthly/annual maximum rainfall, *Rm* monthly/annual minimum rainfall, *RMd* monthly/annual daily maximum rainfall, *H* average relative moisture, *DR* monthly/annual average days with rainfall higher to 1 mm, *DF* monthly/annual average of frosty days, *I* monthly/annual average hours of sunlight

Resistance to slaking depends on many different parameters, commonly cited in the literature as; permeability, porosity, adsorption, mineralogy, microscopic texture, microfabric, presence of microfractures, etc. (Gamble 1971; Franklin and Chandra 1972; Richardson and Long 1987; Taylor 1988; Dick et al. 1994; Dick and Shakoor 1995; Martínez-Bofill et al. 2004; Erguler and Ulusay 2009; Kaufhold et al. 2013; Gautam and Shakoor 2015; Cano and Tomás 2015). This makes the characterization of slaking behavior in rocks using a single parameter extremely complex (Erguler and Ulusay 2009; Gautam and Shakoor 2015).

The slaking susceptibility of clay-bearing rocks is known to influence the stability of heterogeneous slopes, as it may result in differential erosion/degradation. However, not all of these rocks show the same behavior. Considerable differences in the manner of degradation and time taken to degrade may be observed. For this reason, it is extremely important to link behavior quantified in the laboratory to that observed in the field, for example, using weathering patterns and weathering profiles (Cano and Tomás 2015).

Different rock masses show a specific response to each particular combination of weathering-related parameters present in situ, and the intensity and rate of this response determines its susceptibility to weathering, as described by Hack (1998). Results obtained using indices derived from Slake tests are generally useful for predicting the performance of different types of rocks qualitatively, although their use for the prediction of quantitative behavior in field conditions is extremely questionable. However, correlations between slake durability and field performance do exist, as observed by Dick and Shakoor (1995) and Shakoor (1995). In addition, Hack (1998), Hack and Huisman (2002) and Nicholson (2001) have demonstrated that laboratory tests have severe limitations in predicting in situ rock mass performance, especially when discontinuities influence the rock mass behavior. This should be taken into account when using indices obtained in a laboratory environment.

The Slake Durability test (Franklin and Chandra 1972) is the most widely used test worldwide for determining the relationship between slaking and rock durability. This importance is underlined by its endorsement by the International Society for Rock Mechanics (ISRM 1981). Subsequently, the method was standardized by the American Society for Testing and Materials (American Society for testing and Materials (ASTM) 2004), where the second cycle index (Id_2) is denominated the Slake Durability Index, and used to quantify a rock's susceptibility to slaking. However, the test has numerous disadvantages, which will be discussed in subsequent sections of this paper.

The main aim of this paper is to propose a method for characterizing the weathering behavior of carbonate lithologies that outcrop in heterogeneous Flysch-like slopes. The aim of the authors is that the method should be simple enough to be easily used in practice. This will be achieved by analysing the changes in the fragment size distribution of the material retained in the drum during the Slake Durability Test (ASTM 2004). This is important, as the test occasionally does not differentiate between rocks with different slaking behavior, as it only measures the mass of particles smaller than 2 mm lost after each test cycle. This means that any particle larger than 2 mm is retained in the drum, and hence used to calculate the Id index. However, in some cases particles smaller than 2 mm make up a very small percentage of the overall sample mass, but the fraction of the initial sample that is retained in the drum disintegrates into numerous smaller particles. The result is that the Id indices do not correlate with the observed in situ durability.

Erguler and Shakoor (2009) proposed a new method to quantify the nature of rocks. This method quantifies the fragment size distribution of the slaked material, using a "disintegration ratio" calculated for each slaking cycle. This is defined as the ratio between the area under the fragment size distribution curve and the total area encompassing the entire range of the fragment size distribution. Erguler and Ulusay (2009) also suggested a disintegration index that can minimize some of the limitations of slake durability test. The comparison between the disintegration index values measured in the laboratory specimens and those in samples from the same outcrops, exposed to atmospheric conditions for 1 year, showed close agreement. Gautam and Shakoor (2013) proposed a method in which a "disintegration ratio" parameter was calculated from the fragment size distribution curves obtained from samples prepared in a similar manner to those used in the slake durability test, but exposed to natural climatic conditions for 1 year. Subsequently, the same authors (Gautam and Shakoor 2015) compared the laboratory slaking behavior of common clay-bearing rocks to their slaking behavior under natural climatic conditions observed during the aforementioned 1-year experimental study. In both studies the same "disintegration ratio" was used.

Similar to the aforementioned studies, this paper aims to evaluate the degradation potential of a sample by analysing changes in the fragment size distribution curves obtained from the material retained in the drum after each slake durability test cycle, up to a total of five cycles. This means that changes in the type, morphology and number of fragments are accounted for. As such, a single parameter for characterizing slaking behavior is proposed, and the behavior observed in the laboratory is compared with that observed in the field, under natural climatic conditions.

The study area chosen is situated in Southeastern Spain, in the coastal area of the province of Alicante (Figure. 1). From this area, 117 samples were taken, representative of the different carbonate Flysch lithologies present in Alicante. The area has a very abrupt topography, with a high population density and three main transportation corridors, which pass through the Flysch belt. This has resulted in a large number of cuttings, which show a range of instabilities related to the differential degradation of the different lithologies present.

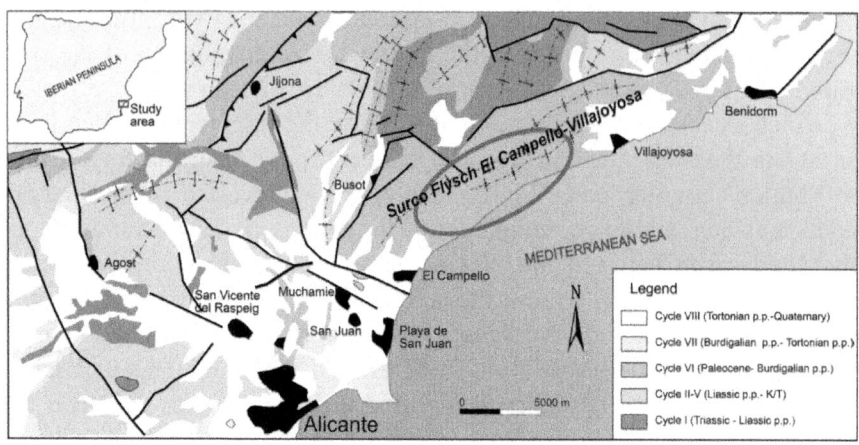

Figure. 1: Location and geological sketch maps of the study area [based on Vera (2004) in Guerrera et al. (2006)]. The *ellipse* indicates the location of the rock exposures in this study.

LITHOLOGICAL SETTING OF THE STUDY AREA

General Framework

The Alicante Flysch sequence (Figure. 1) is composed of pelagic sediments, predominated by sequences of grey marls and thin white marly limestones (hemipelagites) that constitute the rythmite predominated by marls. This sequence may overlap calcarenitic turbiditic episodes. However, the sedimentological complexity of the Flysch formation is even greater because some superposed composite gravitational processes such as mélanges and debrites are also present (Cano and Tomás 2013a).

In this study, five slopes were selected and fully characterized (Figure. 1). 117 intact rock samples were extracted from all of the strata present in the selected slopes and were described in detail in the field. They were geologically classified as:

- thick bedding calcarenites [Grainstone of turbiditic facies of channel (Ta-b)];
- thick bedding calcarenites [grainstone of turbiditic facies of channel (Ta-b) or sheet flood facies (Tb, Tb-c)];
- Thin bedding calcarenites [turbiditic thin beds of fan fringe facies (Tb-c-d)];

- poorly cemented thin bedding calcarenites [turbiditic thin beds of fan fringe facies (Tb-c-d)];
- slightly marly limestones;
- marly limestones;
- silty calcareous marls;
- silty marls;
- calcareous marls–marls;
- sheet silty marls;
- soft marls;
- sheet marls;
- soft calcareous mélanges; and
- calcareous debrites.

Characterization of Rock Mass Jointing

At regional level, the area is fractured into several blocks linked to several main dextran fault systems, such as the Cadiz-Alicante fault (N70E), which generated N20E oriented folding, the Vinalopo fault system (N155E) and the Socovos fault system (N120E) that generated N70E oriented folding (Guerrera et al. 2006).

At local scale, six sets of discontinuities were observed in the slopes in the study, five of which are of tectonic origin (J1–J5). The sixth set corresponds to bedding. These five tectonic joints present a quasi-perpendicular disposition to the bedding, generating prismatic blocks with a variety of sizes and shapes for the different lithologies (Figure. 2a).

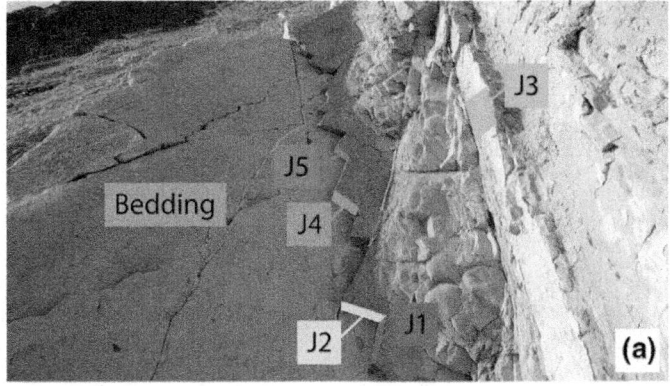

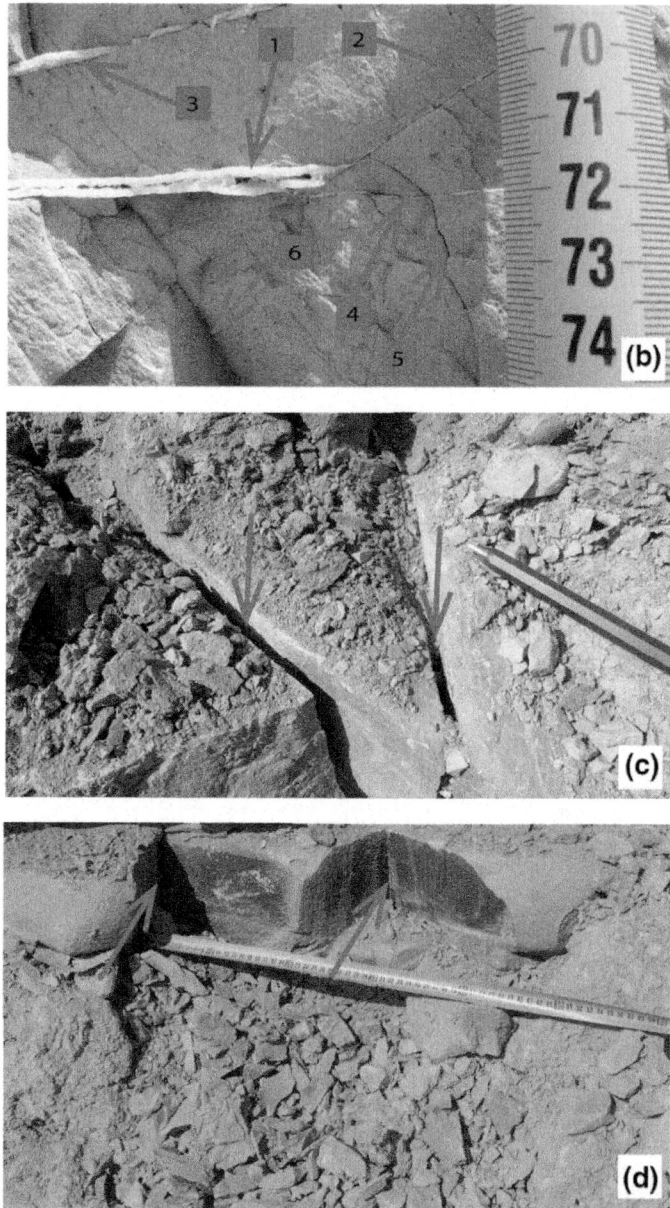

Figure. 2: **a** Plan view of a slope in this study, in which the different discontinuity sets are recognized. **b** Different aperture and infilling cases from the most carbonate lithologies: *1* opening of 4 mm partially filled of calcite, *2* joint of 1 mm partially filled, *3* joint with 1.5 mm of calcite infilling width, *4* joint of 0.5 mm of calcite infilling width, *5* joint with an aperture of 0.5 mm, without filling, *6* joint very tight (<0.1 mm).

c Appearance of a set of thick bedding calcarenites with dislocated blocks because of mechanical excavation. **d** Appearance of a set of calcareous marls-marls with interlocked blocks. Mechanical excavation has not affected these discontinuities.

The two first sets (J1, J2) exhibit a very large persistence (>20 m), and the isolated blocks present a parallelepiped shape. Their average spacings are 28 and 18 cm, respectively. J3 and J4 present a much lower persistence (1.4 and 2.2 m, respectively) and average spacings of 47 and 25 cm, respectively. J5 presents short to very short persistence (1.1 m) and 60 cm of average spacing.

The sixth set of discontinuities corresponds to bedding, which shows a very large persistence. As the studied slopes are heterogeneous, the bedding has been specifically described for each of the lithologies outcropping in the study area. Typically, the thick bedding calcarenites units are 25–120 cm thick and the bedding spacing is 7–30 cm. They are thick bedded blocky, consisting of tabular blocks formed by two to five other intersecting discontinuity sets. The size and shape of these blocks is different to other, less competent lithologies. This is because some discontinuities are well cemented, behaving as if there were no such discontinuity. The slightly marly limestones units are 12–60 cm thick, the bedding spacing is 10–30 cm and they are prismatic blocky. The calcareous mélange units are 50–200 cm thick and exhibit a chaotic structure with a high erratic discontinuity surfaces density that results in heterometric rock blocks (centimetric to decimetric) with different morphologies. The thin bedding calcarenites (C) units are 5–20 cm thick, the bedding spacing is 3–10 cm and they are very blocky with prismatic shape. The thin bedding calcarenites (L) units are 5–30 cm thick and the bedding spacing is 3–10 cm. They are thin bedded blocky consisting of parallelepipedic blocks. The only observed unit of calcareous debrites in the studied slopes is 600 cm thick and exhibits a chaotic structure constituted by blocks and a calcareous matrix with a high erratic discontinuity surfaces density that generates heterometric rock blocks (centimetric to decametric block size) with different morphologies. The only unit of soft calcareous mélange encountered is 20 cm thick and has a chaotic structure. Thin bedding silty calcarenites units are 2–50 cm thick, the bedding spacing is 2–10 cm. They are thin bedded blocky consisting of parallelepipedic blocks. The marly limestones units are 10–80 cm thick and the bedding spacing is 8–30 cm. They are very blocky with prismatic shape, in the same manner as the other lithologies listed below. The silty calcareous marls units are 15–50 cm thick and the bedding spacing is 8–15 cm. The silty marls units are 20–70 cm thick and the bedding spacing is 10–14 cm. The calcareous marls–marls units are 15–160 cm thick and the bedding spacing is 9–30 cm. The sheet silty marls units are 30–160 cm thick and the bedding spacing is 10–

23 cm. Soft marls are 25–40 cm thick and the bedding spacing is 10–15 cm.

Regarding openings, it should be noted that in the studied rock masses the presence of infilling in the openings of discontinuities, which are extension joints (not veins), highly depends on the lithology of the unit. As such, only in the following lithologies, which showed high contents of carbonate, are discontinuities filled by free crystals of calcite: thick bedding calcarenites, slightly marly limestones, calcareous mélange, thin bedding calcarenites (C), calcareous debrites, marly limestones, silty calcareous marls and soft calcareous mélange. In these carbonatic lithologies, the opening of joint J1 varies between 0.5 and 10 mm and is completely or partially filled by calcite. The width of J2 ranges from very tight (<0.1 mm) to open (1 mm) and the filling is also made of calcite that can partially fill the discontinuity space or even not be present, leading to a partially open to open discontinuity. Opening ranges are from 0.5 to 5 mm for J3 and from 0.1 to 3 mm for J4. In both cases, the joint is filled or partially filled by calcite. J5 exhibits 1 to 3 mm calcite infilling width (Figure. 2b). Otherwise, in the marly lithologies, the discontinuities also present an aperture varying between <0.1 mm to 10 mm, but without calcite infill. Finally, the joints that correspond to bedding are mainly very tight or they are cemented in the more calcareous lithologies. Additionally, a patina of iron oxide and manganese oxide species in dendritic form has been commonly observed in all sets of discontinuities, including bedding.

Both intact rock and discontinuities exhibited a weathered state, from the face of the slope to certain depth. The degree of alteration changed according to the lithology and the depth, according to Cano and Tomás (2015).

Generally, both, the tectonic discontinuities and the bedding are planar, mainly rough or sometimes slightly rough in the more carbonatic lithologies and smooth or slightly rough in the marly lithologies. Occasionally, the tectonic discontinuities of the units of thick bedding calcarenites, mainly the thickest ones, show evidence of karstification, which generates more undulated or stepped, and very rough joints, and also greater apertures (20 mm). Occasionally, the bedding surface presents load casts at the bottom of calcarenite units of metric thickness at the contact with marly lithologies, generating very rough joints with large undulation. Sometimes, the bedding presents slickensided surfaces.

Water flow was not observed, although after downpours the discontinuities and the intact rock of marly lithologies were wet.

The effect of the excavation method on the disturbance of a rock mass is a well-known phenomenon (Romana 1993; Hoek et al. 2002). However, a large number of factors can influence the degree of disturbance in the rock mass surrounding an excavation, and it may never be possible to quantify these factors precisely (Hoek et al. 2002). Four of the five studied slopes were excavated using mechanical methods and the fifth slope is a natural hill. Due to the heterogeneous nature of these slopes, when the cuts are excavated by means of mechanical methods, the low competence lithologies (e.g., soft marls, silty marls, etc.) are sheared through the rock matrix. However, in the high competence sets (e.g., thick bedding calcarenites) the blocks are broken off through their discontinuities, dislocating them towards the slope face. This is the reason why in the slope face of these sets, the aperture of the discontinuities is higher than that observed in the inner part of the slope, in which the blocks have not been perturbed by the mechanical action of the excavation (Figure. 2c, d). Another cause of the greater discontinuity opening in the more competent blocks from the slope face is the dislocation of rock blocks due to removing underlying unit effects of erosion and/or differential degradation.

METHODOLOGY

General Overview

The main aim of this paper is to characterize the slaking behavior of the different carbonate lithologies outcropping in the study area, to aid the prediction of their weathering behavior following the excavation of a slope or cutting. To this end, the different lithologies were identified and described in the field and their mineralogical characteristics obtained. 5-cycle slake durability tests were performed on intact rock samples. Fragment size distribution curves were also obtained and their morphology analyzed. A modified parameter (D_{RP}) based on the original "disintegration ratio" (D_R) proposed by Erguler and Shakoor (2009), has also been proposed. From D_{RP}, a novel parameter named potential disintegration index (PDI) based on the change in the D_{RP} ratio between slake cycles has been defined. The combined use of the PDI, together with the analysis of the shape of the particle size distribution curves, and the behavior of retained fragments throughout the five cycles of the slake durability test (changes in size and shape) has allowed a new classification of the slake behavior of these lithologies to be proposed. Additionally, the proposed slake behavior classes were compared with the weathering patterns and weathering profiles observed in the same lithologies in the field (Cano and Tomás 2015) (Figure. 3). The methodology used in this study is described in the following paragraphs.

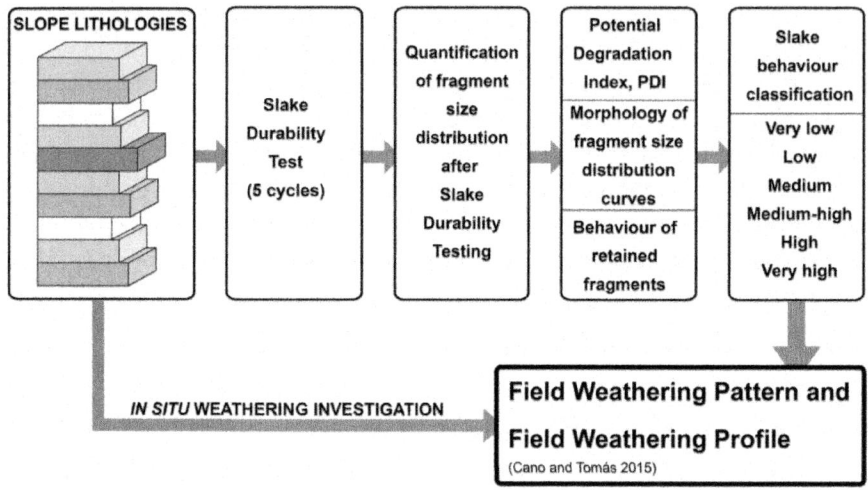

Figure. 3: Conceptual sketch of the slaking and weathering characterization of the Flysch lithologies.

Intact Rock Mineralogy

In this study, the different lithologies were described in the field using a simplified geological classification of rocks based on their genetic classification, structure, composition and grain size (Geological Society of London 1977). Additionally, a mineralogical characterization of the samples by X-ray diffraction was performed. Because some of the samples were of marly composition, they were characterized in two different stages. Firstly, X-ray diffractograms of all the samples were obtained. Secondly, X-ray diffractograms of the oriented aggregate of samples with high phyllosilicate content were obtained, to identify them according to Robert and Tessier's (1974) methodology. Finally, for some representative samples, the carbonate contents obtained from the interpretation of the X-ray diffractograms were compared with those obtained using the Bernard calcimeter test (ASTM 2007a), to validate these results.

Data were collected and interpreted using the XPowder software package, (Martin 2004) whose qualitative search-matching procedure was based on the ICDD-PDF2 database.

Assessment of Fragment Size Distributions after the Slake Durability Test

The Slake Durability Test (SDT) is one of the simplest tests in rock mechanics, and is the most widely used test worldwide for characterizing the environmental weathering resistance of rock. Although originally the slake durability test was developed for testing the weathering potential of shales, mudstones, siltstones, and other clay-bearing rocks (Franklin and Chandra 1972), the slake durability index is also typically used for testing weak rocks such as mudstones, marls, ignimbrites, conglomerates, and poorly cemented sandstones (Sabatakakis et al. 1993; Santi 1998; Czerewko and Crips 2001; Erguler and Ulusay 2009; Miščcević and Vlastelica 2011). Consequently, although in the Flysch formation there are some very competent, hard turbiditic rocks that show very high durability indices, to classify the Flysch lithologies using a uniform weathering potential criteria, the slake durability test was used for testing all of the samples.

The durability of weak rocks is usually assessed using the second-cycle slake durability index (Id_2). Nevertheless, some authors (Gamble 1971; Taylor 1988; Moon and Beattie 1995; Ulusay et al. 1995; Bell et al. 1997; Gökçeoğlu et al. 2000; Erguler and Shakoor 2009; Miščcević and Vlastelica 2011) have suggested that index values taken after three or more cycles of slaking and drying may be useful when evaluating higher durability rocks, such as those in this study.

As part of a previous study, it was observed that within the study area some intact rock samples showed high Id1 and Id2 indices [after Franklin and Chandra (1972) and Gamble (1971)], in contrast with the weathering behavior observed in situ. As such, it was concluded that the observed weathering of the rocks was much higher than that predicted by the SDT indices (Cano and Tomás 2015).

This appeared to be related to the fact that, despite the high Id_2 values obtained, the sample retained in the drum was extremely fragmented and visually appeared highly degraded. However, the fragments were larger than 2 mm, leading to the high Id_2 values obtained (see Figure. 4).

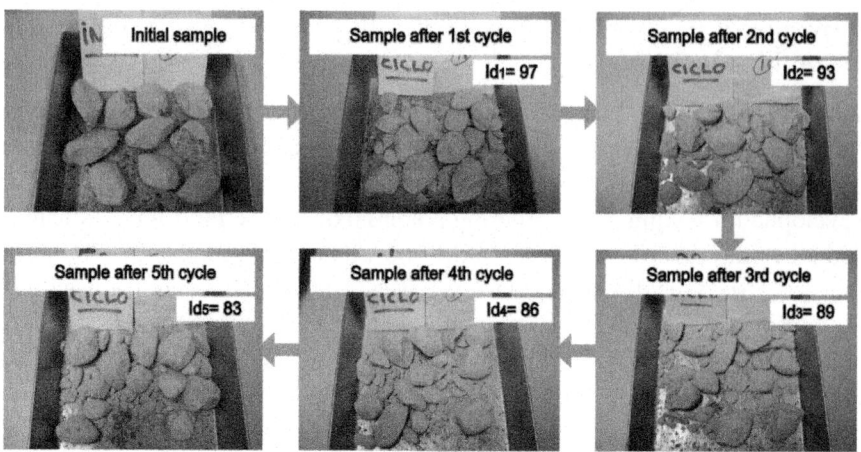

Figure. 4: Example of heavily degraded calcareous marls after successive cycles of slake durability test (SDT) with $Id_2 = 93$. Note that two-cycle (Id_2) SDT results classify this sample as "high durability". The Id values calculated for each cycle are also shown.

The American Society for Testing and Materials requires that as part of the slake durability test, in addition to the Id_2 index, the fragments retained in the drum should be qualitatively categorized as type I material (primarily large fragments), type II material (mixture of large and small fragments), or type III material (primarily small fragments) (American Society for testing and Materials (ASTM) 2004). However, Erguler and Shakoor (2009) demonstrated these categories are insufficiently detailed to give a refined classification. As a consequence, it is obvious that the Id_2 index does not adequately reproduce the real degradation properties of the Flysch lithologies studied, providing optimistic values.

It should be noted that rock specimens found on superficial parts of slopes usually show signs of weathering or even severe degradation. As a consequence, the Flysch rock samples tested correspond to intact rocks that were obtained from the inner part of the slope. Subsequently, the intact rock samples were transported to the laboratory in plastic bags, and maintained at a constant temperature. The time between storage and testing was always less than 1 week. While the tests were performed, the laboratory temperature was also kept constant ($24 \pm 2\,°C$) to conserve humidity and temperature conditions.

The tests were performed according to the ASTM (2004), with five test cycles performed. Five cycles were chosen because of: (a) the need to compare hard and soft lithologies using the same parameter; (b) the existence of some durable rocks which are unaffected by a low number of cycles; (c) the need to study the rocks' long-term weathering behavior; and (d) the need to avoid an excessively long test period.

It is unrealistic to classify carbonate Flysch lithologies according to their in situ weathering behavior after long-term exposure to real conditions based solely on the indices obtained from the slake durability test. As such, many authors have combined (in various different ways) the results of the slake durability test with the analysis of particle size distribution curves (Erguler and Shakoor 2009; Erguler and Ulusay 2009; Gautam and Shakoor 2013, 2015), with the aim of improving the characterization of various lithologies according to their weathering behavior.

The procedure adopted as part of this study was as follows. Firstly, ten 40–60 g pieces of intact rock were taken from the study area, providing a total sample mass of 450–550 g. The samples were dried for 24 h in an oven at 105 °C, sieved, and immediately placed in the slake durability test apparatus. After the first test cycle, the sample retained in the drum was dried, sieved and weighed. This procedure was repeated for a further four cycles, giving a total number of five test cycles. The sieving procedure adopted for determining fragment size distribution curves was the same as that used for soils (ASTM 2007b), using standard sieves whose aperture sizes were 40, 31.5, 25, 20, 12.5, 10, 6.3, 5 and 2 mm. The results were plotted in semi-logarithmic scale, to show the fragment size distribution of samples before and after each test cycle. The curves were plotted on the same graph, to easily observe changes in the samples after each slaking cycle. The sieve apertures used were shown on the x-axis in semi-logarithmic scale, and the percent passing (by weight) on the y-axis. It is important to note that the samples were sieved with extreme care, to avoid further fragmentation of the particles retained in the drum, which could have been mistakenly attributed to the effects of the previous slaking cycle. Erguler and Shakoor (2009) proposed the disintegration ratio (D_R) parameter as the sole indicator of the effects of each slaking cycle on the fragment size distribution curves. It is defined as:

$$D_R = \frac{A_C}{A_T} \qquad (1)$$

where A_C is the area under any size distribution curve and A_T is the total area encompassing the whole range of fragment size distributions. In this study, a similar parameter is proposed. However, as the graphs used show percent passing (by weight), as opposed to percent retained (by weight), when the parameter is equal to 1 ($A_C = A_T$) this represents the maximum degree of degradation possible. To avoid confusion, the proposed new parameter has been named D_{RP} (Figure. 5).

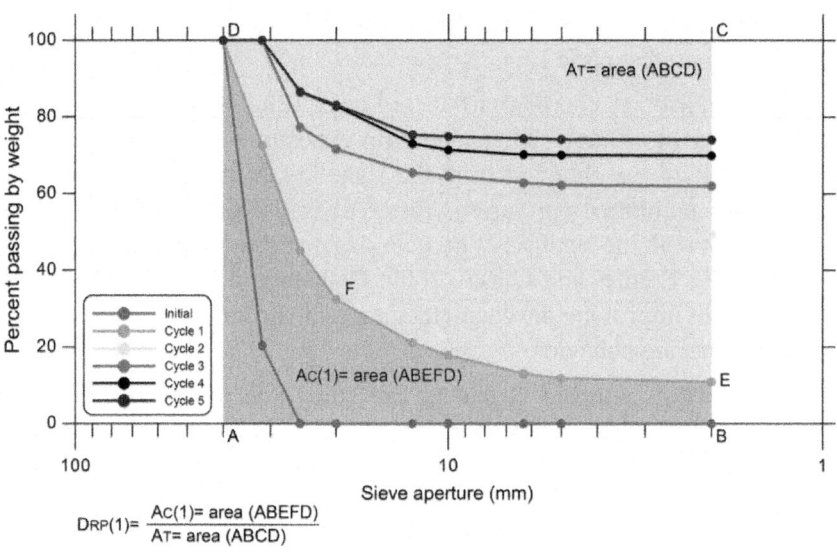

Figure. 5: Example showing the calculation of D_{RP} after each slaking cycle. In this case, after the first cycle $D_{RP}(1)$.

Despite the fact that D_{RP} can be a good indicator of the degradation potential of the weakest rocks, it has been observed that the size and shape of the fragments that are retained in the drum between cycles changes from one cycle to the next, throughout the five slaking cycles. This observation agrees well with in situ observations, where the form and visual condition of lithologies change over time. For this reason, it is proposed that the change in the calculated D_{RP} value between slaking cycles is evaluated. To this end, a logarithmic curve was fitted to the D_{RP} values obtained for each sample and the R^2 and typical error values were calculated (Figure. 6). From this curve, the number of cycles required for a sample to reach 50 % of the maximum possible degradation ($D_{RP} = 1$) could be estimated. This number of cycles is denominated N_{50}. In the example shown in Figure. 6, from the fitting equation:

$$D_{RP} = 0.082 \text{Ln}(Nr) + 0.035 \qquad (2)$$

The Nr value for $D_{RP} = 1$ is $Nr = N_{50} = 290$ cycles. However, owing to the fact that the slaking resistance of the rocks in the study varied greatly, the range of N_{50} values was very large, varying from $N_{50} = 2$ in rocks that were very susceptible to degradation, to $N_{50} = 8.10^{19}$ in rocks which were not. As such, a new parameter was defined to aid the classification of samples—the Potential Degradation Index (PDI). This is calculated as:

$$\text{PDI} = \text{Ln} N_{50} \qquad (3)$$

The values calculated for the samples in this study vary from 0.8 to 46, with a value of PDI = 5.7 calculated for the example in Figure. 6.

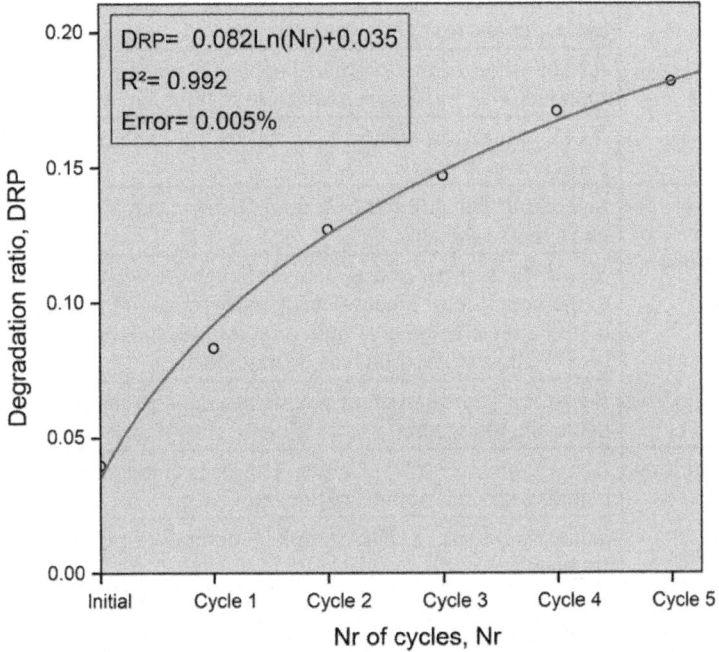

Figure. 6: Example of a logarithmic curve fitted to the plot of the D_{RP} parameter against slake durability test cycles, shown for the sample H15.

As such, using only the Potential Degradation Index, the potential long-term degradation of a given sample from a carbonate lithology may be assessed. However, with the aim of refining the classification limits, to better distinguish between the numerous lithotypes present in the study area, a qualitative study was performed on the fragments retained in the drum during the different cycles of the slake durability test, and a classification was proposed. Three factors are proposed to standardize this classification: roundness, number of fragments, and fragment size (Table 2; Figure. 7).

Table 2: Standardized qualitative factors used to describe the behavior of the fragments retained in the drum throughout the five slaking cycles [adapted from ASTM (2004)]

Roundness	*R1* The edges of the fragments appear similar to before the test cycles, or show slightly blunted edges	
	R2 The edges of the fragments appear completely blunt or lightly rounded. The initial appearance of the fragment may still be noted	
	R3 The fragments show a rounded shape. The initial shape of the particle is not visible	
Variation in number of fragments	*Increasing* The number of fragments increases significantly from one cycle to the next	
	Equal The number of fragments is largely constant from one cycle to the next. Small pieces occasionally break off from the corners of the largest fragments and very occasionally a large fragment breaks into two medium sized fragments	
	Reducing The number of fragments reduces significantly from one cycle to the next	
Fragment size	*Large fragments (L)* The entire sample is composed of large fragments, with occasional smaller fragments	
	Small fragments (S) The sample is composed primarily of small fragments	
	Mixture of large and small fragments (M) The sample is composed of a mixture of fragments of different sizes, which may or may not be rounded	

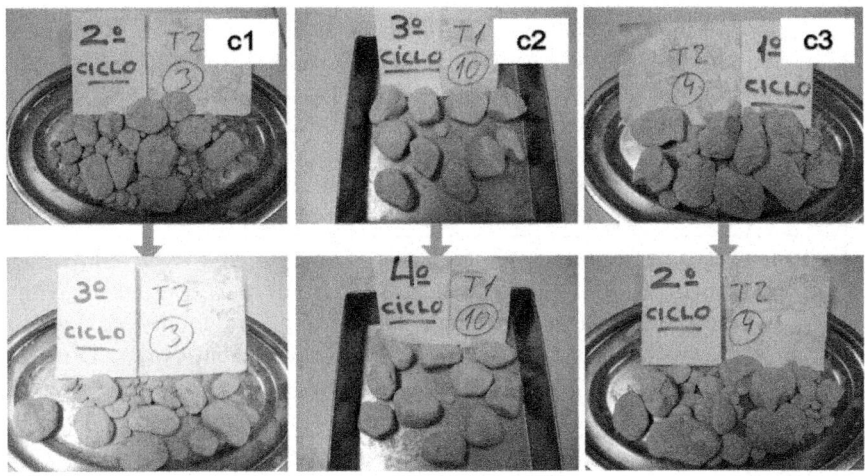

Figure. 7: Samples representative of the three standard factors which have been chosen to describe the behavior of the fragments retained in the drum after each slaking cycle. *a1*, *a2*, and *a3* are examples of roundness types R1, R2 and R3. *b1*, *b2* and *b3* show the different fragment sizes, corresponding to large (L), small (S) and a mixture (M) of fragments, respectively. *c1*, *c2* and *c3* show the variation in number of fragments between two consecutive cycles—reducing, equal and increasing, respectively.

Using these standard factors, 11 slaking behavior patterns were defined based on the changes observed in the fragments. A distinction is made between three different textures present in the lithotypes: compact, laminated and poorly cemented (Table 3).

Table 3: Slaking behavior patterns based on changes observed in the fragments retained in the drum

Compact samples	
Type C1	R3 type roundness of fragments from the first to the last cycle. Initial increase in number of fragments of different sizes (M), maintaining or reducing the number of fragments throughout the test cycles. Fragments of differing size (M) present
Type C2	Initial R2 type roundness, changing to R3 from the second or occasionally the third cycle. Initial increase in number of fragments of differing size (M), with the number of fragments constant or decreasing throughout subsequent cycles. Fragments of differing size (M) present

Type C3	Fragments show R2 roundness until the third cycle, occasionally the second. From the fourth or last cycle the roundness type is R3. In the two first cycles the number of fragments of differing size (M) increases, which is maintained or reduces throughout the following cycles. Habitually during the last or fourth cycle, the smallest fragments tend to disappear, leaving a type L sample
Type C4	R2 type roundness is observed from the first cycle until the fourth, without any increase in the number of fragments, with these being large (L). Occasionally small fragments are generated which tend to disappear in the subsequent cycles. During the fifth cycle the fragments show a more rounded shape (R2 or R3) with very blunt edges
Type C5	Samples retain their initial number of fragments, which have a similar size (L), throughout the five cycles. Occasionally small fragments break off from the corners, or even larger fragments. The roundness of the fragments is type R1 until the third or fourth cycle, and type R2 in the last or last two cycles
Type C6	The samples retain the same number of fragments, and of similar size (L) throughout the five cycles of the Slake Durability Test. Occasionally, small fragments break off from the corners. The roundness of the fragments is type R1 in all cycles, although on occasion R2 roundness is observed during the last cycle
Laminated samples	
These samples show very limited roundness, as they tend to fracture into flat fragments which slide within the drum	
Type L1	Extensive fracturing during the first cycle, into fragments of differing sizes (M), with type R2 roundness observed in the largest fragments and R3 roundness in small- and intermediate-sized fragments. Throughout the following cycles the number of particles of differing sizes (M) is constant or reduces, and the number of fragments with R3 roundness increases
Type L2	During the first cycle very little fracturing is observed, but during the second cycle a greater degree of fracturing is observed, producing fragments of differing size (M), with an R2 type roundness. During subsequent cycles, the number of fragments of differing size is constant or reduces. From the third or fourth cycle some fragments with type R3 roundness are observed

Type L3	Fracturing into fragments of differing sizes (M) is observed during the first two cycles, although the initial large fragments predominate, showing R1 roundness. Throughout subsequent cycles the number of particles of differing sizes (M) is constant or reduces, with the smaller fragments tending to disappear, resulting in a type L sample. From the fourth or fifth cycle R2 roundness is observed
Type L4	Throughout the first two cycles some small fragments are produced, which are maintained or tend to disappear. However, the large initial samples predominate (L–M or L), showing R1 roundness. From the third cycle until the final cycle mixed roundness is observed, with some fragments showing type R1 and others R2
Poorly cemented samples	
The poorly cemented calcarenite samples behaved in a distinct manner, as rounding of the original fragments was predominant as opposed to fragmentation	
Type R	Fragments show type R3 roundness throughout all cycles. The number of large particles (L) is maintained after each cycle, although a reduction in diameter is observed

A qualitative analysis of the changes in morphology in the six fragment size distribution curves obtained for each sample after five slake durability test cycles was performed. The curves showing a similar morphology were grouped.

Using all of the aforementioned parameters (behavior in terms of changes in the retained fragments, changes in the shape of the fragment size distribution curves, and principally the potential degradation index parameter), all of the samples taken as part of the study were classified, and limits between the different classes were established.

However, the procedure required for this classification was somewhat laborious. To improve ease of use, the changes in the morphology of the fragment size distribution curves throughout the five slaking cycles were analysed quantitatively, and a parameter which is simpler to obtain (but correlates well with the potential degradation index) was proposed. For the 117 samples in the study, parameters habitually used in sieve analysis in soil mechanics or sedimentology were calculated, such as the coefficient of uniformity (C_U), curvature coefficient (C_c), diameter for which 50 % of the sample passes (D_{50}), percentage of sample passing a certain sieve size, etc. as well as combinations of these parameters which described the shape of the fragment size distribution curves. Using the IBM SPSS Statistics program, possible correlations between these parameters and the Potential Degradation Index were investigated, with the R^2 and typical error values calculated for each curvilinear approximation. The parameters that showed the best fit were the D_{50} range (i.e. the variation between the initial D_{50} and the value after the fifth slaking cycle) and the sample percentage passing through the 12.5 mm sieve ($P_{12.5}$). These parameters have the additional advantage of quantifying the changes in shape of the fragment size distribution curves (Figure. 8). Of these two parameters, the best correlation was given by $P_{12.5}$. The parameter also has the advantage of allowing the same number of classification groups to be established.

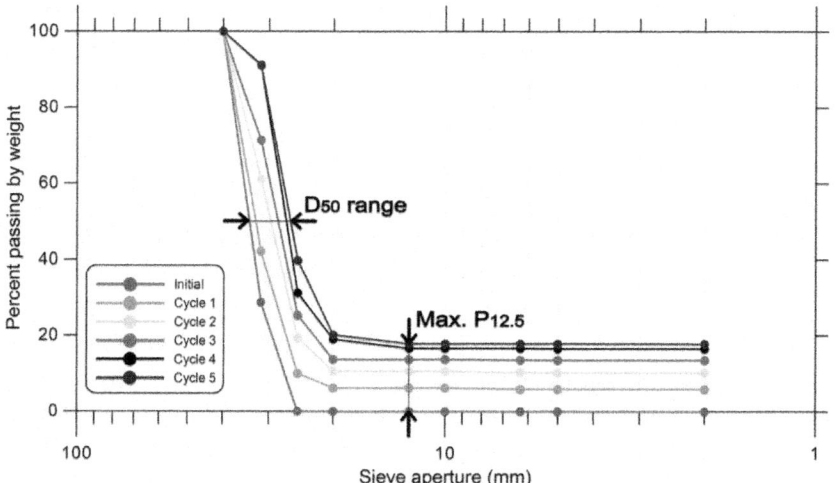

Figure. 8: Calculation of D_{50} range and maximum percent of sample passing through the 12.5 mm sieve ($P_{12.5}$). Note that the higher the $P_{12.5}$ and D_{50} values the worst the slaking behavior of the rock.

Once all of the samples were classified, the behavior of the different lithotypes from which the samples were extracted was studied and compared with the weathering patterns and weathering profiles defined by Cano and Tomás (2015).

SLAKE BEHAVIOR CLASSIFICATION: POTENTIAL DEGRADATION INDEX (PDI)

As discussed in the previous section, changes in the D_{RP} parameter throughout the slake durability test cycles were evaluated for 117 samples (Figure. 6). The R^2 values and typical error were calculated for the logarithmic curves fitted to each sample. All of the curves showed a good fit, with an average R^2 value of $R^2 = 0.977 \pm 0.023$, and an error of $e = 0.05\ \%$.

Using the curves fitted to each sample, the N_{50} value for each was determined, and hence also the PDI parameter (as described in Sect. 3.3), which was used to classify the samples. Six durability classes were established: Very low, Low, Medium, Medium–high, High and Very high (Table 4), which correspond to a specific manner of degradation observed in the fragments retained in the drum (see Figure. 9).

Table 4: Classification of samples based on the Potential Degradation Index (PDI)

Class	PDI	Maximum % Passing 12.5 ($P_{12.5}$)	Slake behavior Patterns (Table 2)	Field weathering profile (Cano and Tomás 2015)	Id_2	Lithotypes
1. Very low	≤1.5	≥50	Type C1	FHIJK	34–61	Sheet marls (1), soft marls (3), calcareous marls–marls (1), sheet silty marls (1)
2. Low	1.5–3	50–23	Type C2	FHIJK	75–93	Sheet silty marls (4), calcareous marls–marls (6), poorly cem. thin bedding calc. (1), Thin bedding silty calc. (1), silty calcareous marls (1)
			Type R	AB (poorly cemented samples)		
			Type L1	LM (laminated samples)		
3. Medium	3–5.5	23–11.5	Type C3	FHIJK	87–95	Thin bedding calcarenites (L) (3), calcareous marls–marls (14), marly limestones (2), soft calcareous mélange (1), silty marls (3), silty calcareous marls (2), thin bedding calcarenites (C) (1), thin bedding silty calc. (1)
				EFG		
			Type C4	AC		
			Type L2	LM (laminated samples)		

4. Medium–high	5.5–8	11.5–7.3	Type C4	EFG	94–97	Calcareous marls–marls (5), calcareous debrites (1), Marly limestones (8), thin bedding calc. (L) (2), thin bedding calc. (C) (1), silty marls (2), thick bedding calc. (1)
				FHIJK		
			Type C5	AC		
			Type L3	LM (laminated samples)		
5. High	8–15	7.3–3.6	Type C5	EFG	97–99	Slightly marly limestones (15), marly limestones (2), silty calcareous marls (1), calcareous mélange (3), thin bedding calc. (L) (1), thin bedding calc. (C) (5), Silty marls (1), Thick bedding calc. (1)
			Type C6	AC		
			Type L4	LM (laminated samples)		
6. Very high	>15	<3.6	Type C6	NW-A	98–100	Slightly marly limestones (10), thin bedding calc. (L) (1), thick bedding calc. (9), thin bedding calc. (C) (1), calcareous mélange (1)
				AC		
				EFG		
				LM		

Additionally, $P_{12.5}$ (which quantifies the changes in shape of the fragment size distribution curves), the slake behavior patterns of fragments during five test cycles, and the field weathering profile are shown

Id_2 is also shown, to compare it with the index proposed in this paper (the number of samples for each lithotype is indicated in brackets)

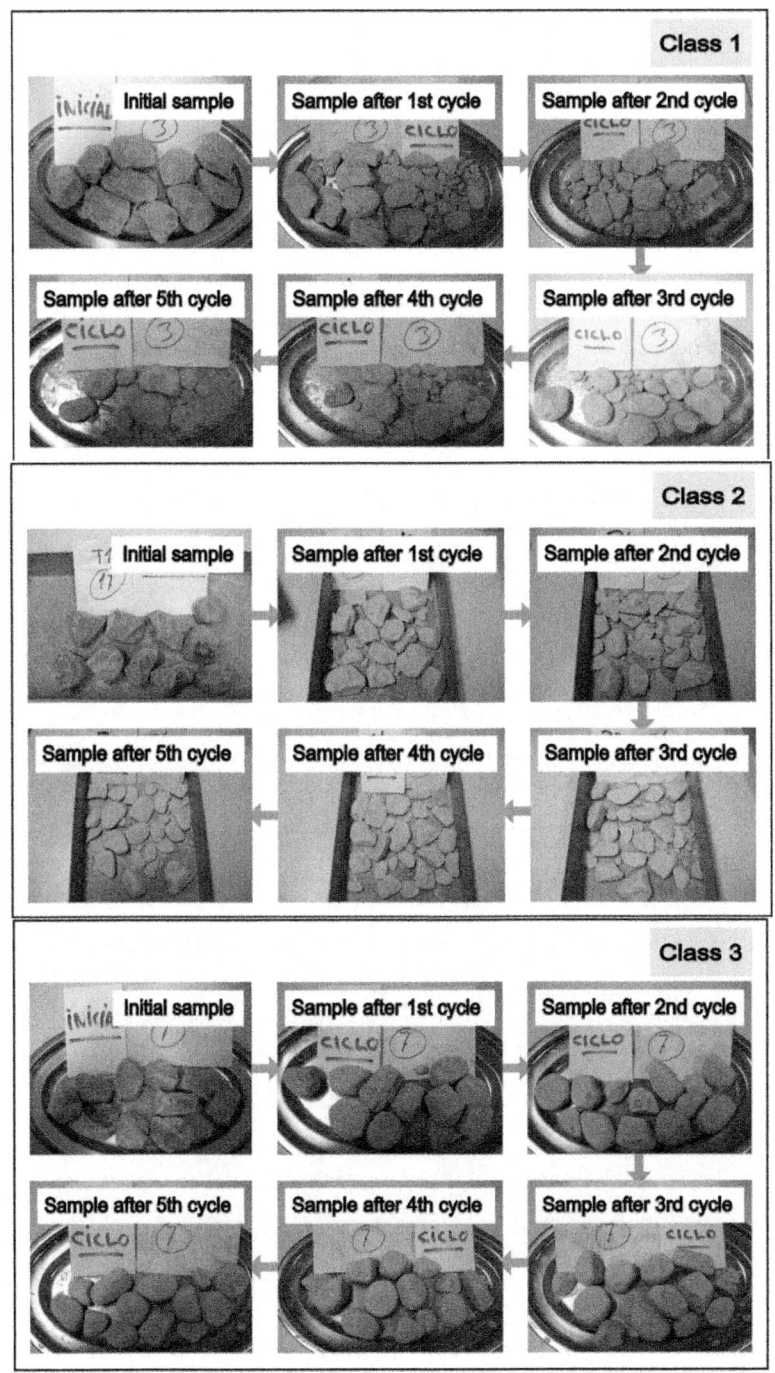

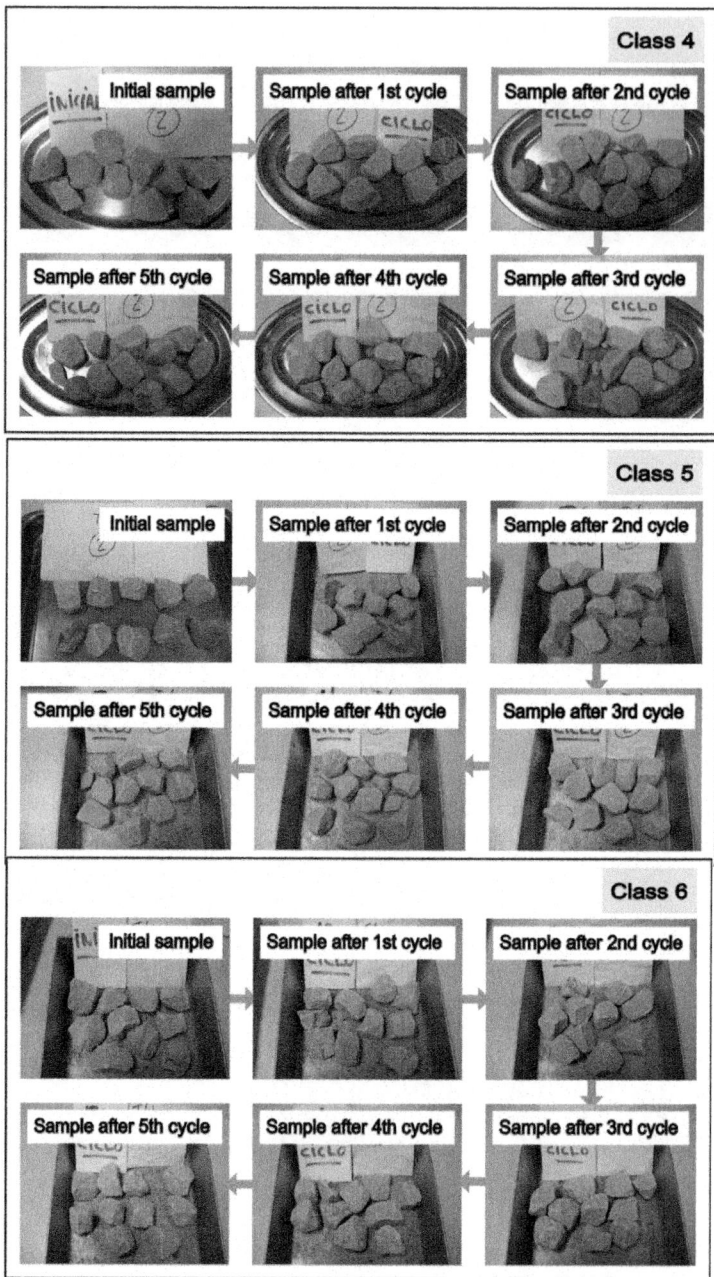

Figure. 9: Models representing each class of slaking behavior, resulting from the analysis of fragments retained in the drum throughout the five slake durability test cycles.

Following qualitative analysis of the change in morphology of the six fragment size distribution curves obtained for each of the five slake durability test cycles, and aided by the quantification of these changes using the D_{50} range, and maximum percent of sample passing through the 12.5 mm sieve $P_{12.5}$ (Figure. 7), similar fragment size distribution curves were grouped together into the same six categories used to group samples by their PDI parameter (Figure. 10).

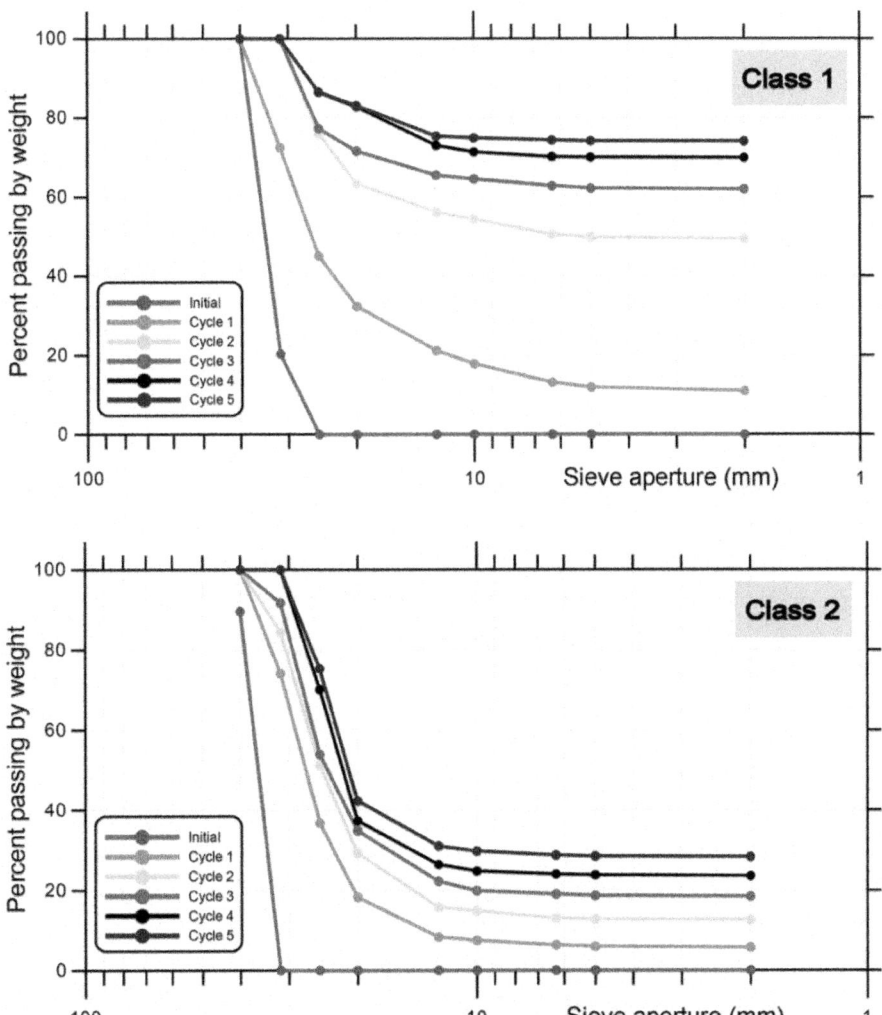

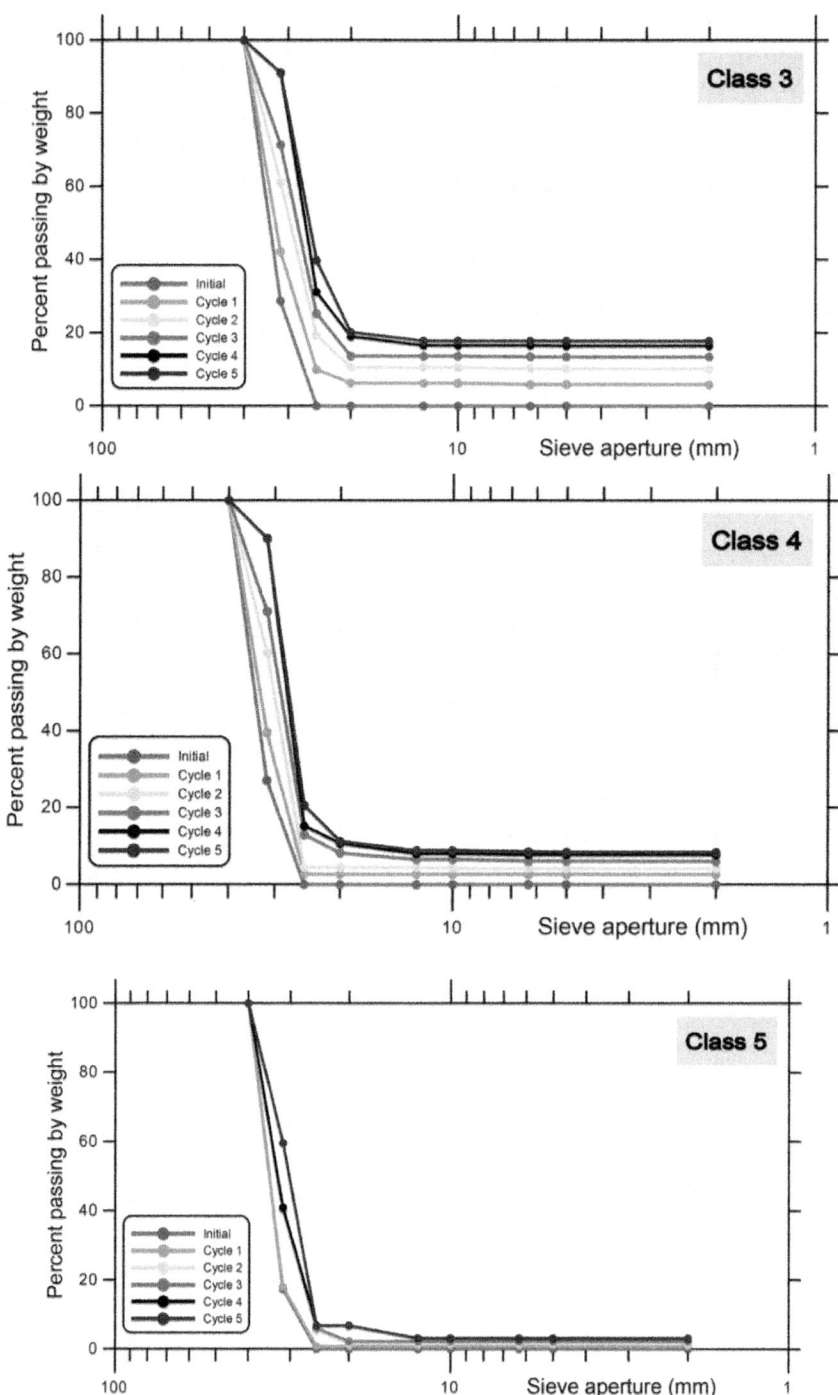

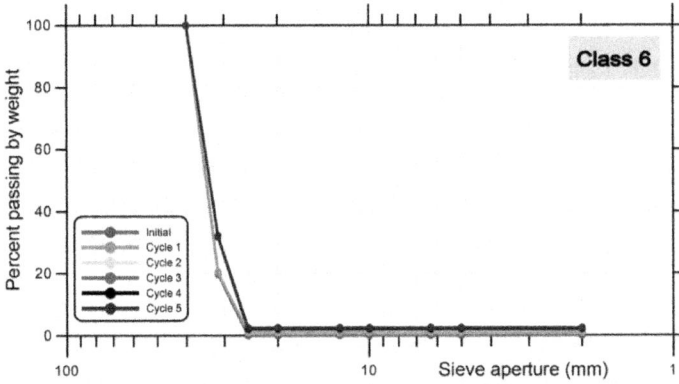

Figure. 10: Models representing each class of slaking behavior, representing the changes in shape in the fragment size distribution curves throughout the five cycles of the slake durability test. Note that the higher the class number the better the slake behavior of the sample.

The calculation of the PDI parameter, while not complex, is somewhat laborious. As such a parameter which is simpler to obtain, but which correlates well with the PDI, is proposed. This is the maximum percent of sample passing through the 12.5 mm sieve ($P_{12.5}$, Figure. 8), which can be fitted to the PDI using a power function (Figure. 11), and used to establish the boundaries of the same six classes (Table 4).

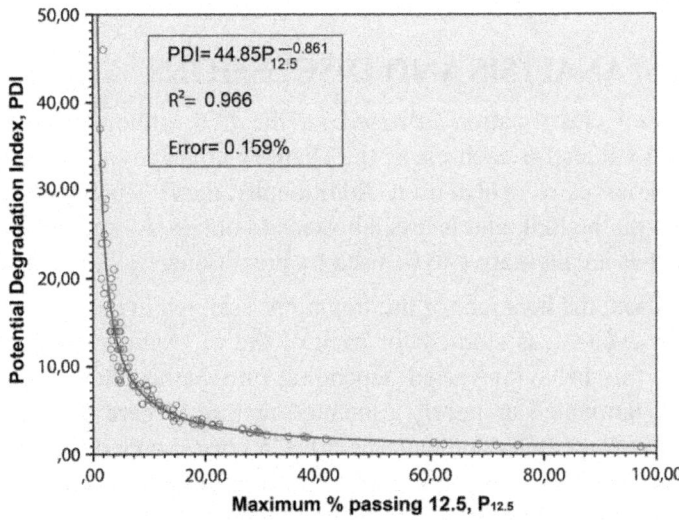

Figure. 11: Correlation between potential degradation index (PDI) and maximum percentage of sample passing through the sieve of 12.5 mm ($P_{12.5}$) for the 117 samples.

FIELD WEATHERING CHARACTERIZATION

The final part of this study consists of comparing the slaking behavior of the different samples, based firstly on the analysis of changes in the fragments retained in the drum (see Tables 2, 3; Figure. 7), secondly on the calculation of the potential degradation index, and finally by attending to the weathering patterns and weathering profiles of the different lithologies, as observed in the field by the authors of this study. The field observations of the different lithologies were made for rock that had been exposed to natural climatic conditions over a long period of time. The weathering patterns are defined as follows [based on Cano and Tomás (2015)]:

Not weathered (NW); slight discoloration (A); reduction by arenization (B); flat weathering front peeling off (C); conchoidal peeling off (D); incipient rounding of blocks formed by tectonic joints (E); ellipsoidal morphology blocks formation (F); cubic centimetre fracturing of ellipsoidal block (G); incipient conchoidal fracture of ellipsoidal blocks and formation of ellipsoidal blocks of minor size (H); total conchoidal exfoliation of ellipsoidal blocks (I); massive fracturing in centimetric pseudocubic blocks (J); residual soil (K); centimetric rhomboidal fracturing in centimetric thickness strata (L); and massive fracturing of centimetric thickness strata (M). The field weathering profile is comprised of the sum of different weathering patterns (e.g. FHIJK, EFG, etc.) and in the field it was observed that weathering profiles depended on the lithological nature of the strata, although some lithologies showed similar weathering profiles (Table 4).

RESULTS, ANALYSIS AND DISCUSSION

The proposed classification is based on the PDI, although to help define the limiting values for each class, the changes in shape of the fragment size distribution curves were also used. Additionally, the $P_{12.5}$ parameter correlated very well with the PDI, and is less laborious to obtain. As such, this parameter is proposed as an alternative to be used by practitioners.

In addition, the behavior of the fragments retained in the drum throughout the slaking cycles was studied for each of the 117 samples. As commented previously, this behavior varied depending on whether the samples showed a compact, laminated or poorly cemented texture. Where the samples were compact, the slake behavior patterns C1–C5 corresponded directly with the first five proposed classes, giving an FHIJK or EFG weathering profile. The laminated samples behaved differently, as the fragments could not become rounded owing to the fact that they slid around the drum. The slake behavior patterns of the laminated samples (L1–L4) corresponded to classes 2–5, respectively. The least durable laminar samples (class 1) behaved in a similar manner to the least durable compact samples (i.e. type C1), and in the same way the most durable laminar samples (class 6) behaved similarly to type C6 samples (Table 4).

The samples with an AC weathering profile (the most durable as observed in the field) had a slaking behavior class equal or greater to "medium" (class 3), and their slake behavior pattern was always one type greater than that which would be expected for rocks of their class, except for class 6, where they coincided. Where the samples were poorly cemented, the behavior during the slake cycles was also unusual as the fragments only degraded by rounding, without fracturing and maintaining the same number of fragments (although reduced in size). These samples corresponded to Class 2. Finally, class 6 samples always showed a C6 type slake behavior pattern, independently of whether they were laminar or compact, and corresponded to the most durable weathering profiles (NW-A, AC, EFG and LM) (Table 4).

It may be observed in Tables 4 and 5 that there is no biunivocal relationship between any one lithotype and durability class, or fragment degradation behavior. This was also reflected in the in situ behavior observed (weathering profile). This is understandable when considering that the samples were identified using field criteria based on a simplified geological classification of rocks based on their genetic category, structure, composition and grain size (Geological Society of London 1977). Although instrument-based techniques were used to identify mineralogy, there does not appear to be a clear relationship between mineralogy and durability. This is probably because durability is affected by other factors, such as the microfabric or microscopic texture of the rock (Martínez-Bofill et al. 2004; Kaufhold et al. 2013; Cano and Tomás 2015). However, it is of greater relevance to note that using this methodology, a particular sample will show a particular PDI value, which allows it to be categorized according to its slaking behavior, and hence allow its susceptibility to long-term weathering in a slope to be evaluated.

Table 5: Slaking classification of Alicante's carbonate Flysch lithotypes

Mineralogy and classes of durability →	Cb (%)	Phy (%)	Qtz (%)	Very low (PDI ≤ 1.5)	Low (PDI: 1.5–3)	Medium (PDI: 3–5.5)	Medium–high (PDI: 5.5–8)	High (PDI: 8–15)	Very high (PDI > 15)
Thick bedding calcarenites	92.5 ± 4.2	5.4 ± 3.0	2.1 ± 1.6				C5	C6	C6
Slightly marly limestones	82.9 ± 3.2	12.5 ± 2.5	4.6 ± 1.2					C5	C6

Lithology	Cb	Qtz	Phy							
Calcareous mélange	87.5 ± 2.4	9.3 ± 1.9	3.2 ± 0.6						C5–C6	C6
Thin bedding calcarenites (C)	78.3 ± 5.4	14.7 ± 3.7	7.0 ± 2.5			C4	–		C5	C6
Thin bedding calcarenites (L)	72.1 ± 9.8	18.5 ± 7.2	3.3 ± 2.3			C3–L2	L3		L4	C6
Calcareous debrites	84.8 ± 0.0	11.2 ± 0.0	4.0 ± 0.0					–		
Marly limestones	79.0 ± 2.6	15.3 ± 2.4	5.7 ± 0.7			C2–C3	C4		C5	
Silty calcareous marls	82.4 ± 3.9	12.7 ± 2.1	5.0 ± 1.9		C2	C3		C4		
Silty marls	74.7 ± 3.1	17.8 ± 2.6	7.6 ± 0.7			–	–		–	
Calcareous marls–marls	75.0 ± 3.2	17.5 ± 2.6	7.5 ± 1.1	C1	C2	C3	C4			
Soft calcareous mélange	84.2 ± 0.0	11.7 ± 0.0	4.2 ± 0.0			C3				
Thin bedding silty calc.	70.0 ± 11.7	20.0 ± 9.7	10.1 ± 2.1		L1	C3				
Sheet silty marls	68.5 ± 6.8	21.4 ± 6.6	10 ± 1.4	C1	C2					
Poorly cem. thin bedding calc.	74.7 ± 0.0	19 ± 0.0	6.4 ± 0.0		R					
Soft marls	75.4 ± 3.8	16.7 ± 3.0	7.7 ± 0.9	C1						
Sheet marls	75.7 ± 3.7	16.9 ± 2.5	6.5 ± 0.0	C1						

Each lithology is associated with a particular slake behavior pattern and different categories of durability based on Potential Degradation Index. Additionally, mineralogy is shown for each lithology

Cb carbonates, dolomite plus calcite, *Qtz* quartz, *Phy* phyllosilicates. *C1, C2, C3, C4, C5, L1, L2, L3, L4* and *R* are the different slaking behavior patterns.

Bold values indicate Very low (PDI ≤ 5) to Very high (PDI > 15) are the six categories of durability based on Potential Degradation Index

If the range of Id_2 values is analysed and compared with the proposed new PDI parameter, it may be clearly observed that the use of Id_2 overstates the slaking resistance of the carbonate samples in this study by at least one class.

The different samples, classified in six categories based on their PDI, have been also associated with their maximum percent of sample passing through the 12.5 mm sieve (P12.5), the slake behavior patterns of fragments during five test cycles, the field weathering profile and Id2 index. Additionally, the lithotypes have been associated with their corresponding classes, including the number of samples and allowing the most representative lithologies corresponding to each class to be shown (Table 4).

If the results are analysed according to lithotypes, a clear relationship may be observed between carbonate content and durability—although this is not the determining factor, for reasons discussed previously. However, there is a general trend in the slaking behavior (classes 1–6) of each lithological group. The thick bedding calcarenites mainly showed a "very high" durability, with a type C6 slake behavior pattern, corresponding to the NW-A field weathering profile. Slightly marly limestones exhibited a high or very high durability, their slake behavior pattern was type C5 and C6 and corresponded to the EFG field weathering profile. Calcareous mélange mainly showed a "high" durability, with a type C5 and C6 slake behavior pattern, corresponding to the AC field weathering profile. Thin bedding calcarenites (compact) mainly exhibited "high" durability, their slake behavior pattern was type C5, corresponding to the AC field weathering profile.

The behavior of the thin bedding calcarenites (laminated) was highly variable, oscillating between "medium" and "very high" durability. Their slake behavior pattern was also variable, although it corresponded exclusively to the LM field weathering profile. When sampling from the slopes in the study area, only one layer of Calcareous debrites was detected, which were classified as "medium–high" durability. No data were available for the slake behavior pattern, and the field weathering profile was AD.

Marly limestones mainly exhibited "medium–high" durability, their slake behavior pattern was type C4, corresponding to the EFG field weathering profile. The behavior of the Silty calcareous marls was very variable, as values indicating both "low" and "high" durability, and intermediate classes, were obtained. Data on the slaking behavior pattern were not available, and the field weathering profile was exclusively FHIJK. The behavior of the silty marls also varied greatly, as they were classed as having "medium" to "high" durability. Data on the slaking behavior pattern were not available, and the field weathering profile was EFG, exclusively. The calcareous marls–marls varied between "low" and "medium–high" durability, with a type C2–C4 slake behavior pattern. However, the majority of the samples exhibited a "medium" durability with a type C3 slake behavior pattern, corresponding exclusively to the FHIJK field weathering profile.

The only sample of soft calcareous mélange detected in this study showed a "medium" durability, slake behavior pattern C3 and corresponded to the LM field weathering profile. The thin bedding silty calcarenites had a "low" to "medium" durability, L1–C3 slake behavior pattern and LM field weathering profile. Sheet silty marls mainly exhibited a low durability. Their slake behavior pattern was type C2, which corresponded to the FHIJK field weathering profile. The only sample of poorly cemented thin bedding calcarenites detected in the study area showed a low durability. The slake behavior pattern was type R, corresponding to the AB field weathering profile. Finally, soft marls and sheet marls were the least durable lithotypes, with a "very low" durability, and a type C1 slake behavior pattern which corresponded to the FHIJK field weathering profile (Table 5).

CONCLUSIONS

The classification of carbonate rocks according to their slaking behavior using the Id_1 indices (Franklin and Chandra 1972; Gamble 1971) does not accurately reflect the weathering behavior observed in situ. The use of only one sieve size (2 mm) in the laboratory does not adequately explain slaking behavior, since the study of the fragments retained in the drum shows evidence of degradation which is not reflected in the results given by the slake durability test.

A new index based on changes in the disintegration ratio (D_{RP}; Eq. (1), Figure. 5) throughout five slaking cycles—the potential degradation index [PDI; Eq. (3)] has been proposed as an alternative to the Id_2 index. This index takes into account the degradation of the sample fragments retained in the 2 mm drum, as well as changes in these fragments throughout the experimental process. This has allowed six durability classes to be established to describe slaking behavior: very low, low, medium, medium–high, high and very high durability. These classes also agree more closely with weathering behavior patterns observed in the field.

The results indicate that the Id_2 index overstates the slaking resistance of the carbonate rock samples tested by at least one class.

This methodology has also allowed the samples to be grouped according to slake behavior patterns, which take into account changes in morphology in the fragments retained in the drum after each cycle. Additionally, these parameters have been related to the field weathering profiles defined by Cano and Tomás (2015), which allowed the results of laboratory slake tests to be compared with weathering resulting from long-term exposure to natural climactic conditions observed in the field. Additionally, a correlation has been proposed between the potential degradation index (PDI) and the maximum percentage of sample passing through the 12.5 mm sieve ($P_{12.5}$), which is a less laborious parameter to obtain, and hence is considered to be more adequate for practitioners.

ACKNOWLEDGMENTS

This work was funded by the University of Alicante under the project GRE14-04, the Ministerio de Economía y Competitividad del Gobierno de España in the framework of the projects TIN2014-55413-C2-2-P and TEC2011-28201-C02-02. The authors wish to express their sincere thanks to Dr. T. Miranda for comments made during Dr. M. Cano's stay at the University of Minho (funded by the University of Alicante), which have substantially improved this work.

REFERENCES

1. Agencia Estatal de Meteorología (AEMET) (2012) Guía resumida del clima en España 1981-2010, Madrid. Available in http://www.aemet.es/es/conocermas/publicaciones. Accessed 3 Nov 2015
2. American Society for testing and Materials (ASTM) (2004) Standard test method for slake durability of shales and similar weak rocks (D4644-04), Philadelphia
3. American Society for Testing and Materials (ASTM) (2007a) Standard test method for particle-size analysis of soils (D422–63(2007) e2. West Conshohocken, PA
4. American Society for testing and Materials (ASTM) (2007b) Standard test method for rapid determination of carbonate content of soils [ASTM D4373—02(2007)], Philadelphia
5. Bell FG, Entwisle DC, Culshaw MG (1997) A geotechnical survey of some British Coal measures mudstones, with particular emphasis on durability. Eng Geol 46:115–129
6. Cano M, Tomás R (2013a) Characterization of the instability mechanisms affecting slopes on carbonatic Flysch: Alicante (SE Spain), case study. Eng Geol 156:68–91
7. Cano M, Tomás R (2013b) Assessment of corrective measures for alleviating slope instabilities in carbonatic Flysch formations: Alicante (SE of Spain) case study. Bull Eng Geol Environ 72(3–4):509–522
8. Cano M, Tomás R (2015) An approach for characterising the weathering behavior of Flysch slopes applied to the carbonatic Flysch of Alicante (Spain). Bull Eng Geol Environ 74:443–463
9. Czerewko MA, Crips JC (2001) Assessing the durability of mudrocks using the modified jar slake index test. Q J Eng Geol Hydrogeol 34(2):153–163
10. Dick JC, Shakoor A (1995) Characterizing durability of mudrocks for

slope stability purposes. Geol Soc Am Rev Eng Geol X 5:121–130

11. Dick JC, Shakoor A, Wells NA (1994) A geological approach toward developing a mudrock durability classification system. Can Geotech J 31(5):17–27

12. Erguler ZA, Shakoor A (2009) Quantification of fragment size distribution of clay-bearing rocks after slake durability testing. Environ Eng Geosci 15(2):81–89

13. Erguler ZA, Ulusay R (2009) Assessment of physical disintegration characteristics of clay-bearing rocks: disintegration index test and a new durability classification chart. Eng Geol 105:11–19

14. Franklin JA, Chandra A (1972) The slake-durability test. Int J Rock Mech Mining Sc. 9:325–341

15. Gamble JC (1971) Durability-Plasticity Clasification of Shales and other Argillaceous Rocks. Ph. D. Thesis. University of Illinois

16. Gautam TP, Shakoor A (2013) Slaking behavior of clay-bearing rocks during a one-year exposure to natural climatic conditions. Eng Geol 166:17–25

17. Gautam TP, Shakoor A (2015) Comparing the Slaking of Clay-Bearing Rocks Under Laboratory Conditions to Slaking Under Natural Climatic Conditions. Rock Mech Rock Eng. doi:10.1007/s00603-015-0729-7

18. Geological Society of London (1977) The description of rock masses for engineering purposes: report by the Geological Society Engineering Group Working Party. Q J Eng Geol Hydrogeol 10:55–388

19. Gökçeoğlu C, Ulusay R, Sönmez H (2000) Factors affecting the durability of selected weak and claybearing rocks from Turkey, with particular emphasis on the influence of the number of drying and wetting cycles. Eng Geol 57:215–237

20. Guerrera F, Estévez A, López-Arcos M, Martín-Martín M, Martín-Pérez JA, Serrano F (2006) Paleogene tectono-sedimentary evolution of the Alicante Trough (External Betic Zone, SE Spain) and its bearing on the timing of the deformation of the South-Iberian Margin. Geodin Acta 19(2):87–101

21. Hack HRGK (1998) Slope Stability Probability Classification, 2nd ed.: Ph.D. Thesis, ITC, Delft, The Netherlands

22. Hack HRGK, Huisman M (2002) Estimating the intact rock strength of a rock mass by simple means. In: van Rooy JL, Jermy CA (eds) Engineering geology for developing countries—proceedings of 9th congress of the international association for engineering geology and the environment:

south african institute of engineering geologists (SAIEG). Durban, South Africa, pp 1971–1977
23. Hoek E, Carranza-Torres C, Corkum B (2002) Hoek-Brown failure criterion-2002 edition. Proceedings of NARMS-Tac 1:267–273
24. INGEMISA (Investigaciones geológicas y mineras, S. A.), Auernheimer C (1991) Mapa geocientífico de la provincia de Alicante. Conselleria d'Administració Pública, Valencia, Agència de Medi Ambient, p 93
25. ISRM (1981) Rock characterization. In: Brown ET (ed) Testing and monitoring—ISRM suggested methods. Pergamon press, Oxford, p 211
26. Kaufhold A, Gräsle W, Plischke I, Dohrmann R, Siegesmund S (2013) Influence of carbonatic content and micro fabrics on the failure strength of the sandy facies of the Opalinus Clay from Mont Terri (underground rock laboratory). Eng Geol 156:111–112
27. Martin JD (2004) Using XPowder: A software package for Powder X-Ray diffraction analysis
28. Martínez-Bofill J, Corominas J, Soler A (2004) Behavior of the weak rocks cutslopes and their characterization using the results of the Slake Durability Test. In: Lecture notes in Earth Sciences. 104. Engineering Geology for Infrastructure Planing in Europe
29. Miščcević P, Vlastelica G (2011) Durability characterization of marls from the region of Dalamtia. Croatia. Geotech Geol Eng 29:771–781
30. Miščcević P, Vlastelica G (2014) Impact of weathering on slope stability in soft rock mass. J Rock Mech Geotech Eng 6:240–250
31. Moon VG, Beattie AG (1995) Textural and microstructural influences on the durability of Waikato coal measures mudrocks. Q J Eng Geol 14:255–279
32. Nicholson DT (2001) Deterioration of Excavated Rockslopes: mechanisms, morphology and assessment: Ph.D. Thesis, University of Leeds, Leeds, UK
33. Santi PM (1998) Improving de jar slake, slake index, and slake durability tests for shales. Environ Eng GeoSci IV(3):385–396
34. Richardson DN, Long JD (1987) The sieve slake durability test. Bull Ass Eng Geol 2(5):247–258
35. Robert M, Tessier D (1974) Méthode de préparation des argiles des sols pour des études minéralogiques. Annal Agronom 25(6):859–882
36. Romana M (1993) A geomechanical classification for slopes: slope mass rating. Comp Rock Eng 3:575–599
37. Sabatakakis N, Tsiambaos G, Koukis G (1993) Index properties of soft

marly rocks of the Athens basin, Greece. Geotech Eng Hard Soils-Soft Rocks. Anagnostopoulos A (eds) Rotterdam, Balkema, pp 275–279
38. Shakoor A (1995) Slope stability considerations in differentially weathered mudrocks. Geol Soc Am Rev Eng Geol X 1995(5):131–138
39. Taylor RK (1988) Coal Measures mudrocks: composition, classification and weathering processes. Q J Eng Geol 21:85–99
40. Ulusay R, Arikan F, Yoleri MF, Caglan D (1995) Engineering geological characterization of coal mine waste material and an evaluation in the context of back-analysis of spoil pile instabilities in a strip mine SW Turkey. Eng Geol 40:77–101
41. Vera JA (2004) Geología de España. Sociedad Geológica de España, Instituto Geológico y Minero de España

Chapter 8

ROCK MASS HYDRAULIC CONDUCTIVITY ESTIMATED BY TWO EMPIRICAL MODELS

Shih-Meng Hsu[1], Hung-Chieh Lo[1], Shue-Yeong Chi[1] and Cheng-Yu Ku[2]

[1]Sinotech Engineering Consultants, Inc
[2]National Taiwan Ocean University, Taiwan

INTRODUCTION

Undertaking engineering tasks such as tunnel construction, dam construction, mine development, the abstraction of petroleum, and slope stabilization require the estimation of hydraulic conductivity for fractured rock mass. The understanding of hydraulic properties of fractured rock mass, which involves the fluid flow behavior in fractured consolidated media, is a critical step in support of these tasks. To obtain hydraulic properties of fractured rock mass, double packer systems can be adopted (NRC 1996). They can be used to determine the hydraulic conductivity in a portion of borehole using two inflatable packers. Although this type of test can directly measure the hydraulic parameter, costs of the testing are fairly high. Several studies (Snow, 1970; Louis, 1974; Carlsson & Olsson, 1977; Burgess, 1977; Black, 1987; Wei et al., 1995;) have proposed the estimation of rock mass hydraulic conductivity using different empirical equations, which were based on the concept that rock mass permeability decreases with depth, as shown in Table 1. These empirical equations provide a great feature for characterizing rock mass hydraulic properties quickly and easily. However, the applicability of these equations is very limited because depth is not the only significant variable on the prediction of rock mass permeability. Hydraulic properties of rock mass may vary with geostatic stress, lithology and fracture properties, including fracture aperture and frequency, fracture length, fracture orientation and angle, fracture interconnectivity, filling materials, and fracture plane features (Lee & Farmer, 1993; Sahimi, 1995; Foyo et al., 2005; Hamm et al., 2007). Thus, a more applicable empirical equation for estimating hydraulic conductivity of rock mass possibly must include the aforementioned factors.

This chapter proposes two empirical models to estimate hydraulic conductivity of fractured rock mass. The first empirical model was based on the rock mass classification concept. The study developed a new rock mass classification scheme for estimating hydraulic conductivity of fractured rocks. The new rock mass classification system called as "HCsystem" based on the following four parameters: rock quality designation (RQD), depth index (DI), gouge content designation (GCD), and lithology permeability index (LPI). HCvalues can be calculated from borehole image data and rock core data. The second empirical model was simply based on results of borehole televiewer logging, flowmeter logging and packer hydraulic tests. Three borehole prospecting techniques for fractured rock mass hydrogeologic investigation were performed to explore various hydrogeologic characteristics, such as fracture width, fracture angle, flow velocity and hydraulic conductivity. By adopting a correlation analysis, the dependence between hydraulic conductivity and other prospecting data can be identified. The consequence results can be used to determine rock mass hydraulic conductivity.

Table 1. Diverse approximations for estimating rock mass hydraulic conductivity

Equation	Reference
$k = az^{-b}$	Black (1987) a and b are constants. z is the vertical depth below the groundwater surface.
$\log K = -8.9 - 1.671 \log Z$	Snow (1970) K (ft²) is the permeability. z (ft) is the depth.
$K = 10^{-(1.6\log z + 4)}$	Carlson and Olsson (1977) K (m/s) is the hydraulic conductivity. z (m) is the depth.
$K = K_s e^{(-Ah)}$	Louis (1974) K (m/s) is the hydraulic conductivity. K_s is the hydraulic conductivity near ground surface. H (m) is the depth. A is the hydraulic gradient.
$\log K = 5.57 + 0.352 \log Z$ $-0.978(\log Z)^2 + 0.167(\log Z)^3$	Burgess (1977) K (m/s) is the hydraulic conductivity. Z (m) is the depth.
$K = K_i[1 - Z/(58.0 + 1.02Z)]^3$	Wei et al. (1995) Z is the depth. K is the hydraulic conductivity. K_i (m/s) is the hydraulic conductivity near ground surface.

THE FIRST EMPIRICAL MODEL

The classical rock mass classification systems have gained wide attention and are frequently used as powerful design aids in rock engineering. A great feature of the existing systems is that the characterization of rock mass properties for specific engineering purposes can be quickly obtained at a relatively low cost. There are six common systems used for engineering purposes, including Rock load (Terzaghi, 1946), Stand-up time (Lauffer, 1958), RQD (Deere et al., 1967), RSR (Wickham et al., 1972), RMR (Bieniawski, 1973), and Q-system (Barton et al., 1974). The above systems or other available systems were designed on the geomechanical assessment of rock mass. However, there is very limited study on rock mass permeability assessments (Bieniawski, 1989). Because permeability of rock mass is related to groundwater seepage into excavations for tunnels, mines, and other construction sites, Gates (1997) proposed the hydro-potential (HP) value as a new rock mass classification, semiquantitative technique employed to evaluate the potential for developing groundwater in bedrock. The HP-value technique is a modification of the engineering rock mass quality designation (Q) originally developed for evaluation of rock competency in tunnel design and seismic rock fall susceptibility. To reduce the cost of estimating the hydraulic parameter, the objective of the study presented herein is to propose a new application of the rock mass classification concept on the estimation of hydraulic conductivity of fractured rocks. The new rock mass classification system will be verified by in situ hydraulic test data from two hydrogeological investigation programs in three boreholes to demonstrate its rationality in predication of hydraulic conductivity of fractured rocks. Besides, the model verification using another borehole data with four additional in-situ hydraulic tests from similar geologic rocks was also conducted to further verify the feasibility of the proposed empirical HC model.

Components of New Rock Mass Classification System

Prior to describing the new rock mass classification system, potential factors, including rock quality designation (RQD), depth index (DI), gouge content designation (GCD), and lithology permeability index (LPI), that may affect the degree of permeability should be considered. In addition, the rating approach for each factor that represents the magnitude of permeability is also described as follows.

Rock Quality Designation (RQD)

In rock engineering, from the mechanical point of view, the degree of fracturing stands for rock quality. This provides a simple index to judge the engineering quality of the rock. From the hydrogeological point of view, fractures reflect the ability of a formation to transmit water through fractures themselves. Thus, the degree of fracturing may be regarded as a factor in evaluating rock mass permeability.

To assess the influence of the fracture characteristic on permeability, the rock quality designation (RQD) index, which was developed by Deere et al. (1967), can be adopted. The RQD index was introduced over 40 years ago as an indicator of rock mass conditions. The RQD value is defined as the cumulative length of core pieces longer than 100 mm in a run (R_s) divided by the total length of the core run (R_T) and can be obtained from the following equation.

$$RQD = \frac{\sum Length\ of\ Intact\ and\ Sound\ Core\ Pieces > 100\ mm}{Total\ Length\ of\ Core\ Run,\ mm} \times 100\% = \frac{R_s}{R_T} \times 100\% \quad (1)$$

In this study, a core run for calculating a RQD value is herein defined as a selected zone of a hydraulic test. Equation (1) may be utilized to identify rock mass permeability.

Depth Index (DI)

Many researchers (for example Lee & Farmer, 1993; Singhal & Gupta, 1999) pointed out that rock mass permeability may decrease systematically with depth. The decrease in permeability with depth in fractured rocks is usually attributed to reduction in fracture aperture and fracture spacing. The reduction is due to the effect of geostatic stresses, and thereby the permeability of fractured rocks will reduce. The depth may be considered as a factor in evaluating rock mass permeability.

To assess the influence of the depth on permeability, a depth index called as DI was defined as the following equation.

$$DI = 1 - \frac{L_c}{L_T} \quad (2)$$

in which L_T is the total length of a borehole; L_c is a depth which is located at the middle of a double packer test interval in the borehole. The value of DI is always greater than zero and less than one. The greater the DI value, the higher the permeability.

Gouge Content Designation (GCD)

In general, the permeability of clay-rich gouges has extremely low values (Singhal & Gupta, 1999). The RQD value may decrease by an increase of fractures in a core run. If the fractures contain infillings such as gouges, permeability of the fractures will reduce. To assess the influence of the gouge materials on permeability, a gouge content designation index called as GCD was defined as the following equation.

$$GCD = \frac{R_G}{R_T - R_S} \quad (3)$$

in which RG is the total length of gouge content. The value of GCD is always greater than zero and less than one. The greater GCD value stands for the more gouge content in a core run, and thereby it will reduce the permeability of the core run.

Lithology Permeability Index (LPI)

Lithology is the individual character of a rock in terms of mineral composition, grain size, texture, color, and so forth. For an intact rock, the magnitude of permeability depends largely on the individual character of the rock. It may be affected by the average size of the pores, which in turn is related to the distribution of particle sizes and particle shape. In sedimentary formations grain-size characteristics are most important because coarse grained and well-sorted material will have high permeability as compared with fine-grained sediments like silt and clay. Thus, the lithology may be regarded as a factor in evaluating rock mass permeability. To assess the influence of lithology on permeability, a lithology permeability index called as LPI was defined as Table 2.

Rock Mass Permeability System

As stated in Section 2.1, the rock mass permeability may be dependent on the following four parameters: rock quality designation (RQD), depth index (DI), gouge content designation (GCD), and lithology permeability index (LPI). However, the permeability is not simply affected by only one factor. It is possibly affected by any two factors, three factors or even all four factors. Thus, a rock mass classification scheme was applied to establish the rock mass permeability system. The new rock mass classification scheme is the product of the four parameters. It can account for the synthetic effect from the four parameters on permeability. The new rock mass classification system called as "HC-system" can be given by the following equation:

$$HC = \left(1 - \frac{RQD}{100}\right)(DI)(1\text{-}GCD)(LPI) \quad (4)$$

The value of each parenthesis at the right hand side of Equation (4) is always greater than zero and less than one depending on the values assigned to the four parameters. The greater the value of each parenthesis, the higher the permeability. Thus, the system performs a numerical assessment of the rock mass permeability using the four parameters. However, it should be noted that if $(1-RQD)$ is zero, the value of 0.01 in the term of $(1-RQD)$ is suggested to avoid the HC-value to be zero. Currently, the study took the same weight for each factor in Equation (4). While collecting more observed data, a further study can be considered to assign a different weight for each factor to giver better correlation between the hydraulic conductivity and HC-value. The rationality of Equation (4) must be verified by observed data through in situ hydraulic tests.

Table 2. Description and ratings for lithology permeability index

Lithology	Hydraulic conductivity (m/s)				Range of rating	Suggested Rating
	Reference[1]	Reference[2]	Reference[3]	$K_{average}$		
Sandstone	$10^{-6} \sim 10^{-9}$	$10^{-7} \sim 10^{-9}$	$10^{-7} \sim 10^{-9}$	$10^{-7.5}$	0.8-1.0	1.00
Silty Sandstone	-	-	-	-	0.9-1.0	0.95
Argillaceous Sandstone	-	-	-	-	0.8-0.9	0.85
S.S. interbedded with some Sh.	-	-	-	-	0.7-0.8	0.75
Alternations of S.S & Sh.	-	-	-	-	0.6-0.7	0.65
Sh. interbedded with some S.S.	-	-	-	-	0.5-0.7	0.60
Alternations of S.S & Mudstone	-	-	-	-	0.5-0.6	0.55
Dolomite	$10^{-6} \sim 10^{-10.5}$	$10^{-7} \sim 10^{-10.5}$	$10^{-9} \sim 10^{-10}$	10^{-8}	0.6-0.8	0.70
Limestone	$10^{-6} \sim 10^{-10.5}$	$10^{-7} \sim 10^{-9}$	$10^{-9} \sim 10^{-10}$	10^{-8}	0.6-0.8	0.70
Shale	$10^{-10} \sim 10^{-12}$	$10^{-10} \sim 10^{-13}$	-	$10^{-10.5}$	0.4-0.6	0.50
Sandy Shale	-	-	-	-	0.5-0.6	0.60
Siltstone	$10^{-10} \sim 10^{-12}$	-	-	10^{-11}	0.2-0.4	0.30

Sandy Siltstone	-	-	-	-	0.3-0.4	0.40
Argillaceous Siltstone	-	-	-	-	0.2-0.3	0.20
Claystone	-	$10^{-9} \sim 10^{-13}$	-	10^{-11}	0.2-0.4	0.30
Mudstone	-	-	-	-	0.2-0.4	0.20
Sandy Mudstone	-	-	-	-	0.3-0.4	0.40
Silty Mudstone	-	-	-	-	0.2-0.3	0.30
Granite	-	-	$10^{-11} \sim 10^{-12}$	$10^{-11.5}$	0.1-0.2	0.15
Basalt	$10^{-6} \sim 10^{-10.5}$	$10^{-10} \sim 10^{-13}$	-	$10^{-11.5}$	0.1-0.2	0.15

[1] B.B.S. Singhal & R.P. Gupta (1999)
[2] Karlheinz Spitz & Joanna Moreno (1996)
[3] Bear (1972)

Study on Correlation of Hydraulic Conductivity and HC-System

To verify rationality of Equation (4), the study collected in situ hydraulic test data to determine a relationship between hydraulic conductivity and HC-values. The needed data include rock core logs, BHTV (Borehole Acoustic Televiewer) image logs, locations of hydraulic tests, and hydraulic conductivity results. The hydraulic test results of boreholes were used to perform the dependence of HC on hydraulic conductivity.

Description of Study Area, Boreholes, BHTV Investigation and Hydraulic Test Data

The study area is located in Taiwan. Taiwan's strata are distributed in long and narrow strips, almost parallel to the island's axis. Metamorphic rock lies under the Central and Snow Mountain Ranges. Sedimentary rock forms part of the island-wide piedmonts and coastal plains as well as the Coastal Mountain Range. The island of Taiwan has three geological zones divided by longitudinal faults: the Central Range, Western Piedmont and Eastern Coastal Mountain Range zones (Figure 1).

About 26 hydraulic conductivity measurements and borehole image scanning using BHTV were conducted in four boreholes in Western Piedmont, primarily at three sites: Da-Keng, Shang-Ming, and Caoling (Figure 1) in which borehole HB-94-01 is in the Da-Keng site, boreholes HB-95-01 and HB-95-02 are in the Shang-Ming site, and borehole CH-04 is in the Caoling site. Besides, the Da-Keng and Caoling sites are in central Taiwan and the ShangMing site is in southern Taiwan. The dominant rock strata of the Shang-Ming site include Miocene sedimentary rock with layers of sandstone or shale or their alternation. The major structures consist of a series of parallel easterly inclined thrust faults and folds, which often form local fractured zones, including geological structures such as the Pingshi fault, the Biauhu fault, and the Chin-Shan fault. In addition, borehole HB-94-01 in the Da-Keng site and borehole CH-04 in the Caoling site also have similar rock strata but without geological structures. Based on the loggings and geological analysis, HB-95-01 and HB-95-02 are strongly influenced by the faults; nevertheless, HB-94-01 and CH-04 are not. Rock core photos (Figure 2(a)) indicated soft and cohesive gouges are extensive in borehole HB-95-02; while the borehole CH-04 is not influenced by the faults (Figure 2(b)).

The depth of the boreholeHB-94-01 is 110 m. The principal lithologic units for the borehole are sandstone and siltstone. The interval of 36 m to 44 m is a fractured zone compared to other depths in the borehole. A total of 8 hydraulic tests using a double packer system were carried out to determine hydraulic conductivity (Sinotech Engineering Consultants, LTD., 2005). The strategy of the test design from the drilling work was to determine hydraulic properties from different geological structures such as no fracture, a single fracture, or multiple fractures at different depths. The drilling depths of HB-95-01 and HB-95-02 are 250 m and 350 m, respectively. The principal lithologic units for HB-95-01 are sandstone, argillaceous sandstone, and sandy mudstone. The principal lithologic units for HB-95-02 are sandstone, argillaceous sandstone, and sandstone mixed with some mudstone. HB-95-01 and HB-95-02 are close to the Biauhu fault and Pingshi fault, respectively. Rock core data indicated soft and cohesive gouges are extensive in both boreholes (Figure 2) in which the hydraulic properties of fault-related rocks can be studied. The study completed 3 and 14 hydraulic tests in HB-95-01 and HB-95-02, respectively (Sinotech Engineering Consultants, LTD., 2007). The strategy of the test design was to determine hydraulic conductivity in more permeable zones and clay-rich gouge zones. Besides, the drilling length of borehole CH-04 is 120 m. The principal lithologic units of the borehole CH-04 are mainly sandstone, shale, and sandstone with some thin shale. Four different intervals were sealed by double packers for conducting the hydraulic tests. Those hydraulic test data were used for the model verification and it is described in Section 2.3.3.

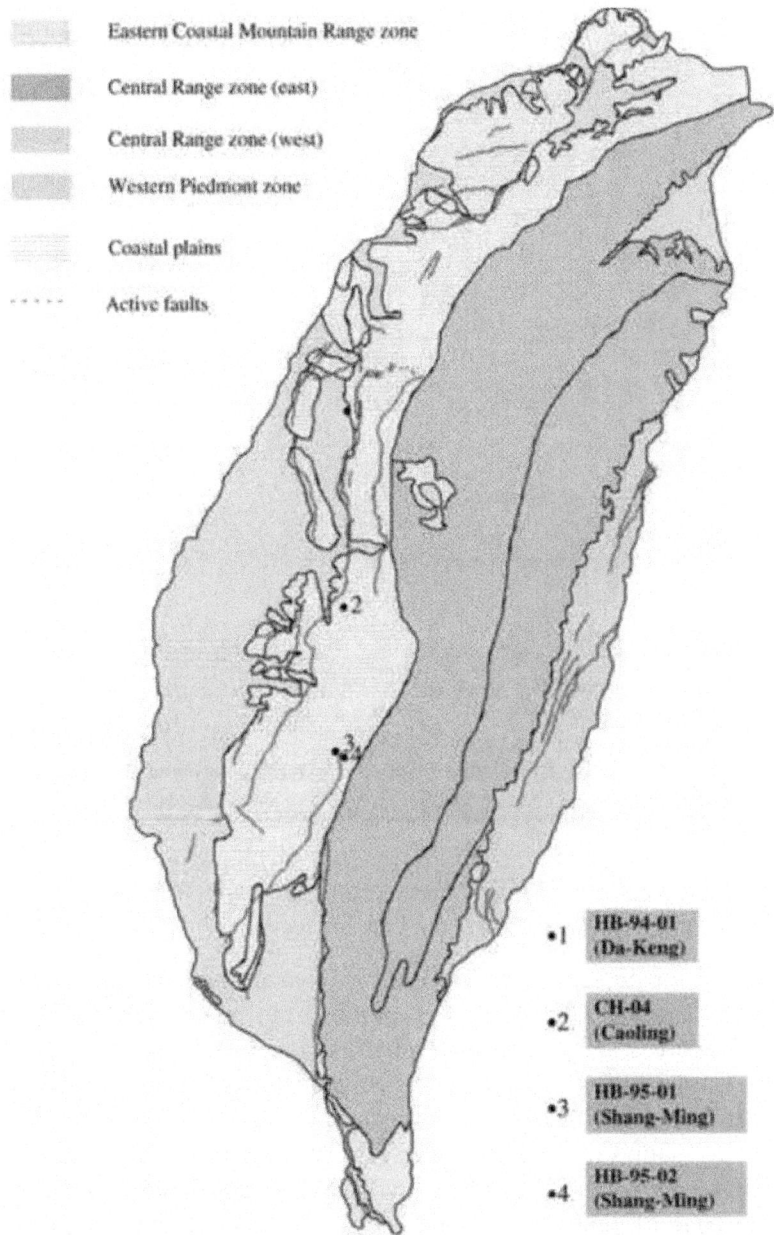

Figure. 1. Location of major faults and four boreholes for this study in Taiwan.

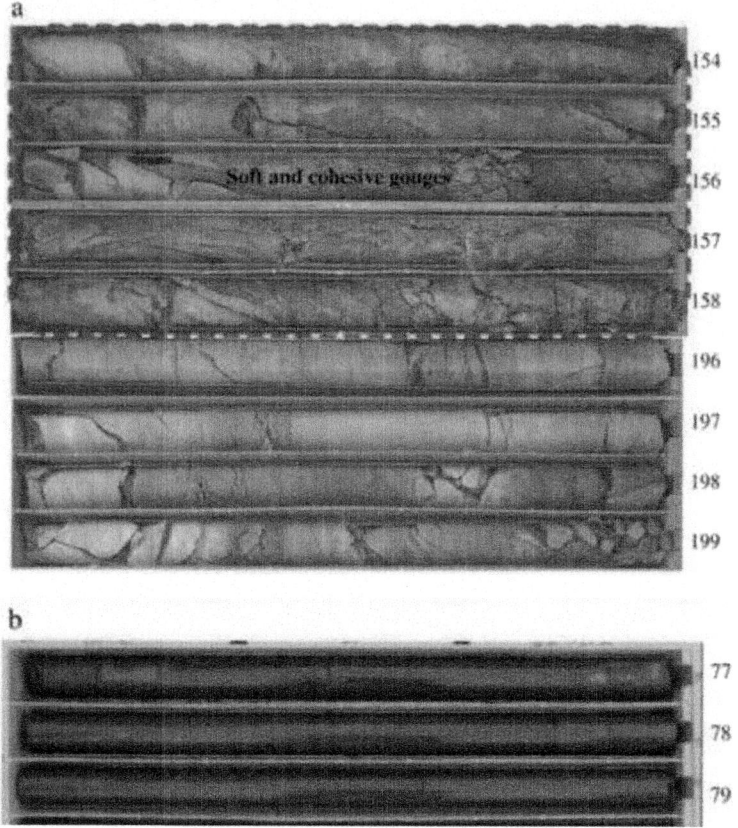

Figure. 2. (a). Rock core photos of borehole HB-95-02 with fault influence; (b). Rock core photos of borehole CH-04 without fault influence.

BHTV image logs were gathered to characterize lithology, gouge and fractures for each borehole and essential to the proper design of hydraulic testing. Figure 3 shows two hydraulic test zones and corresponding BHTV images from different boreholes. It is obvious that fractures or shear band can be identified clearly using the high-resolution BHTV. The BHTV image in Figure 3(a) shows multiple fractures. This information with a double packer system (Figure 4) can be used to investigate hydraulic property of a water-bearing zone of subsurface. The BHTV image in Figure 3(b) shows shear band. This information can be utilized to investigate hydraulic property of fault-related rocks.

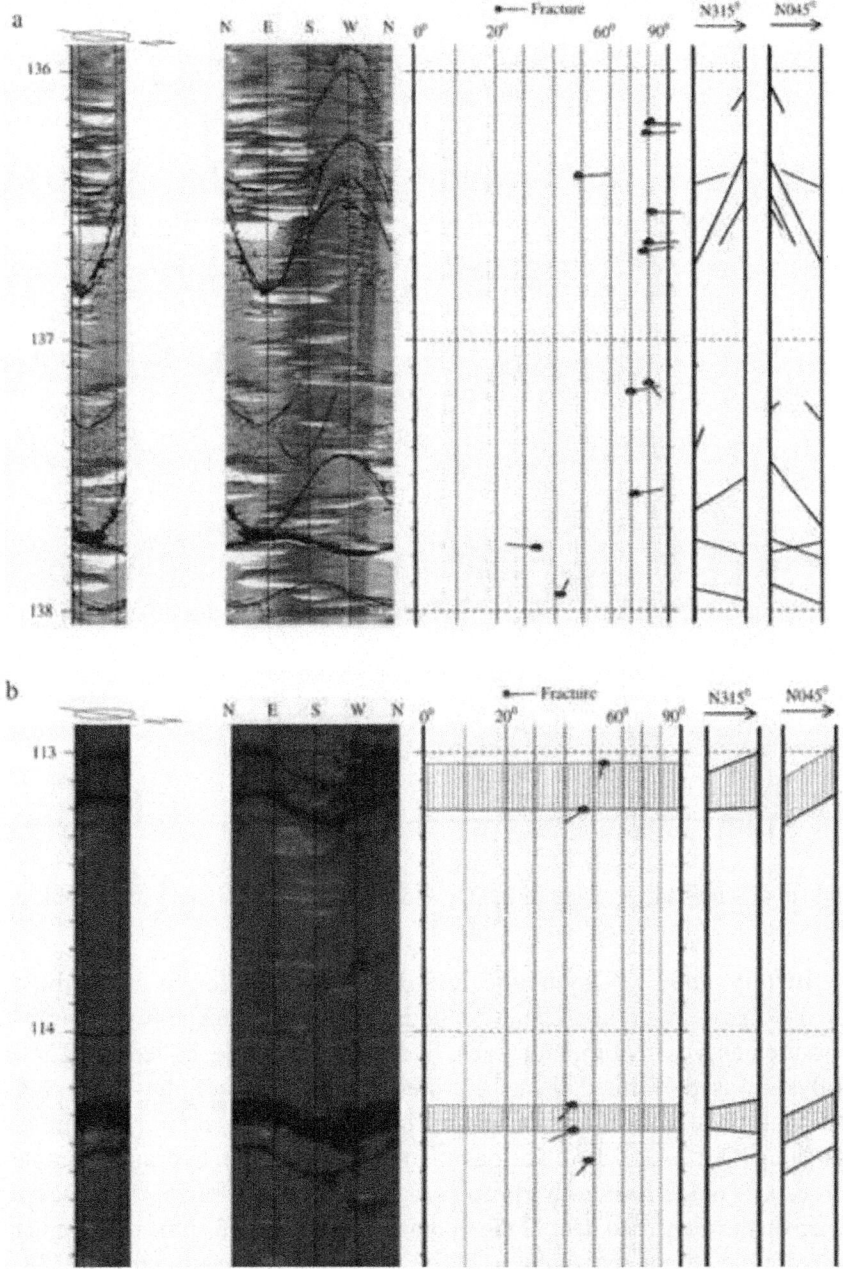

Figure. 3. (a). hydraulic test interval and corresponding BHTV image (depth 136 m-138 m, HB- 95-01); (b). hydraulic test interval and corresponding BHTV image (depth 113 m-115 m, HB- 95-02).

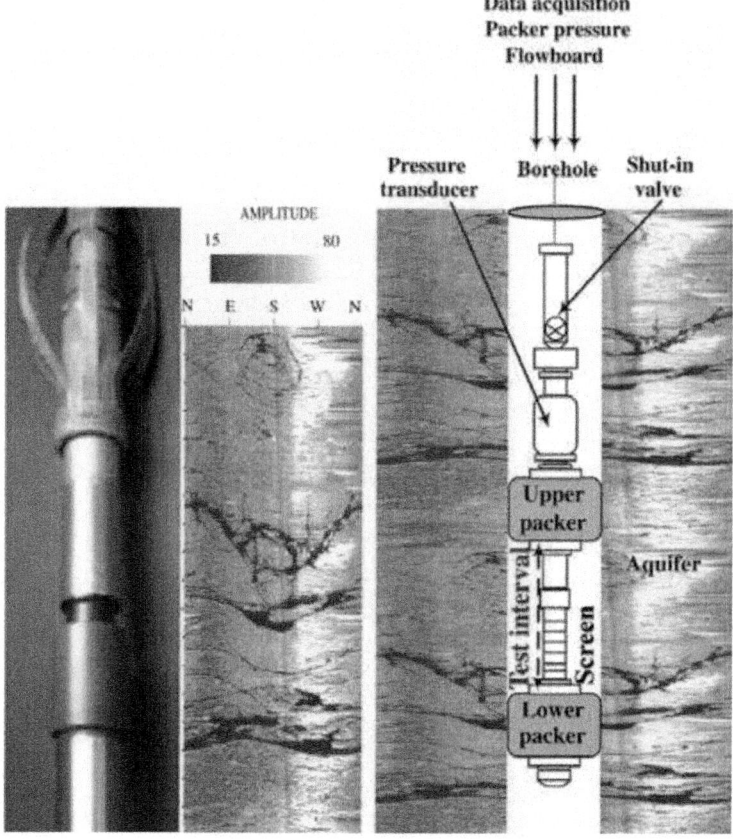

Figure. 4. Schematic drawing of BHTV, acoustic image of borehole, and double packer system.

In this study, 26 hydraulic test data collected during hydraulic tests can be analyzed by analytical methods. Water pressure and discharge rate measurements with time for each hydraulic test were collected. The data analysis was performed using a professional version of the AQTESOLVE test analysis software, which enables both virtual and automatic type curve matching (Duffield, 2004). The quantitative evaluation of hydraulic parameters was carried out as an iterative process of the best-fit theoretical response curves based on the measured data of the hydraulic test. Figure 5 shows the evaluation of hydraulic parameters using AQTESOLVE. For the test interval of 118.5 m to 121.7 m, although three fractures and a fracture zone of approximately 7.25 cm thickness were seen on the borehole image, lack of interconnectivity of fractures and soft and cohesive gouges existing at the fractures may reduce the permeability of rock masses.

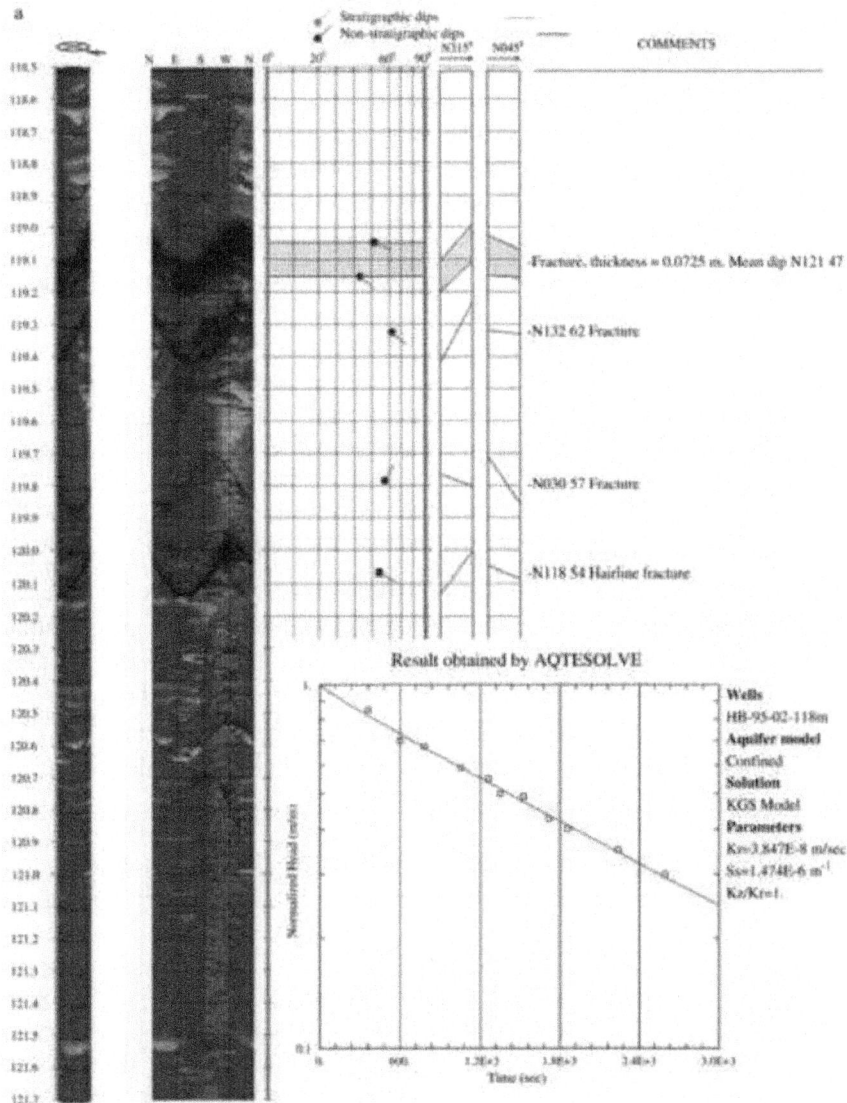

Figure. 5. Evaluation of hydraulic parameters using AQTESOLVE and BHTV images at pack-off zones of 118.5 m to 121.7 m in borehole HB-95-02.

Relationship between Hydraulic Conductivity and HC-System

Regression analysis was performed to estimate the dependence of HC on hydraulic conductivity. A total of 22 hydraulic test data (borehole CH-04 data not included) were applied to the study. HC-values for the hydraulic tests can be computed from borehole image data and rock core data, in which the values of RQD and GCD at each test interval can be calculated from borehole image data and rock core data with Equations (1) and (3), respectively. The value of DI can be calculated using Equation (2). The value of LPI for each test zone can be obtained from rock core data and Table 2. Table 3 shows the calculated results for the HC-system based on the verified data. The regression results indicated that a power law relationship exists between the hydraulic conductivity and HC values with a coefficient of determination of 0.866 as shown in Figure 6. The empirical HC model is obtained as shown in Equation 5.

$$K = 2.93 \times 10^{-6} \times (HC)^{1.380}, R^2 = 0.866 \qquad (5)$$

If only HB-94-01 testing data were adopted, a better correlation with the coefficient of determination of 0.905 can be obtained as shown in Equation (6).

$$K = 2.31 \times 10^{-6} \times (HC)^{1.342}, R^2 = 0.905 \qquad (6)$$

It should be noted that the values of (1−GCD) in HB-94-01 borehole are all equal to 1. The results of Equation (6) demonstrate that the empirical HC model may also be more accurate for the estimation of the rock mass hydraulic conductivity if the fractures do not contain infillings. There are a few limitations that need to be noted for the use of Equation (5). The data used to develop the equation are limited in number and in the lithologies represented. From the definition of DI, DI cannot be determined for inclined boreholes because the data collected were from vertical boreholes.

Table 3. The calculated results for HC-system based on 22 hydraulic test data

Boreholes	Test intervals (m)	1-RQD/100	DI	1-GCD	LPI	HC	K (m/s)
HB-94-01	34.7-36.3	0.094	0.677	1.000	1.000	0.0635	7.06E-08
	36.4-38.0	0.438	0.662	1.000	1.000	0.2895	1.64E-06
	56.7-58.3	0.063	0.477	1.000	0.950	0.0283	1.53E-08
	74.6-76.2	0.500	0.315	1.000	0.400	0.0629	5.30E-08
	77.2-78.8	0.010	0.291	1.000	0.400	0.0012	4.22E-10
	82.6-84.2	0.125	0.242	1.000	0.400	0.0121	2.31E-09
	90.2-91.8	0.010	0.173	1.000	0.400	0.0007	2.86E-10
	94.2-95.8	0.500	0.136	1.000	0.400	0.0273	4.53E-09
HB-95-01	99.0-101.9	0.345	0.598	0.200	0.400	0.0165	9.80E-09
	117.2-120.1	0.690	0.526	1.000	0.850	0.3081	9.76E-07
	133.2-136.1	0.724	0.461	0.286	1.000	0.0954	4.68E-08
HB-95-02	88.6-91.4	0.071	0.743	1.000	0.600	0.0318	1.56E-07
	96.0-99.2	0.031	0.721	1.000	0.600	0.0135	2.42E-08
	118.5-121.7	0.219	0.657	0.071	0.700	0.0072	1.36E-09
	134.8-138.0	0.344	0.610	0.727	0.700	0.1068	1.17E-07
	154.8-158.0	0.938	0.553	0.103	0.700	0.0376	1.99E-08
	173.0-176.2	0.938	0.501	0.103	0.700	0.0340	9.08E-09
	189.8-193.0	0.594	0.453	1.000	0.700	0.1883	1.01E-06
	196.6-199.8	0.563	0.434	0.500	1.000	0.1220	6.00E-08
	213.2-216.0	0.679	0.387	1.000	1.000	0.2625	4.54E-07
	249.0-251.8	0.393	0.285	0.091	0.700	0.0071	4.03E-09
	272.0-274.8	0.214	0.219	1.000	0.700	0.0328	3.36E-08

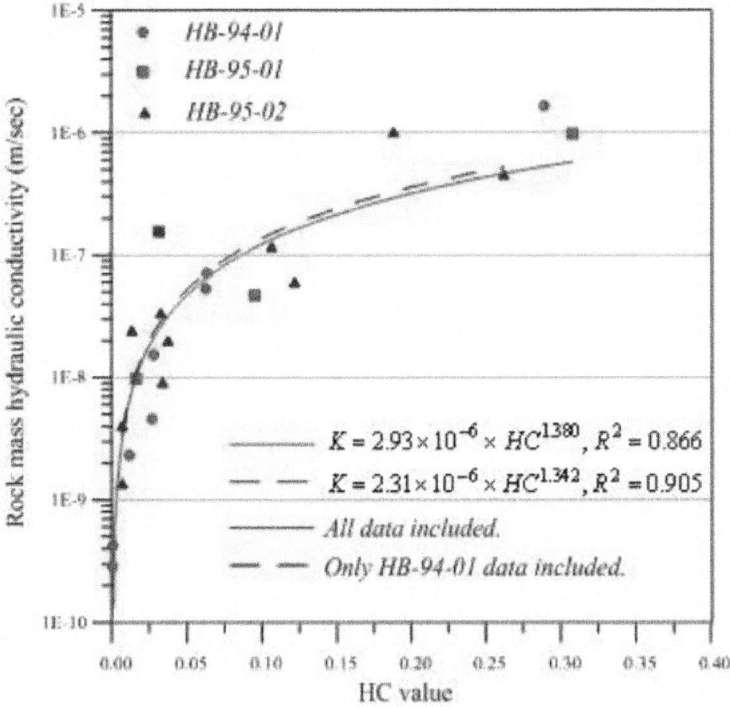

Figure. 6. Relationship between hydraulic conductivity and HC values.

The Empirical HC Model Verification

In order to further verify the feasibility of the proposed empirical HC model, the model verification is conducted. Borehole CH-04 data was adopted to verify the empirical HC model. The depth from 24.5 m to 26.6 m, 32.5 m to 34.1 m, 65.7 m to 67.8 m, and 77.8 m to 79.9 m were sealed by double packers for conducting the hydraulic tests. The quantitative evaluation of hydraulic parameters was then performed using AQTESOLVE. Table 4 shows four hydraulic test data for the model verification, in which $K_{HC-model}$ and $K_{in-situ}$ represent K obtained by Equation (5) and the in situ hydraulic test, respectively. Figure 7 shows that the comparison of the rock mass hydraulic conductivity obtained by in situ test and that from the estimation of the empirical HC model. Very good correlation can be found (Figure 7). This verification example demonstrates that the empirical HC model is able to determine the rock mass hydraulic conductivity for different sites in which the lithologic conditions are similar.

Table 3. Four hydraulic test data for the model verification

Test intervals (m)	RQD (%)	DI	1-GCD	LPI	HC	$K_{HC\text{-model}}$	$K_{in\text{-situ}}$
24.5-26.6	81.0	0.787	0.952	0.55	0.0785	9.06E-08	7.14E-08
32.5-34.1	43.8	0.723	0.975	0.55	0.2179	3.69E-07	1.11E-06
65.7-67.8	47.6	0.444	0.976	0.55	0.1248	1.71E-07	9.95E-08
77.8-79.9	95.2	0.343	1.000	0.55	0.0090	4.59E-09	9.09E-09

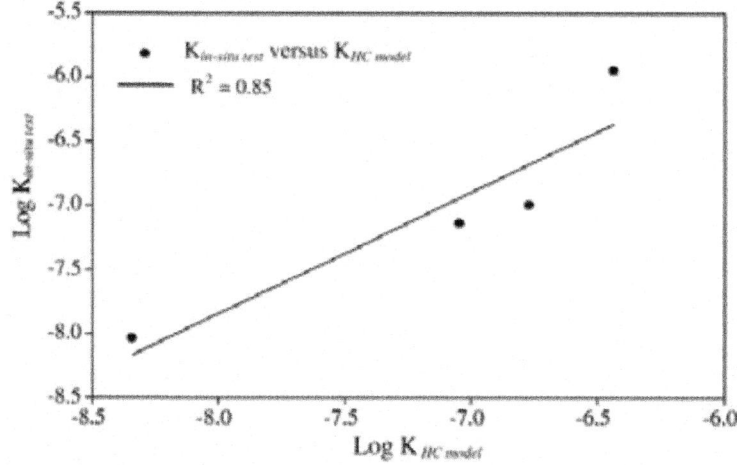

Figure. 7. Correlation between $K_{HC\text{-model}}$ and $K_{in\text{-situ}}$.

Application of HC Model

According to the above study results, the high correlation between hydraulic conductivity and HC implies that the new rock mass classification in the presented study is reasonable. It may provide two important applications in hydrogeology. The first application is that the regression equation (Equation (5) or Equation (6)) is capable of providing a useful tool to predict hydraulic conductivity of fractured rocks based on measured HC-values. By using this approach, hydraulic conductivity data in a given site can be directly acquired, which removes the cost on hydraulic testing. Secondly, for in-situ aquifer tests the HC-system is a valuable new rock mass classification system for preliminary assessment of the degree of permeability in a pack-off interval of a borehole. It is beneficial to the hydraulic test design.

THE SECOND EMPIRICAL MODEL

In recent years, many borehole hydrogeological investigations have been conducted to understand the relationship between hydraulic conductivity and fracture properties. For example, Tanaka and Miyakawa (1992) reported that some hydraulically conductive regions within a borehole existed in high-density fracture zone. Their results were supported by data acquired through packer tests and borehole televiewer logging. In addition, Gustafson et al., (1991) described that 44 to 61% of fractures in granite are non-conductive based on the field hydrological tests. By performing similar approaches, Hamm et al., (2007) demonstrated that the fracture aperture has stronger relationship to hydraulic conductivity than fracture frequency. They also proposed that the cubic fracture aperture model has close relationship to transmissivity with the highest correlation coefficient of 0.88.

In this study, a hydrogeologic investigation employing a series of subsurface exploration technologies was conducted at three active landslide sites in southern Taiwan. Each site was initially investigated with borehole televiewer logging to identify potentially significant fracture features and fracturing degree at depth and its hydrogeologic implications. Flowmeter logging was then performed to measure high permeability zones and fracture hydraulic connectivity along the borehole. Based on the prospecting results, test sections of hydraulic tests can be arranged. The hydraulic packer tests were carried out to further characterize the hydrogeologic system of the site and quantitatively determine the hydraulic properties of major hydrogeologic units or different geological structures. Finally, the borehole data were used in correlation and regression analyses to define the dependence and establish models between the fracture properties and hydraulic conductivity. The rationality of the regression results was carefully assessed in predicting fractured rock hydraulic conductivity.

Investigation Technologies

A comprehensive hydrogeologic investigation on slopeland may include surficial geology investigations, borehole drilling, testing of soil and rock properties from rock cores, landslide mapping with light detecting and ranging (LIDAR), resistivity image profiling (RIP), double-ring infiltration tests, borehole televiewer logging, electrical logging, flowmeter logging, and packer tests. This study focuses on describing borehole televiewer logging, flowmeter logging, and packer tests with the purpose of obtaining relationships between fracture properties and their corresponding hydraulic conductivity. The techniques are described as follows.

Borehole Televiewer Logging

Using borehole televiewers to characterize fractured-rock properties has been adopted for many years. The tool acquires continuous, 360-degree images of the borehole wall while the probe moves along the length of the borehole. The results provide relevant geological and structural information needed to hydraulically analyze the subsurface, such as the location, orientation and angle of the fractures, fracture width, infilling material of fractures, and structural planar features. In addition, the borehole image is capable of clarifying the uncertainties of the traditional rock core drilling technique, including those derived from human errors for misplacing rock core samples from its original place, or interpretation for missing intervals (Williams & Johnson, 2004; Hsu et al., 2007). Generally, borehole televiewers are of two types, including: the acoustic televiewer (ACTV) and the optical televiewer (OPTV) (Figure 8). The ACTV uses a fixed acoustic transducer and rotating acoustic mirror to scan the borehole wall with a focused ultrasound beam. The amplitude and travel time of the reflected acoustic signal are recorded simultaneously as separate image logs. The OPTV system consists a ring of lights, a hyperboloidal mirror, and video camera housing in the transparent window. The OPTV is capable of providing real-time borehole images (Williams & Johnson, 2004). Each probe has its own suitable prospecting environments that depend on specific groundwater conditions. Therefore, both acoustic and optical televiewers were used in this study.

Heat-Pulse Flowmeter Logging

The heat-pulse flowmeter (HPFM) system consists of a wire-grid heating element and two sensitive thermistors (heat sensors) located above and below the wire-grid. The wire-grid generates a sheet of heat in the water and the heat migrates towards one of the thermistors, depending on direction and rate of groundwater flow (Sloto & Grazul, 1989). The direction and rate of flow can be computed once the peak temperature has been detected by the thermistor (Figure 9). When the stationary measurements are conducted at several depths along a borehole, the distribution of groundwater velocity can be obtained. This not only provides useful information to characterize the aquifer permeability, but also brings good indications to identify the location of the flow path, water-producing/receiving zones, and fracture connectivity (Paillet, 1986; Miyakawa et al., 2000; Williams & Paillet, 2002). In addition, Miyakawa et al. (2000) pointed out that the HPFM is usually carried out by pumping water into or extracting water from the borehole, because it is difficult to detect hydraulic pathways in a natural state due to low groundwater velocities.

Rock Mass Hydraulic Conductivity Estimated by Two Empirical Models

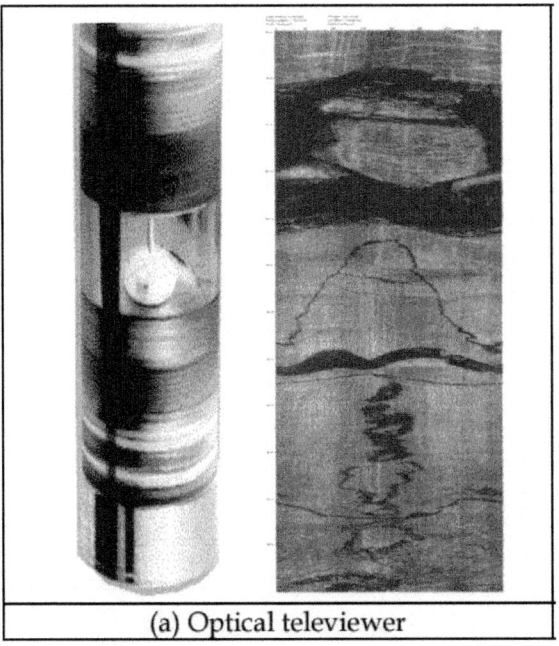

(a) Optical televiewer

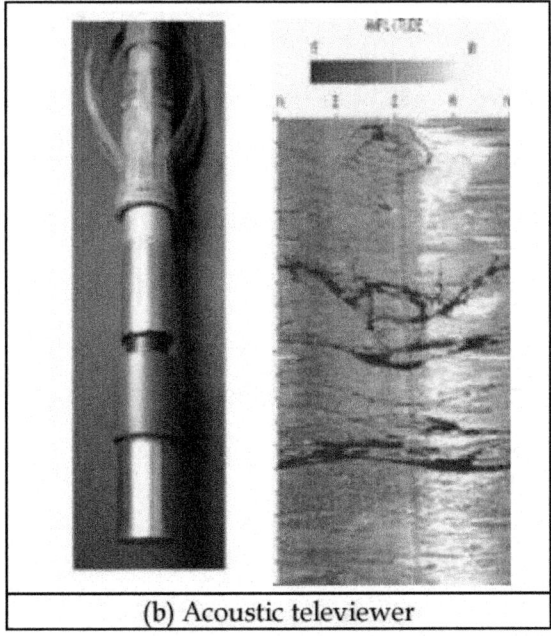

(b) Acoustic televiewer

Figure. 8. The televiewer and scanned borhole images.

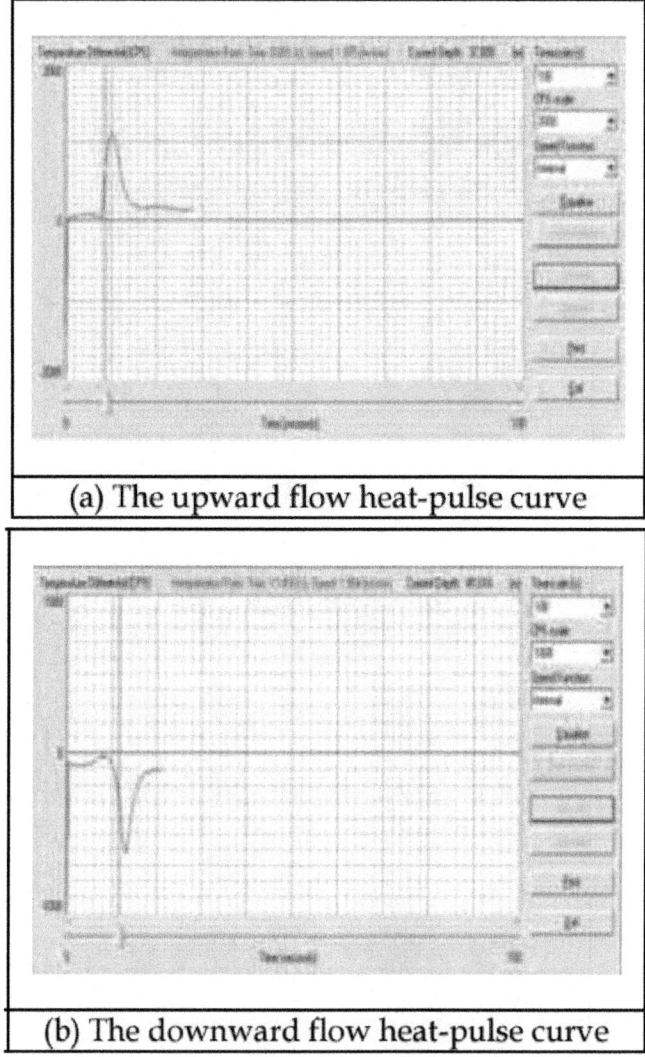

(a) The upward flow heat-pulse curve

(b) The downward flow heat-pulse curve

Figure. 9. The heat-pulse curve (the elapsed time in X-axis and the temperature difference measured by the thermistor in Y-axis).

Double Packer Hydraulic Test

The double packer hydraulic test is one of the most common approached applied to determining the hydraulic conductivity and storage coefficient along discrete sections of a borehole. It is now recognized that this approach is capable of investigating the variability of a borehole as it intersects various hydrogeological units.

The double packer hydraulic test was conducted by isolating a section of borehole with a set of packers and measuring the rate of flow and/or pressure over a period of time. The system adopted in this study contained two inflatable rubber packers, a shut-in valve, flow meters, a submersible pump, and three transducers for measuring the piezometric pressure in the isolated interval and the areas above and below the packers. The rubber packers were inflated using nitrogen delivered through a polyethylene airline. The shut-in valve was used to open and close the hydraulic connection between the pipe string and the test section. The pumping or injection rate was measured at the land surface with a flow meter.

Four types of hydraulic tests can be applied to the double packer system including the pump test, injection test, slug test, and pressure pulse test. A pump test involves pumping at a constant or variable rate and measuring changes in water levels during pumping. In an injection test, fluid is injected into the test interval at a constant head. In the slug test, a known amount of water is delivered to the test interval and the changes in pressure are monitored as equilibrium conditions return. In a pressure head test, an increment of pressure is applied to the test interval and the pressure decay is monitored over time. Typically, the selection for type of test is based on the expected permeability of strata, the volume of rock to be sampled, and the availability of time and equipment. The data collected during the hydraulic test can be analyzed by analytical methods using professional software AQTESOLVE as stated in Section 2.3.1.

Case Study and Prospecting Results

Borehole logging was conducted at three landslide sites in the south central portion of Taiwan (Figure 10). The purpose of the investigation was to determine the hydraulic properties from various geological structures, such as the degree of fracturing and hydraulic conductivity, to test hypotheses related to the causes of the landslides. The logging was conducted in six boreholes that ranged in depth from 70 to 80 m. Two boreholes from each of the following three sites were investigated: Tung-Chi (borehole FH-13 and FH-15), BaoLong (FH-03 and FH-05), and Gi-Lu (borehole FH-21 and FH-23). The Tung-Chi and BaoLong sites are located in Kaohsiung County, southern Taiwan. The principle lithologic units of the two sites are weathering slate with clay-rich gouges and fresh shale with thin layered sandstone, respectively. The Gi-Lu site is located in Pingtung County, southern Taiwan, where the dominated geologic unit composed of fresh slate with a minor amount of quartz and metamorphic sandstone.

The investigation first identified the position and degree of the fracturing using borehole televiewer logging. Second, the angle and width of fracture or fracture zone were calculated by adopting the post processing software. Lastly, flowmeter logging was used to determine the strata permeability and fracture connectivity. Based on the results of these preliminary logging data to determine favorable hydraulic conditions, test sections were selected for the double packer hydraulic test, and the results were used to determine the hydraulic conductivity for different geological structures (Figure 11). A total of 18 double packer hydraulic test sections were obtained in this study. Table 4 summarizes the results of the average fracture angle, fracture width, average flow velocity, hydraulic conductivity, and product of fracture width and flow velocity for each test section. These calculated data were used in correlation and regression analyses to define the dependence and establish models between the fracture properties and hydraulic conductivity.

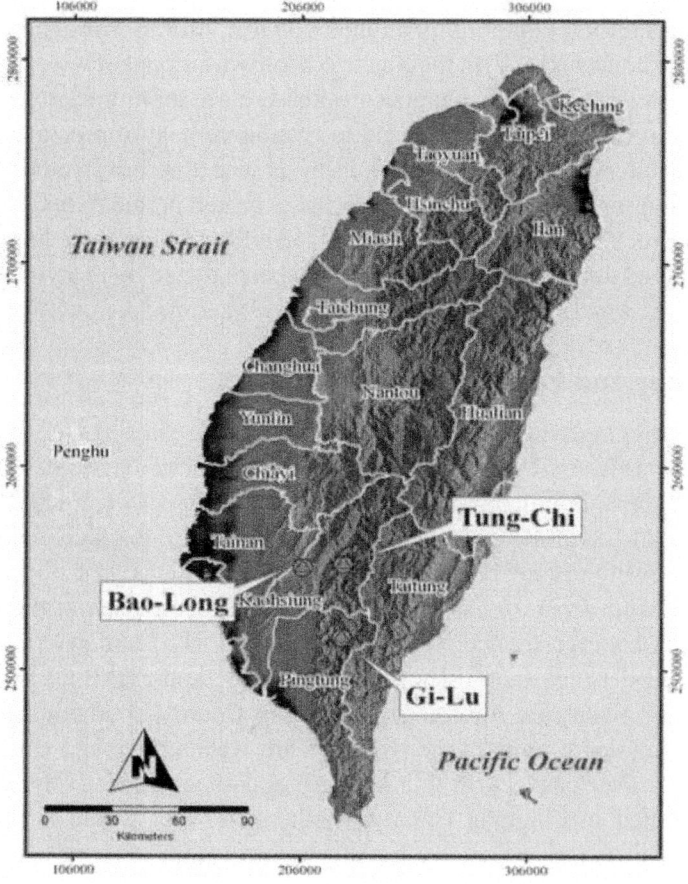

Figure. 10. Location of three active landslide sites.

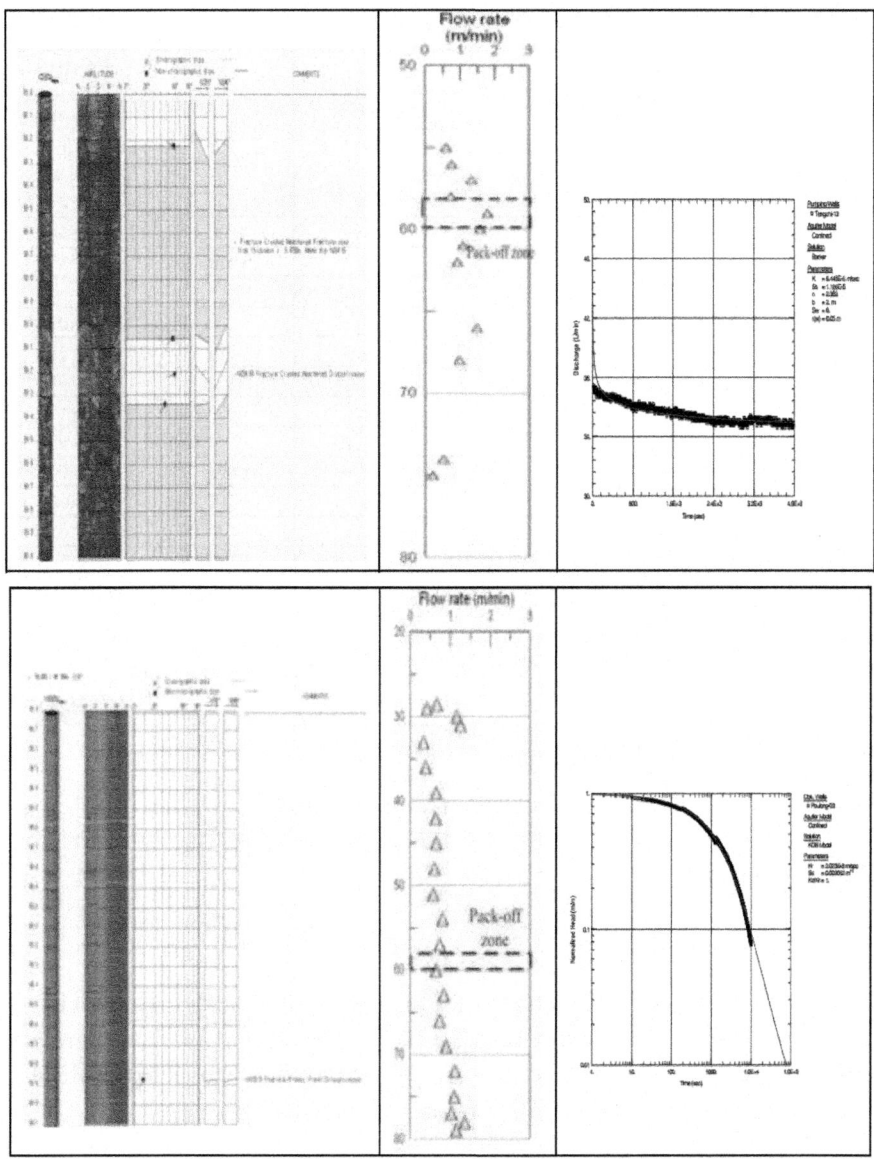

Figure. 11. The computed angle and width for single fracture or fracture zone, the measured groundwater velocity and hydraulic conductivity obtained from AQTE-SOLVE (from left to right) at 58.0-60.0m pack-off zones in Tung-Chi landslide site (above); and at 58.0-60.1m pack-off zones in Bao-Long landslide site (below).

Table 4. The prospecting results

Borehole	Test Intervals (m)	Average Fracture Angle, FA (deg)	Fracture (zone) Width, FW (m)	Average Flow Velocity, V (m/min)	Hydraulic Conductivity, K (m/sec)	Product of Fracture Width and Flow Velocity (FW×V)
FH-13	46.5-48.5	42.0	2.000	NA	3.59E-06	NA
	58.0-60.0	57.0	1.485	1.25	6.45E-06	1.856
	61.0-63.0	35.0	0.845	1.01	9.57E-06	0.845
	68.0-70.0	32.6	0.015	0.96	3.25E-07	0.014
FH-15	45.5-47.5	29.5	1.430	2.40	1.59E-05	3.432
	56.6-58.6	29.0	2.000	2.47	2.91E-05	4.940
	65.8-67.8	32.3	0.080	2.45	1.67E-06	0.196
FH-03	29.5-31.6	19.0	0.005	1.20	1.13E-08	0.006
	39.0-41.1	11.0	0.005	0.97	1.98E-10	0.005
	58.0-60.1	09.0	0.005	0.70	3.20E-09	0.004
FH-05	34.0-36.1	38.0	0.055	0.52	1.68E-07	0.029
	50.0-52.1	31.2	0.035	1.15	4.10E-07	0.040
FH-21	31.9-33.4	34.0	0.445	NA	8.40E-06	NA
	35.3-37.0	41.7	0.220	NA	2.54E-06	NA
	51.8-53.3	38.0	0.430	0.10	3.19E-07	0.043
	63.8-65.3	40.0	0.525	0.49	4.88E-07	0.257
FH-23	50.5-52.0	54.3	0.130	0.10	2.67E-08	0.013
	57.0-58.5	38.2	0.105	0.33	1.12E-07	0.035

NA: Non-detectable when test section above groundwater table.

Data Analysis

Correlation Analysis

To define the dependence between the rock mass hydraulic conductivity and fracture angle, fracture width, and the flow velocity, a univariate correlation analysis was performed. The correlation coefficient between the different values can be computed by following equation:

$$r_{xy} = \frac{\sum_{i=1}^{n}(x_i - \bar{x})(y_i - \bar{y})}{S_x S_y} \quad (7)$$

where x and y are the sample means of x and y, Sx and Sy are the sample deviations of x and y, which are defined as:

$$S_x = \sqrt{\frac{\sum_{i=1}^{n}(x_i - \bar{x})}{n}} \qquad (8)$$

$$S_y = \sqrt{\frac{\sum_{i=1}^{n}(y_i - \bar{y})}{n}} \qquad (9)$$

The results of the correlation analysis are shown in Table 5. The table shows that the fracture width and flow velocity have a good correlation with hydraulic conductivity, with correlation coefficient of 0.75 and 0.69, respectively. In addition, the analysis indicates that the fracture angle and hydraulic conductivity were independent. Most noticeably, by multiplying fracture width and flow velocity, the products of two values shown strong positive correlation to hydraulic conductivity with correlation coefficient of 0.93. Since the borehole flow velocity is usually related to fracture connectivity, a strong correlation to hydraulic conductivity can be found when considering for both fractures width and connectivity.

Table 5. The results of correlation analysis

	Correlation Coefficient			
	Average Fracture Angle, FA (deg)	Fracture (zone) Width, FW (m)	Average Flow Velocity, V (m/min)	Product of Fracture Width and Flow Velocity (FW×V)
Hydraulic Conductivity, K (m/sec)	-0.01	0.75	0.69	0.93

Regression Analysis

A regression analysis was used to establish the relationship between hydraulic conductivity and fracture angle, fracture width, flow velocity and the product of fracture width and flow velocity. Although the fracture angle and flow velocity were difficult to establish good relationship with hydraulic conductivity as shown in Table 6, a power low relationship exists between the hydraulic conductivity and fracture width in the semi-log scale with a coefficient of determination of 0.73. Furthermore, a better regression result, which the coefficient of determination of 0.83 was established between hydraulic conductivity and the product of fracture width and flow velocity in log-log scale (Figure 12) Therefore, it can be concluded that the rock mass hydraulic conductivity not only related to the fracture width, but also possesses strong relationship when their corresponding connectivity took into account. This regression results is desired to predict hydraulic conductivity of the study area based on borehole televiewer and flowmeter logging results, which removes cost of packer test for additional test intervals or boreholes.

Table 6. The results of regression analysis

Factors Related to Hydraulic Conductivity	Best Regression Equation	Coefficient of Determination (R^2)
Average Fracture Angle, FA (deg)	$K = 1.79\text{E-}13(FA)^{4.27}$	0.42
Fracture (zone) Width, FW (m)	$K = 5.42\text{E-}6(FW)^{1.30}$	0.73
Average Flow Velocity, V (m/min)	$K = 2.65\text{E-}8\exp(2.18V)$	0.28
Product of Fracture Width and Flow Velocity (FW×V)	$K = 6.29\text{E-}6(FW \times V)^{1.25}$	0.83

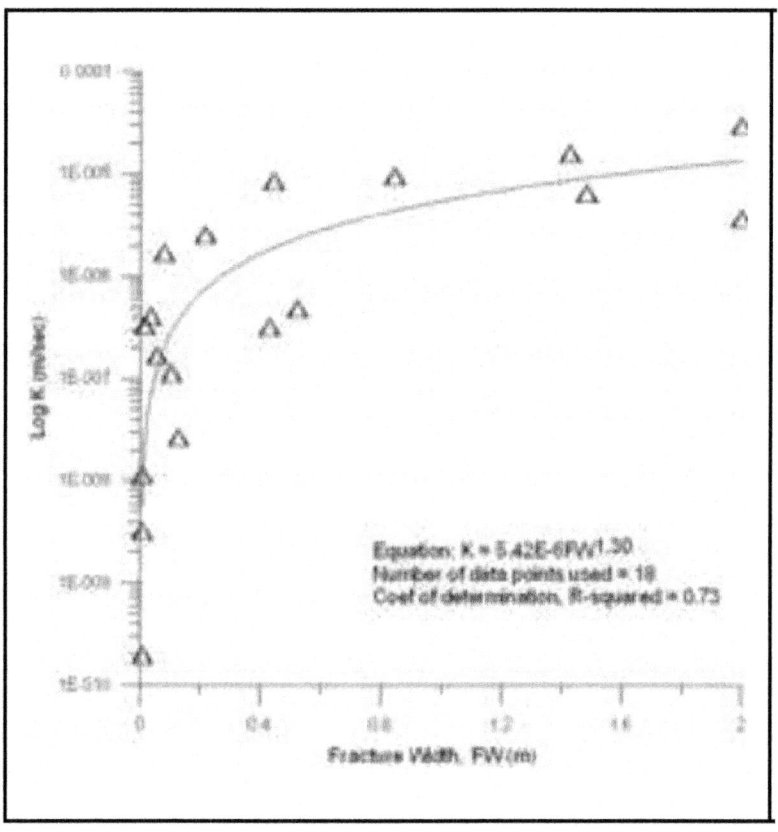

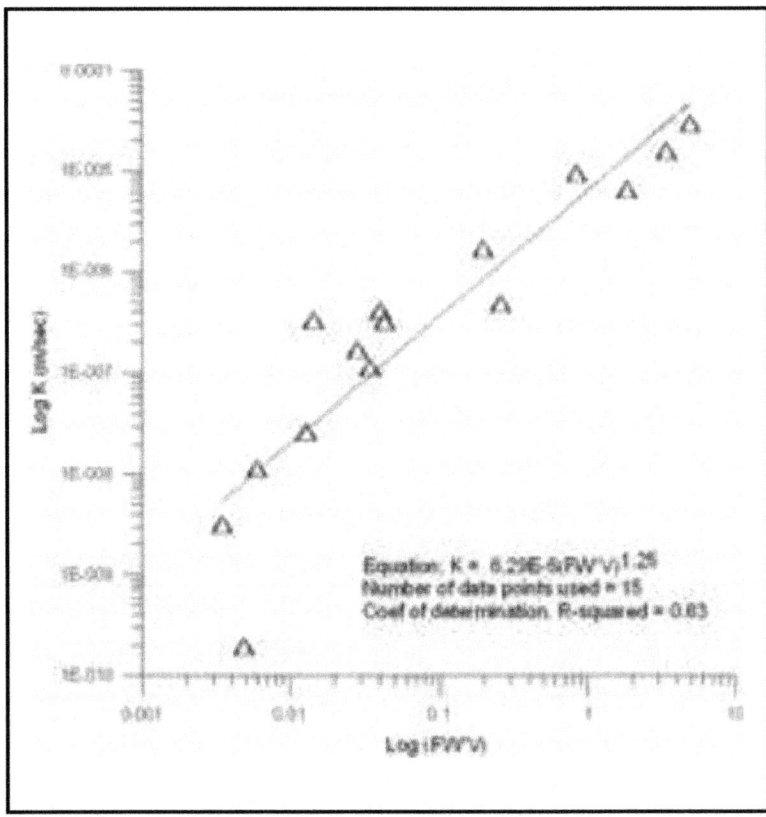

Figure. 12. Regression results of Fracture width (FW) with hydraulic conductivity (left); and the product of fracture width and flow velocity (FW×V) with hydraulic conductivity (right).

Model Verification

In order to further verify the applicability of the regression result, the verification was conducted using another two in-situ borehole prospecting data obtained in a nuclear power plant. The geologic unit of the site is mainly composed of shale or shale with thin layered sandstone. Both televiewer and flowmeter loggings were performed, and five intervals were selected for packer test. The quantitative evaluation of hydraulic conductivity was then applied using AQTESOLV based on the data of packer test. Table 7 shows five test data for the model verification, in which K_{FW} and $K_{FW \times V}$ represent K obtained by corresponding equations in Table 6; $K_{in\text{-}situ}$ represents K obtained by the in situ hydraulic test. Figure 13 shows the comparison of rock mass hydraulic conductivity obtained from in-situ test data and from regression results. The coefficient of determination between in-situ hydraulic conductivity and that predicted from the product of fracture width and velocity is 0.74, which is higher than the hydraulic conductivity predicted from fracture width only. Although the prediction still possesses some deviations when compared to real data (a predicted value is an order of magnitude lower than the measured value), the verification result demonstrated that the regression equation is capable of estimating the fractured rock hydraulic conductivity without packer testing, especially for the site with similar lithologic environments.

Table 7. Five test data for model verification

Borehole	Test Intervals (m)	Fracture (zone) Width, FW (m)	Average Flow Velocity, V (m/min)	Product of Fracture Width and Flow Velocity (FW×V)	Predicted Hydraulic Conductivity, K_{FW}(m/sec)	Predicted Hydraulic Conductivity, $K_{FW×V}$(m/sec)	Measured Hydraulic conductivity, $K_{in-situ}$ (m/sec)
BH-37	65.5-67.0	0.08	2.99	0.24	2.03E-07	1.05E-06	1.17E-06
	73.5-75.0	0.07	2.22	0.14	1.55E-07	5.59E-07	2.31E-07
BH-43	28.0-29.5	0.17	3.22	0.55	5.41E-07	2.96E-06	1.16E-05
	43.0-44.5	0.08	2.24	0.18	2.03E-07	7.33E-07	2.10E-07
	54.0-55.5	2.00	2.60	5.20	1.33E-05	4.94E-05	1.81E-05

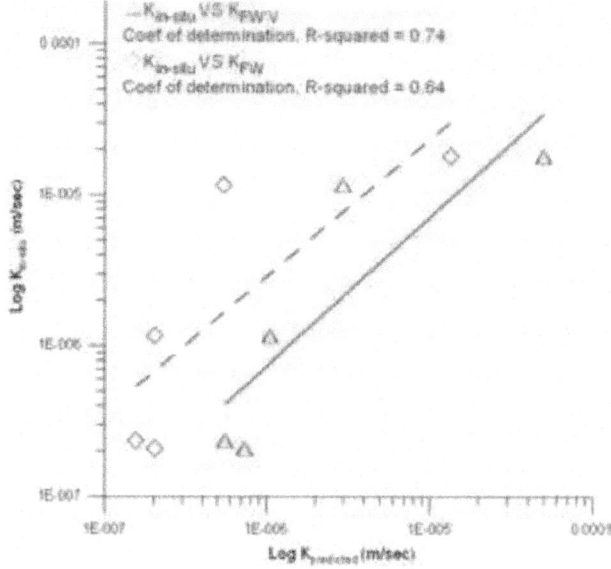

Figure 13. Verification of regression results: hydraulic conductivity predicted from the product of fracture width and flow velocity (solid line); hydraulic conductivity predicted from fracture width only (dash line).

CONCLUSIONS

The estimation of rock mass hydraulic conductivity using feasible empirical equations possesses great advantages. This chapter proposes two empirical models to estimate hydraulic conductivity of fractured rock mass with the features of high efficiency and low cost.

The first empirical model was based on the rock mass classification concept. The study develops a new rock mass classification scheme for estimating hydraulic conductivity of fractured rocks. The new rock mass classification system called as "HC-system" based on the following four parameters: rock quality designation (RQD), depth index (DI), gouge content designation (GCD), and lithology permeability index (LPI). HC-values can be calculated from borehole image data and rock core data. To verify rationality of the defined HC-system, the study collected data from the results of two hydrogeological investigation programs in three boreholes with 22 in-situ hydraulic tests to determine a relationship between hydraulic conductivity and HC. Regression analysis was performed to estimate the dependence of HC on hydraulic conductivity. The field results indicated that the rock mass in the study area has the conductivity between the order 10^{-10} and 10^{-6} m/s at the depth between 34 m and 275 m below ground surface. The regression results demonstrated that a power law relationship exists between the two variables with a coefficient of determination of 0.866. Besides, the model verification using another borehole data with four additional insitu hydraulic tests was also conducted to further verify the feasibility of the proposed empirical HC model. The regression equation provides a useful tool to predict hydraulic conductivity of fractured rocks based on measured HC-values. By using this regression equation, hydraulic conductivity data in a given site can be directly acquired, which removes the cost on hydraulic tests. For in-situ aquifer tests, the HC-system is a valuable new rock mass classification system for preliminary assessment of the degree of permeability in a packed-off interval of a borehole.

The second empirical model was simply based on the results of borehole televiewer logging, flowmeter logging and packer hydraulic tests at three active landslide sites in southern Taiwan. Three borehole prospecting techniques for hydrogeologic investigation of fractured rock mass were performed to explore various hydrogeologic characteristics, such as fracture width, fracture angle, flow velocity and hydraulic conductivity. By adopting a correlation analysis, the dependence between hydraulic conductivity and other prospecting data was identified. While the analysis revealed that the fracture width and flow velocity showed good correlation with hydraulic conductivity, the fracture angle and hydraulic conductivity were uncorrelated. In addition, by multiplying fracture width and flow velocity, the product of two values strongly correlated to hydraulic conductivity with the correlation coefficient of 0.93. The regression analysis also indicated that a power law relationship with a coefficient of determination of 0.83 existed between the hydraulic conductivity and the product of fracture width and flow velocity. Furthermore, the regression equation was verified using other borehole prospecting data. The results demonstrated that the regression equation is capable of predicting the hydraulic conductivity of fractured rock based on borehole televiewer and flowmeter logging results at the site with similar lithologic conditions. The study also demonstrated that such an approach is very constructive for a subsurface hydrogeologic assessment, particularly in the absence of packer test data due to a limited budget.

REFERENCES

1. Barton, N.; Lien, R. & Lunde, J. (1974). Engineering classification of rock masses for the design of tunnel support, Rock Mech., Vol. 6, 183-236.
2. Bear, J. (1972). Dynamics of Fluids in Porous Media, American Elsevier Publication Co., New York.
3. Bieniawski, Z. T. (1973). Engineering classification of jointed rock masses, Trans. S. Afr. Inst. Civ. Eng., Vol. 15, 335-344.
4. Bieniawski, Z. T. (1989). Engineering Rock Mass Classifications-A Complete Mannual for Engineers and Geologists in Minging, Civil, and Petroleum Engineering, John Wiley & Sons, New York.
5. Black, J. H. (1987). Flow and flow mechanisms in crystalline rock, in Fluid Flow in Sedimentary Basins and Aquifers, Geol. Soc. Special Publication No. 34, pp. 186-200.
6. Burgess, A. (1977). Groundwater Movements Around a Repository—Regional Groundwater Analysis, Kaernbraenslesaekerhet, Stockholm, Sweden.
7. Carlsson, A. & Olsson, T. (1977). Hydraulic properties of Swedish crystalline rocks-hydraulic conductivity and its relation to depth, Bulletin of the Geological Institute, University of Uppsala, 71–84.
8. Deere, D. U.; Hendron, A. J.; Patton, F. D. & Cording, E. J. (1967). Design of surface and near surface construction in rock, Proceedings of 8th U.S. Symposium. Rock Mechanics. AIME, pp. 237-302, New York.
9. Duffield, G. M. (2004). AQTESOLVE Version 4 User's Guide, Developer of AQTESOLV HydroDOLVE, Inc., Reston, VA, USA.

10. Foyo, A.; Sa'nchez, M. A.; & Tomillo, C. (2005). A proposal for secondary permeability index obtained from water pressure test in dam foundations, J. Geo. Eng., Vol. 77, 69-82.
11. Gates, W .C. B. (1997). The hydro-potential(HP) value: a rock classification technique for evaluation of the ground-water potential in fractured bedrock, Environmental & Engineering Geoscience, Vol. 3, No. 2, 251-267.
12. Gustafson, G.; Rhen, I. & Stanfor, R. (1991). Evaluation and conceptual modeling based on the pre-investigations 1986-1990, Aspo Hard Rock Laboratory Technical Report 91-22, Swedish Nuclear and Waste Management.
13. Hamm, S.; Kim, M.; Cheong, J.; Kim, J.; Son, M. & Kim, T. (2007). Relationship between hydraulic conductivity and fracture properties estimated from packer tests and borehole data in a fractured granite, Engineering Geology, 92, 73-87.
14. Hsu, S.; Chung, M.; Ku, C.; Tan, C. & Weng, W. (2007). An application of acoustic televiewer and double packer system to the study of the hydraulic properties of fractured rocks, 60th Canadian Geotechnical Conference & 8th joint CGS/IAH-CNC Groundwater Conference, pp. 415-422, V1, Ottawa, Canada.
15. Lauffer, H. (1958). Gebirgsklassifizierung für den stollenbau, Geo. Bauwesen, Vol. 74, 46-51.
16. Lee, C. H. & Farmer, I. (1993). Fluid flow in discontinuous rocks, Chapman&Hall, London, UK.
17. Louis, C., (1974). Rock Hydraulics in Rock Mechanics (ed. L. Muller), Springer Verlag, Vienna.
18. Miyakawa, K.; Tanaka, K.; Hirata, Y. & Kanauchi, M. (2000). Detecting of hydraulic pathways in fractured rock masses and estimation of conductivity by a newly developed TV equipped flowmeter, Engineering Geology, Vol. 56, 19-27.
19. National Research Council. (1996). Rock fractures and fluid flow: contemporary understanding and applications, National Academy Press, Washington DC, USA.
20. Paillet, F. L. & Hess, A. E. (1986). Geophysical well-log analysis of fractured crystalline rocks at east Bull Lake, Ontario, Canada: U.S. Geological Survey Water Resources Investigations Report 86-4052.
21. Sahimi, M. (1995). Flow and transport in porous media and fractured rock, Wiley-VCH.

22. Singhal, B. B. S. & Gupta, R.P. (1999). Applied hydrogeology of fractured rocks, Kluwer Academic Publishers, The Netherlands, 400 p.
23. Sinotech Engineering Consultants, LTD. (2005). "Tseng-Wen transbasin diversion tunnel project-supplemental geology investigation," Southern Water Resources Office, Water Resources Agency, Ministry of Economic Affairs, Taiwan.
24. Sloto, R. A. & Grazul, K. E. (1989). Results of borehole geophysical logging and hydraulic tests conducted in area D supply wells, former U.S. Naval Air Warfare Center; Warminster, Pennsylvania, Water Resource Investigation Report 01-4263, U.S. Geological Survey.
25. Snow, D. T. (1969). Anisotropic permeability of fractured media, Water Resources Research, Vol. 5, No. 6, 1273-1289.
26. Spitz, K., and Moreno, J. (1996). "A practical guide to groundwater and solute transport modeling," John Wiley, New York, 480 p. Tanaka, K. & Miyakawa, K. (1992). Application of borehole television system to deep underground survey, Jpn. Soc. Eng. Geol., Vol. 32, 289-303.
27. Terzaghi, K. (1946). Rock defects and loads on tunnel support, Rock Tunneling with steel supports, ed. R. V. Proctor and T. White, Commercial Shearing Co., Youngstown, OH, 15-99.
28. Wickham, G.E., Tiedemann, H.R., and Skinner, E.H. (1972). "Support determineation based on geologic predictions," Proc. Rapid Excav. Tunnelinf Conference, AIME, New Yprk: 43-64.
29. Hess, A.E. (1986). "Identifying hydraulically conductive fractures with a slow velocity borehole flowmeter." Canadian Geotechnical Journal 23: 69-78.
30. Sloto, R.A. and Grazul, K.E. (1989). "Results of borehole geophysical logging and hydraulic tests conducted in area D supply wells, former U.S. Naval Air Warfare Center; Warminster, Pennsylvania" Water Resource Investigation Report 01-4263, U.S. Geological Survey.
31. Tanaka, K. and Miyakawa, K. (1992). "Application of borehole television system to deep underground survey." Jpn. Soc. Eng. Geol. 32: 289-303.
32. Wei, Z. Q.; Egger, P. & Descoeudres, F. (1995). Permeability predictions for jointed rock masses, International Journal of Rock Mechanics, Mineral Science and Geomechanics, Vol. 32, 251-261.
33. Wickham, G. E.; Tiedemann, H. R. & Skinner, E. H. (1972). Support determination based on geologic predictions, Proc. Rapid Excavation Tunnel, AIME, pp. 43-64, New York.

34. Williams, J. H. & Paillet, F. L. (2002). Using flowmeter pulse to define hydraulic connections in the subsurface: a fractured shale example, Journal of Hydrology, Vol. 265, 100-117.
35. Williams, J. H. & Johnson, C. D. (2004). Acoustic and optical borehole-wall imaging for fractured-rock aquifer studies, Journal of Applied Geophysics, Vol. 55, 151-159.

Chapter 9

APPLICATION OF BASE FORCE ELEMENT METHOD ON COMPLEMENTARY ENERGY PRINCIPLE TO ROCK MECHANICS PROBLEMS

Yijiang Peng, Qing Guo, Zhaofeng Zhang, and Yanyan Shan

The Key Laboratory of Urban Security and Disaster Engineering, Ministry of Education, Beijing University of Technology, Beijing 100124, China

ABSTRACT

The four-mid-node plane model of base force element method (BFEM) on complementary energy principle is used to analyze the rock mechanics problems. The method to simulate the crack propagation using the BFEM is proposed. And the calculation method of safety factor for rock mass stability was presented for the BFEM on complementary energy principle. The numerical researches show that the results of the BFEM are consistent with the results of conventional quadrilateral isoparametric element and quadrilateral reduced integration element, and the nonlinear BFEM has some advantages in dealing crack propagation and calculating safety factor of stability.

INTRODUCTION

The finite element method (FEM) has been playing a very important role in solving various problems in engineering and science. However, the conventional finite element method (FEM) based on the displacement model has some shortcomings, such as large deformation, treatment of incompressible materials, bending of thin plates, and moving boundary problems. In the past decades, numerous efforts techniques have been proposed for developing finite element models which are robust and insensitive to mesh distortion, such as the hybrid stress method [1–4], the equilibrium models [5, 6], the mixed approach [7], the integrated force method [8–11], the incompatible displacement modes [12, 13], the assumed strain method [14–17], the enhanced strain modes [18,

19], the selectively reduced integration scheme [20], the quasiconforming element method [21], the generalized conforming method [22], the Alpha finite element method [23], the new spline finite element method [24, 25], the unsymmetric method [26–29], the new natural coordinate methods [30–33], the smoothed finite element method [34], and the base force element method [35–43].

In recent years, some scholars are studying other types of numerical analysis methods, such as boundary element method [44, 45] and meshless method [46, 47]. And some scholars still adhere to explore the finite element method based on complementary energy principle [48–51]. However, these methods have not been widely applied in engineering.

In this paper, the base force element method (BFEM) on complementary energy principle is used to analyze the engineering problems of rock mechanics. The "base forces" was introduced by Gao [52], who used the concept to replace various stress tensors for the description of the stress state at a point. These base forces can be directly obtained from the strain energy. For large deformation problems, when the base forces were adopted, the derivation of basic formulae was simplified by Gao [53] and Gao et al. [54–56]. Based on the concept of the base forces, precise expressions for stiffness and compliance matrices for the FEM were obtained by Gao [52]. The applications of the stiffness matrix to the plane problems of elasticity using the plane quadrilateral element and the polygonal element were researched by Peng et al. [37]. Using the concept of base forces as state variables, a three-dimensional formulation of base force element method (BFEM) on complementary energy principle was proposed by Peng and Liu [35] for geometrically nonlinear problems. And the new finite element method based on the concept of base forces was called as the Base Force Element Method (BFEM) by Peng and Liu [35]. A three-dimensional model of base force element method (BFEM) on complementary energy principle was proposed by Liu and Peng [36] for elasticity problems. A 4-mid-node plane element model of the BFEM on complementary energy principle was proposed by Peng et al. [38] for geometrically nonlinear problem, which is derived by assuming that the stress is uniformly distributed on each edges of a plane element. In the paper [39], an arbitrary convex polygonal element model of the BFEM on complementary energy principle was proposed for geometrically nonlinear problem. In the paper [43], a 4-mid-node plane model of BFEM on complementary energy principle was researched, and its computational performance was studied. The convex polygonal element model of BFEM on complementary energy principle was given by Peng et al. [40] for arbitrary mesh problems. In the paper [41], the concave polygonal element model of BFEM on complementary energy principle was proposed for the concave polygonal mesh problems. In the paper [42], the BFEM on potential energy principle was used to analyze recycled aggregate concrete (RAC) on mesolevel, in which the model of BFEM with triangular element was derived, and the simulation results of the BFEM agree with the test results of recycled aggregate concrete. In recently, the BFEM on damage mechanics has been used to analyze the compressive strength, the size effects of compressive strength, and fracture process of concrete at mesolevel, and the analysis method is the new way for investigating fracture mechanism and numerical simulation of mechanical properties for concrete.

The purpose of this paper is to survey the base forces element method on complementary energy principle for large-scale computing problems in rock engineering problems.

MODEL OF THE BFEM

Compliance Matrix

Consider a 4-mid-node plane element as shown in Figure 1; the compliance matrix of a base force element can be obtained as [43]

$$\mathbf{C}_{IJ} = \frac{1+\nu}{EA}\left(r_{IJ}\mathbf{U} - \frac{\nu}{1+\nu}\mathbf{r}_I \otimes \mathbf{r}_J\right), \quad (I, J = 1, 2, 3, 4) \tag{1}$$

in which E is Young's modulus, ν is Poisson's ratio, A is the area of an element, U is the unit tensor, and r_{IJ} is the dot product of radius vectors r_I and r_J at points I and J.

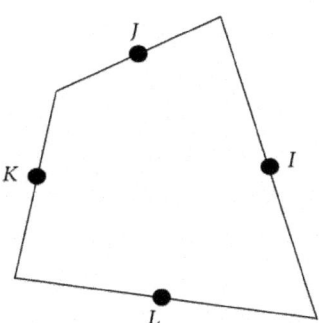

Figure 1: Four-mid-node plane element.

For a plane rectangular coordinate system, the radius vectors r_I and r_J of points I and J can be written as

$$\mathbf{r}_I = x_I \mathbf{e}_x + y_I \mathbf{e}_y, \qquad \mathbf{r}_J = x_J \mathbf{e}_x + y_J \mathbf{e}_y \tag{2}$$

in which $\mathbf{e}_x$, $\mathbf{e}_y$ are the unit vectors.

Further, the compliance matrix of an element can be reduced as follows:

$$\begin{aligned}\mathbf{C}_{IJ} = \frac{1+v}{EA} \Bigl[& \Bigl(\frac{1}{1+v} x_I x_J + y_I y_J \Bigr) \mathbf{e}_x \otimes \mathbf{e}_x \\ & - \frac{v}{1+v} x_I y_J \mathbf{e}_x \otimes \mathbf{e}_y - \frac{v}{1+v} y_I x_J \mathbf{e}_y \otimes \mathbf{e}_x \\ & + \Bigl(x_I x_J + \frac{1}{1+v} y_I y_J \Bigr) \mathbf{e}_y \otimes \mathbf{e}_y \Bigr]. \end{aligned} \tag{3}$$

For a plane strain problem, it is necessary to replace E by $E/(1-v^2)$ and v by $v/(1-v)$ in (1) and (3).

The characteristics of the BFEM on complementary energy principle are that the model does not introduce an interpolating function and is not necessary to introduce the Gauss integral for calculating the compliance coefficient at a point.

Governing Equations

The total complementary energy of the elastic system which has n elements can be written as

$$\Pi_C = \sum_n \left(W_C^e - \bar{\mathbf{u}}_I \cdot \mathbf{T}^I \right), \tag{4}$$

where W_C^e is the complementary energy of an element $\mathbf{T}^I$ and $\bar{\mathbf{u}}_I$ are the resultant force vectors and the given displacement acting on the center node I of the edge I, respectively.

The equilibrium conditions can be released by the Lagrange multiplier method, and a new complementary energy function for an element can be introduced as follows:

$$\Pi_C^{e*}(\mathbf{T}, \boldsymbol{\lambda}, \lambda_\theta) = \Pi_C^e(\mathbf{T}) + \boldsymbol{\lambda} \left(\sum_{I=1}^4 \mathbf{T}^I \right) + \lambda_\theta \left(\mathbf{T}^I \times \mathbf{r}_I \right), \tag{5}$$

where $\boldsymbol{\lambda} = \lambda_x \mathbf{e}_x + \lambda_y \mathbf{e}_y$ and λ_θ are the Lagrange multipliers.

For the elastic system, the modified total complementary energy function of the elastic system which contains n elements should meet the following equation by means of the modified complementary energy principle:

$$\delta \Pi_C^* = \sum_n \left[\delta \Pi_C^{e\,*} (\mathbf{T}, \lambda, \lambda_\theta) \right] = 0. \tag{6}$$

Further, (6) can be expressed as

$$\frac{\partial \Pi_C^* (\mathbf{T}, \lambda, \lambda_\theta)}{\partial \mathbf{T}} = 0, \quad \frac{\partial \Pi_C^* (\mathbf{T}, \lambda, \lambda_\theta)}{\partial \lambda} = 0,$$

$$\frac{\partial \Pi_C^* (\mathbf{T}, \lambda, \lambda_\theta)}{\partial \lambda_\theta} = 0. \tag{7}$$

The first of (7) is the compatibility equations and displacement boundary conditions for the elastic system. According to this equation, the displacement boundary conditions in this paper can be implemented in the BFEM. The second of (7) is the force equilibrium equation of each element. The third of (7) is the moment equilibrium equation of each element. These are the governing equations of the BFEM. From the equations, we can obtain the resultant forces acting on the center points of the edges of all elements.

Stress Tensor of an Element

Consider the 4-mid-node plane element as shown in Figure 1; the real stress σ of the element can be replaced by the average stress $\bar{\sigma}$ if the element is small enough. According to the definitions of Cauchy stress tensors, the stress expressions of an element can be obtained as

$$\sigma = \frac{1}{A} \mathbf{T}^I \otimes \mathbf{r}_I, \tag{8}$$

where $\otimes$ is the dyadic symbol, $\mathbf{T}^I$ and $\mathbf{r}_I$ are the resultant force vectors acting on the center node I of the edge I and the radius vector of the node I, respectively, and the summation rule is implied.

For a plane rectangular coordinate system, the force vectors $\mathbf{T}^I$ of the node I can be written as

$$\mathbf{T}^I = T_x^I \mathbf{e}_x + T_y^I \mathbf{e}_y, \tag{9}$$

where T_x^I and T_y^I are the components of the force vector $\mathbf{T}^I$ along coordinates x and y, respectively

Further, the stress tensors of an element can be reduced as follows:

$$\sigma = \frac{1}{A}\sum_{I=1}^{4}\left[T_x^I x_I \mathbf{e}_x \otimes \mathbf{e}_x + T_x^I y_I \mathbf{e}_x \otimes \mathbf{e}_y\right.$$

$$\left. + T_y^I x_I \mathbf{e}_y \otimes \mathbf{e}_x + T_y^I y_I \mathbf{e}_y \otimes \mathbf{e}_y\right]. \tag{10}$$

Displacement Vector of Nodes

According to the governing equation of an element of the BFEM, the explicit expression of displacement can be obtained as

$$\delta_I = \mathbf{C}_{IJ} \cdot \mathbf{T}^J + \lambda + \lambda_\theta \varepsilon \cdot \mathbf{r}_I, \tag{11}$$

in which ε is the alternating tensor, λ and λ_θ are the Lagrange multipliers [43], and ε and λ can be expressed as

$$\varepsilon = \mathbf{e}_x \otimes \mathbf{e}_y - \mathbf{e}_y \otimes \mathbf{e}_x,$$

$$\lambda = \lambda_x \mathbf{e}_x + \lambda_y \mathbf{e}_y. \tag{12}$$

Further, the displacement vectors of an element can be reduced as follows [43]:

$$\delta_I = \left(C_{IxJx}T_x^J + C_{IxJy}T_y^J + \lambda_x + \lambda_\theta y_I\right)\mathbf{e}_x$$

$$+ \left(C_{IyJx}T_x^J + C_{IyJy}T_y^J + \lambda_y - \lambda_\theta x_I\right)\mathbf{e}_y. \tag{13}$$

SIMULATION OF GRAVITY AND MATERIAL

In engineering problems, regardless of the dam, rock slope, or other structure of rock mass, the gravity of structure should be considered in the numerical calculation. We did not take into account the gravity problem when we previously prepared the program of BFEM on complementary energy principle. In order to consider the gravity of structure, three problems must be solved, including the calculation problem of equivalent node loads in the BFEM on complementary energy principle, the problem of exerting gravity in the software of the BFEM and the calculation problem of stress tensor in an element when the gravity is added.

Equivalent Node Loads of Gravity

For the BFEM on complementary energy principle, the equivalent node loads of gravity in an element can be calculated according to the principle of virtual work, as shown in Figure 2, and the expression can be given as

$$\{Q\}^e = -\frac{\rho g A t}{4} \begin{bmatrix} 0 & 1 & 0 & 1 & 0 & 1 & 0 & 1 \end{bmatrix}^T \quad (14)$$

or

$$\mathbf{Q}^I = -\frac{\rho g A t}{4} \mathbf{e}_y, \quad (I = 1, 2, 3, 4), \quad (15)$$

where t is the thickness of an element, ρ is the density of material, and g is acceleration of gravity.

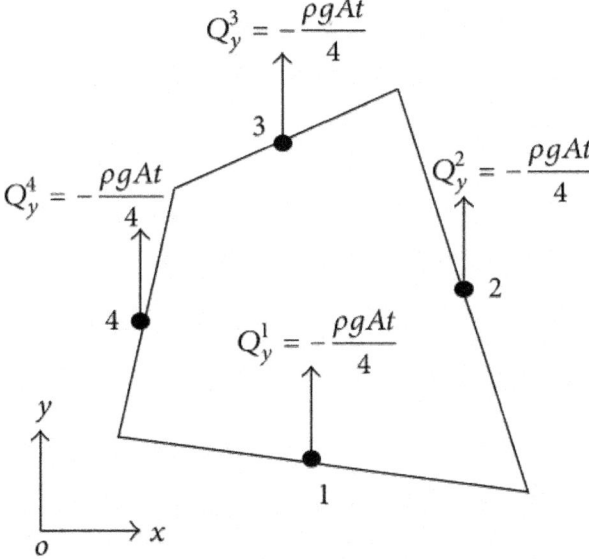

Figure 2: Equivalent node loads of gravity in an element.

Stress Calculation of an Element

When the gravity of an element is not taken into account, as shown in Figure 3, we calculate the force acting on the edges of an element first. Then, the stress tensor of the element can be calculated by (8) or (10).

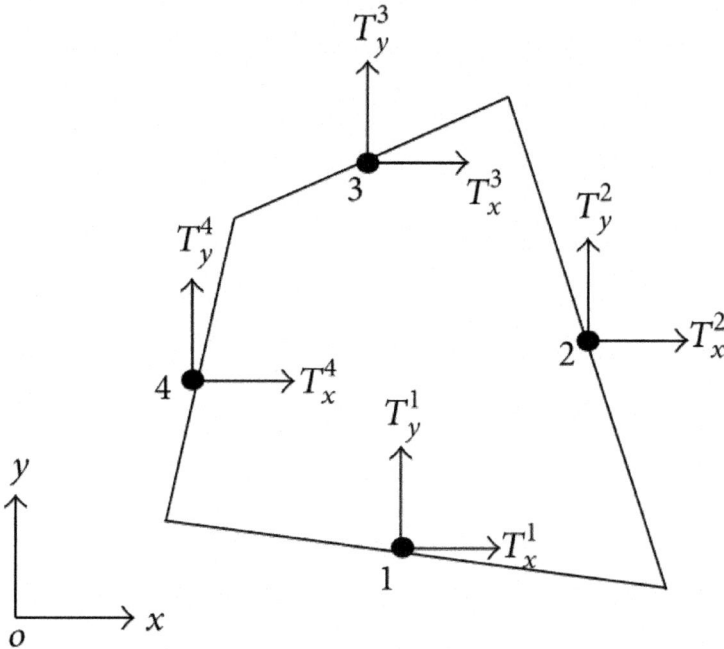

Figure 3: An element without gravity.

Further, the stress tensors of an element can also be reduced as follows:

$$\bar{\sigma} = \frac{1}{A}\left(\mathbf{T}^1 \otimes \mathbf{r}_1 + \mathbf{T}^2 \otimes \mathbf{r}_2 + \mathbf{T}^3 \otimes \mathbf{r}_3 + \mathbf{T}^4 \otimes \mathbf{r}_4\right), \tag{16}$$

where $\mathbf{T}^I = T_x^I \mathbf{e}_x + T_y^I \mathbf{e}_y$, $\mathbf{r}^I = x_I \mathbf{e}_x + y_I \mathbf{e}_y$, ($I$ = 1, 2, 3, 4), x_I and y_I are the coordinates of node I, respectively.

When the gravity of an element is taken into account and there is a free boundary, as shown in Figure 4, the stress tensor of the element can be calculated as

$$\bar{\sigma} = \frac{1}{A}\Big[\left(\mathbf{T}^1 + \mathbf{Q}^1\right) \otimes \mathbf{r}_1 + \mathbf{Q}^2 \otimes \mathbf{r}_2$$
$$+ \left(\mathbf{T}^3 + \mathbf{Q}^3\right) \otimes \mathbf{r}_3 + \left(\mathbf{T}^4 + \mathbf{Q}^4\right) \otimes \mathbf{r}_4\Big]. \tag{17}$$

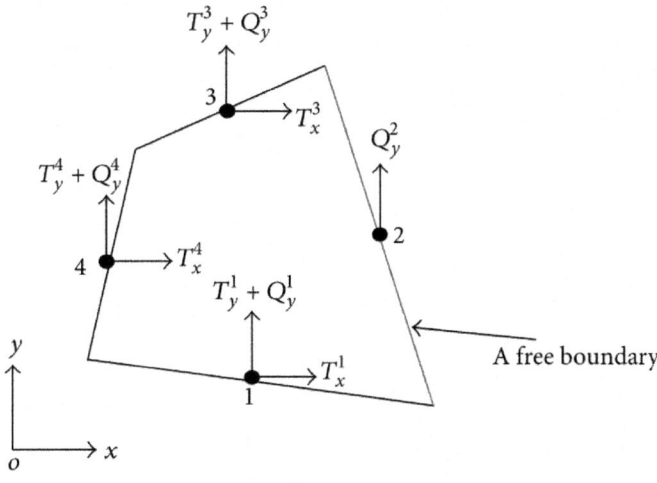

Figure 4: Considering gravity and a free boundary.

When the gravity of an element is taken into account and there is a force boundary condition, as shown in Figure 5, the stress tensor of the element can be calculated as

$$\overline{\sigma} = \frac{1}{A}\left[\left(\mathbf{T}^1 + \mathbf{Q}^1\right) \otimes \mathbf{r}_1 + \left(\mathbf{F} + \mathbf{Q}^2\right) \otimes \mathbf{r}_2 \right.$$
$$\left. + \left(\mathbf{T}^3 + \mathbf{Q}^3\right) \otimes \mathbf{r}_3 + \left(\mathbf{T}^4 + \mathbf{Q}^4\right) \otimes \mathbf{r}_4\right] \quad (18)$$

in which $\mathbf{F} = F_x \mathbf{e}_x + F_y \mathbf{e}_y$.

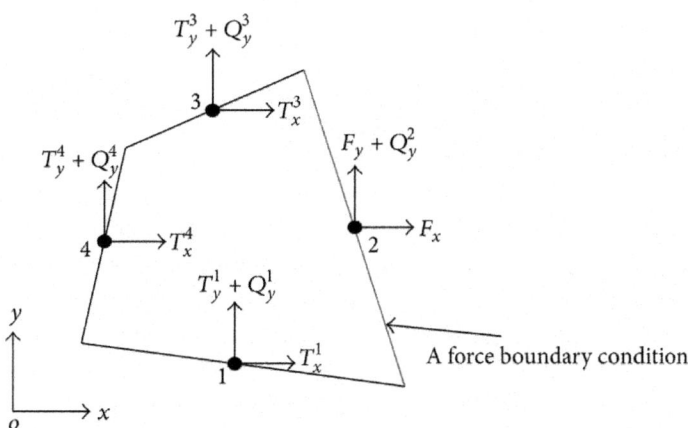

Figure 5: Considering gravity and a force boundary condition.

When the gravity of an element is not taken into account and there is a force boundary condition, as shown in Figure 6, the stress tensor of the element can be calculated as

$$\overline{\sigma} = \frac{1}{A}\left(\mathbf{T}^1 \otimes \mathbf{r}_1 + \mathbf{F} \otimes \mathbf{r}_2 + \mathbf{T}^3 \otimes \mathbf{r}_3 + \mathbf{T}^4 \otimes \mathbf{r}_4\right). \tag{19}$$

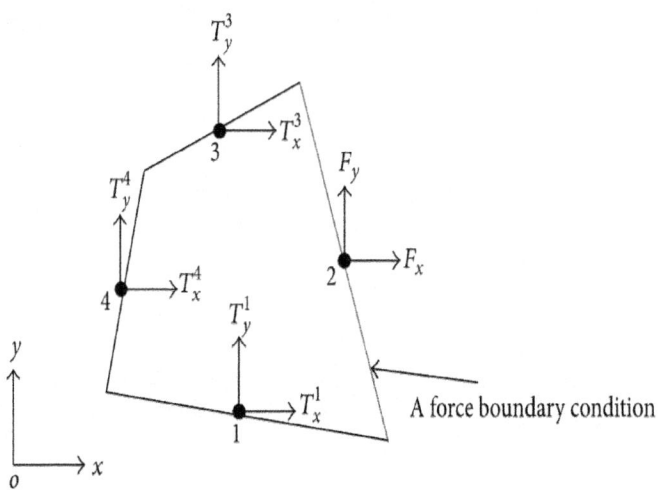

Figure 6: Considering a force boundary condition and no gravity.

Simulation of Different Materials

We adopt two one-dimensional array variables to reflect the change of elastic modulus and Poisson's ratio of different materials, respectively.

THE NONLINEAR MODEL OF BFEM FOR CRACK PROPAGATION PROBLEMS

The conventional displacement model of FEM requires the mesh reconstruction for the crack propagation problems. Therefore, the conventional FEM has deficiencies for the crack propagation problems. Because the contact forces between each element are used, the BFEM on complementary energy principle has advantage in the simulation of crack propagation problems. In the BFEM on complementary energy principle, we only need to deal with the constraint conditions and compatibility equations of displacement.

Failure Criteria of the Contact Interface between Two Elements

According to the control equations of the BFEM on complementary energy principle, the forces acting on the edge on an element can easily be calculated. According to the failure criteria expressed by the forces acting on the edge of an element, we can judge whether the interface of elements is cracking. If there is cracking of element interface, the nonlinear processing must be carried out.

Condition of Elastic State at Contact Interface

When $T_n < 0$ and $|T_s| < -f \cdot T_n + c \cdot l$ or $0 < T_n < t_0$ and $|T_s| < c \cdot l$, the contact interface is not cracked and in the elastic state. Here, T_n and T_s are the normal interface force and tangential interface force at contact interface between two elements, respectively. c and f are the cohesion and the internal friction coefficient of the contact interface, respectively. l is the length of an element. t_0 is the tensile strength of an element, and it is positive in tension.

Condition of Positive Slip at Contact Interface

When $T_n < 0$ and $T_s \geq -f \cdot T_n + c \cdot l$, the contact interface of the two elements began to slip. We need to get rid of the tangential displacement constraint at the contact interface between the two elements that it has been cracked and put the frictional force as load that is used to an initial force condition. The new load acting on the contact interface between the two elements in the second loop of calculation can be written as follows:

$$T_{s0} = f \cdot T_s. \qquad (20)$$

When $0 < T_n < t_0$ and $T_s \geq c \cdot l$, the contact interface of the two elements begins to slip. We need to get rid of the tangential displacement constraint and the normal displacement constraint at the contact interface of the two elements. And the contact interface will crack after slipping.

Condition of Negative Slip at Contact Interface

When $T_n < 0$ and $T_s \leq f \cdot T_n - c \cdot l$, the contact interface of the two elements began to slip. We need to get rid of the tangential displacement constraint at the contact interface between the two elements that it has been cracked, and put the frictional force as load that is an initial force condition. The new load acting on the contact interface between the two elements in the second loop of calculation is

$$T_{s0} = -f \cdot T_s. \qquad (21)$$

When $0 < T_n < t_0$ and $T_s \leq -c \cdot l$, the contact interface of the two elements begins to slip. We need to get rid of the tangential displacement constraint and the normal displacement constraint at the contact interfaces of elements. And the contact interface will crack after slipping.

Condition of Pull Cracking at Contact Interface

When $T_n \geq t_0$, the contact interface of the two elements begins to crack. We need to get rid of the tangential displacement constraint and the normal displacement constraint at the contact interfaces of elements. After the above checks, the computer program of BFEM uses the new loads and constraint conditions to solve the governing equations of BFEM on complementary energy principle and obtains the new forces acting on the contact interfaces of elements. The program repeats the above checks until there are no new cracking at the contact interfaces or the solutions of nonlinear equations cannot be convergence since the interface cracks are too long.

Flow Chart of the Nonlinear BEFM for Crack Propagation Problems

The flow chart of the nonlinear BEFM of crack propagation problems can be shown in Figure 7.

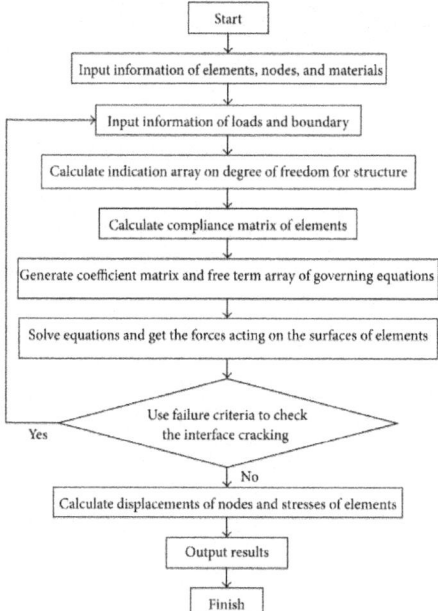

Figure 7: Flow chart of the nonlinear BEFM for crack propagation problems.

CALCULATION METHOD ON SAFETY FACTOR OF STABILITY IN THE BFEM

There are many methods to calculate the safety factor in engineering. The traditional finite element method usually used the rock joint elements to calculate factor of safety along the joint path. When the base force element method is used to analyze the stability of the rock mass in order to get the safety factor along the joint path, it is very easy. First, we calculate the surface forces of all elements according to the different load combinations. Then, we accumulate the sliding resistances and the sliding forces along the sliding path, respectively. Further, the safety factor of stability along the sliding path can obtained by the following equation:

$$K = \frac{\sum_{i=1}^{n}(-T_{ni}f_i + c_i l_i)}{\sum_{i=1}^{n} T_{si}}, \qquad (22)$$

where T_{ni} and T_{si} are the normal interface force and tangential interface force at contact interface between two elements along the sliding path, respectively. c_i and f_i are the cohesion and the internal friction coefficient of the contact interface between two elements along the sliding path, respectively. l_i is the length of the interface in the element along the sliding path.

NUMERICAL EXAMPLES

Example 1: A Rock Pillar under the Action of Gravity

Consider a thick pillar of rock under the action of gravity shown in Figure 8. And its width is 5 m, its height is 10 m, modulus of elasticity $E = 1 \times 10^8$ Pa, Poisson ratio $v = 0.3$, density of rock $\rho = 2.45$ t/m³, and acceleration of gravity $g = 9.8$ m/s². The calculation is considered into the plane stress problem.

The values of stress components and displacement components of the rock pillar are listed in Tables 1–4, respectively. Comparisons of the results from the conventional quadrilateral isoparametric element (Q4 model) and quadrilateral reduced integration element (Q4R model) are also given in Tables 1–4, respectively. The numerical results of the present model are consistent with those of the Q4 model and Q4R model and have shown good computational stability.

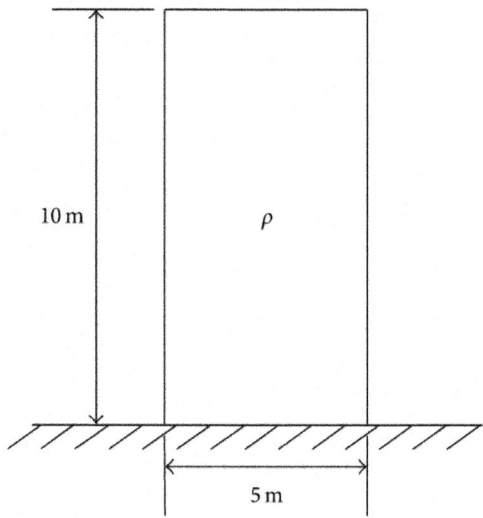

Figure 8: A rock pillar under the action of gravity.

The calculation is done using three different element meshes with the center nodes of edges of elements as shown in Figure 9.

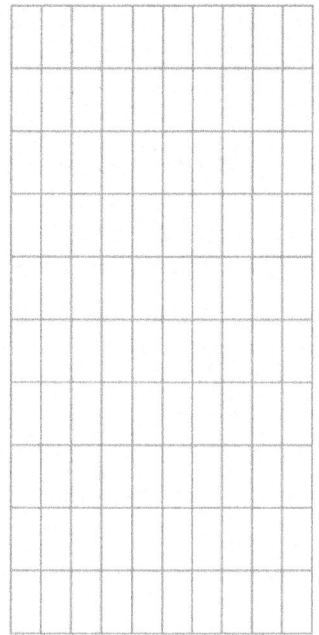

(a) 100 elements, 220 nodes

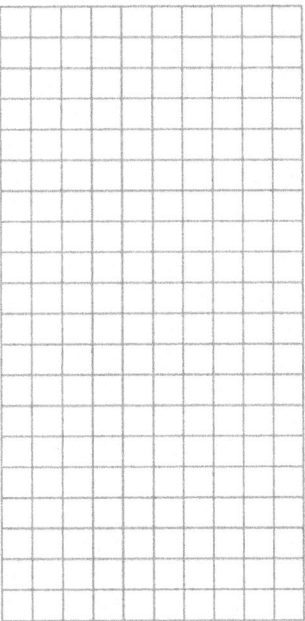

(b) 200 elements, 430 nodes

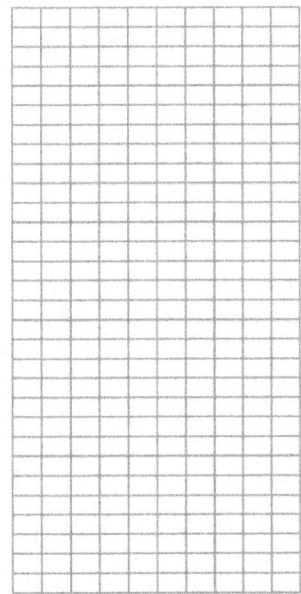

(c) 300 elements, 640 nodes

Figure 9: Three kinds of meshes for a rock pillar.

Table 1: Displacement u_y at the top of the pillar

Meshes	BFEM model ($\times 10^{-2}$ m)	Q4 model ($\times 10^{-2}$ m)	Q4R model ($\times 10^{-2}$ m)
10 × 10	−1.1939	−1.1917	−1.2048
10 × 20	−1.1940	−1.1931	−1.1964
10 × 30	−1.1940	−1.1934	−1.1940

Table 2: Displacement u_y at $h=5$ m of the pillar

Meshes	BFEM ($\times 10^{-3}$ m)	Q4 model ($\times 10^{-3}$ m)	Q4R model ($\times 10^{-3}$ m)
10 × 10	−8.7592	−8.6975	−8.8045
10 × 20	−8.7630	−8.7118	−8.6726
10 × 30	−8.7637	−8.7149	−8.7363

Table 3: Stress σ_x at center of the elements at top of the pillar.

Meshes	BFEM (kPa)	Q4 model (kPa)	Q4R model (kPa)
10 × 10	−12.016	−12.244	−12.062
10 × 20	−6.005	−6.0773	−6.013
10 × 30	−4.003	−4.0382	−4.002

Table 4: Stress σ_y at center of the elements at bottom of the pillar.

Meshes	BFEM (kPa)	Q4 model (kPa)	Q4R model (kPa)
10 × 10	−244.194	−242.945	−244.349
10 × 20	−262.976	−260.116	−263.387
10 × 30	−272.102	−268.002	−272.552

Example 2: Analysis for a Rock Pillar with Four Materials under Pure Shear Effect

Consider a rock pillar with four materials under pure shear effect shown in Figure 10. And its modulus of elasticity $E_1 = 10^9$, $E_2 = 10^8$, $E_3 = 10^7$, and $E_4 = 10^6$, respectively. And Poisson ratio $v = 0.3$, the shear stress on surface of the structure $\tau=1$. The calculation is considered into the plane stress problem and the dimensionless values.

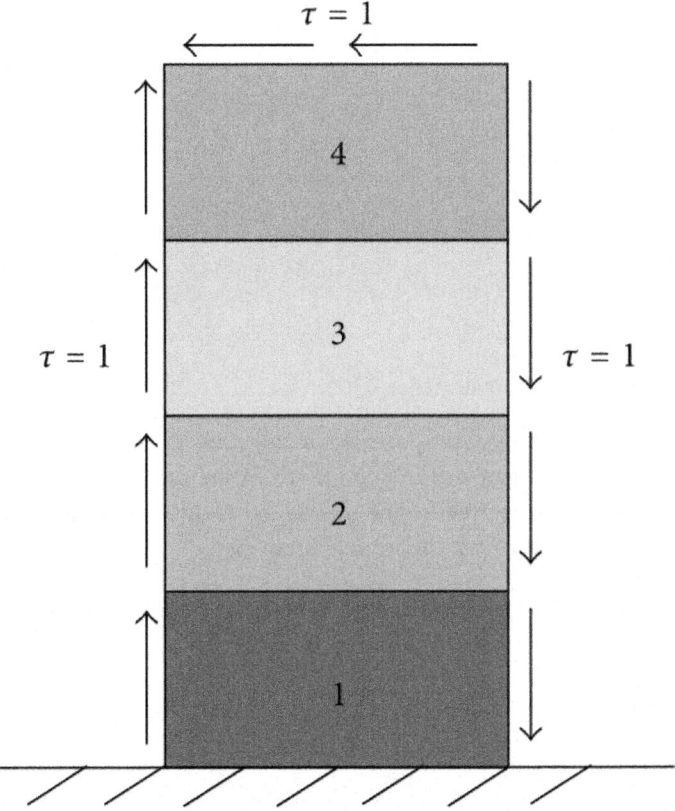

Figure 10: A rock pillar with four kinds of materials.

The calculation is done using the mesh with the center nodes of edges of elements as shown in Figure 11.

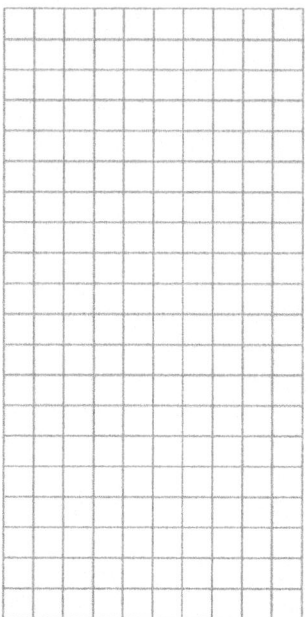

Figure 11: Mesh with 200 elements, 430 nodes.

The values of stress components of the rock pillar are listed in Table 5, respectively. Comparisons of the results from the theoretical analysis are also given in Table 5, respectively. The numerical results of the present model are consistent with those of the theoretical analysis.

Table 5: Stress solution of the pillar with four kinds of materials under uniform shearing forces.

	σ_x	τ_{xy}	σ_y
BFEM	0.000	1.000	0.000
Exact	0.000	1.000	0.000

Example 3: A Rock Pillar under Water Pressure and Gravity

Consider a rock pillar under water pressure and gravity shown in Figure 12. And its modulus of elasticity $E = 10^8$ Pa, Poisson' ratio $v = 0.3$, density of rock $\rho_1 = 2.4$ t/m³, density of water $\rho_2 = 1.0$ t/m³, and acceleration of gravity $g = 9.8$ m/s². The calculation is considered into the plane strain problem.

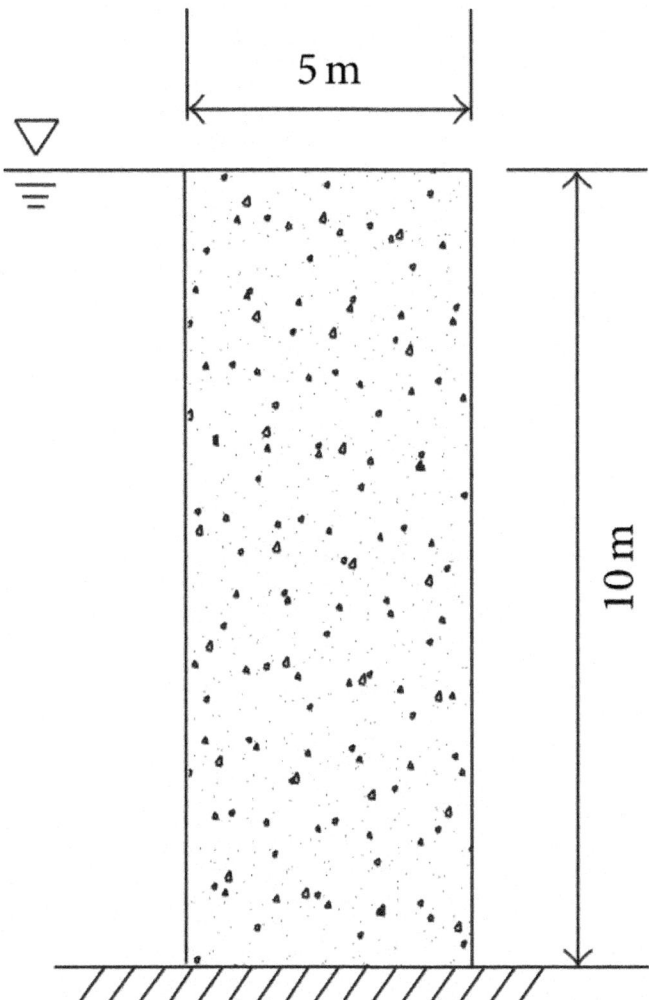

Figure 12: A rock pillar subjected to water pressure and gravity.

The calculation is done using four different element meshes with the center nodes of edges of elements as shown in Figure 13.

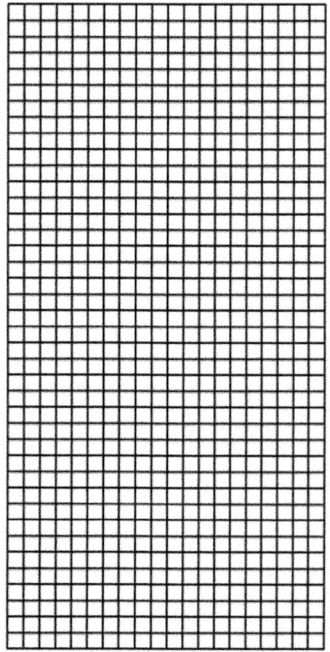

(a) 800 elements, 1660 nodes

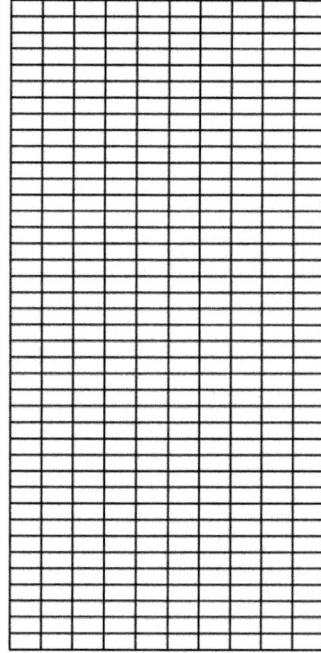

(b) 400 elements, 850 nodes

Application of Base Force Element Method on Complementary ... 229

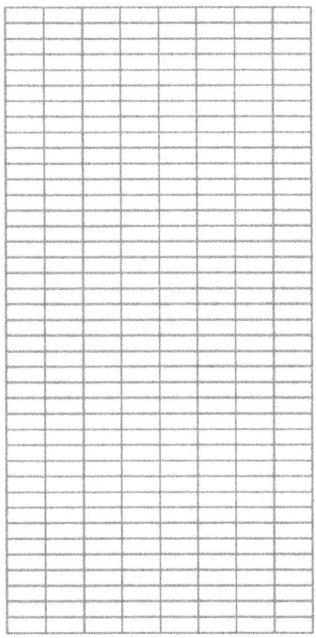

(c) 320 elements, 688 nodes

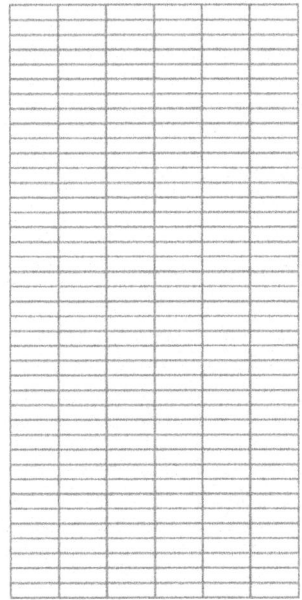

(d) 240 elements, 526 nodes

Figure 13: Four kinds of meshes for a rock pillar.

The values of stress components and displacement components of the rock pillar are listed in Tables 6–12, respectively. Comparisons of the results from the conventional quadrilateral isoparametric element (Q4 model) and quadrilateral reduced integration element (Q4R model) are also given in Tables 6–12, respectively. The numerical results of the present model are consistent with those of Q4R model, and the 4-mid-node element of BFEM has given good performance compared with Q4 model for the large aspect ratio of elements. Due to the different method calculating the equivalent node loads of water pressure between the base force elements and the element of traditional FEM, there is a slight error about the calculation results of the horizontal stresses σ_x of the element in lower right corner of rock pillar between the BFEM and the Q4R model, as shown in Table 10.

Table 6: Displacement u_x at $x = 2.5$ m of the pillar top.

Meshes	BFEM ($\times 10^{-2}$ m)	Q4 model ($\times 10^{-2}$ m)	Q4R model ($\times 10^{-2}$ m)
20 × 40	4.0219	4.0056	4.0190
10 × 40	4.0415	4.0094	4.0398
8 × 40	4.0563	4.0135	4.0532
6 × 40	4.0891	4.0239	4.0863

Table 7: Displacement u_x at $h=5$ m on the right of the pillar

Meshes	BFEM ($\times 10^{-2}$ m)	Q4 model ($\times 10^{-2}$ m)	Q4R model ($\times 10^{-2}$ m)
20 × 40	2.0837	2.0758	2.0830
10 × 40	2.0816	2.0667	2.0815
8 × 40	2.0820	2.0626	2.0823
6 × 40	2.0849	2.0566	2.0862

Table 8: Displacement u_y at $x = 2.5$ m of the pillar top.

Meshes	BFEM ($\times 10^{-3}$ m)	Q4 model ($\times 10^{-3}$ m)	Q4R model ($\times 10^{-3}$ m)
20 × 40	−9.8891	−9.8869	−9.8895
10 × 40	−9.8888	−9.8856	−9.8898
8 × 40	−9.8889	−9.8850	−9.8943
6 × 40	−9.8889	−9.8840	−9.8926

Table 9: Displacement u_y at $h=5$ m on the right of the pillar

Meshes	BFEM model ($\times 10^{-2}$ m)	Q4 model ($\times 10^{-2}$ m)	Q4R model ($\times 10^{-2}$ m)
20 × 40	−1.5918	−1.5878	−1.5913
10 × 40	−1.5488	−1.5412	−1.5483
8 × 40	−1.5290	−1.51894	−1.5284
6 × 40	−1.4983	−1.4832	−1.4975

Table 10: Stress σ_x at center of the element at lower right of pillar

Meshes	BFEM (kPa)	Q4 model (kPa)	Q4R model (kPa)
20 × 40	−128.963	−179.706	−132.974
10 × 40	−142.729	−191.962	−150.839
8 × 40	−141.996	−193.919	−152.815
6 × 40	−134.954	−194.249	−150.941

Table 11: Stress τ_{xy} at center of the element at lower right of pillar

Meshes	BFEM (kPa)	Q4 model (kPa)	Q4R model (kPa)
20 × 40	152.442	172.480	153.366
10 × 40	153.439	167.708	154.637
8 × 40	151.262	163.966	152.612
6 × 40	146.664	157.624	148.288

Table 12: Stress σ_y at center of the element at lower right of pillar.

Meshes	BFEM (kPa)	Q4 model (kPa)	Q4R model (kPa)
20 × 40	−802.328	−796.173	−802.741
10 × 40	−707.932	−700.016	−708.582
8 × 40	−675.867	−666.975	−676.616
6 × 40	−634.179	−623.753	−635.039

Example 4: Stress Analysis of Concrete Gravity Dam

Consider a concrete gravity dam shown in Figure 14. And height of the dam is 65 m, bottom width is 49 m, the water level is 60 m, the elastic modulus of concrete E_1 = 15 GPa, Poisson ratio of concrete v_1 = 0.2, the elastic modulus of rock E_2 = 30 GPa, Poisson ratio of rock v_2 = 0.3, density of concrete is 2.45 t/m^3, density of water is 1 t/m^3, and acceleration of gravity g = 9.8 m/s^2. The calculation is considered into the plane strain problem and considered the effect of rock foundation of the dam.

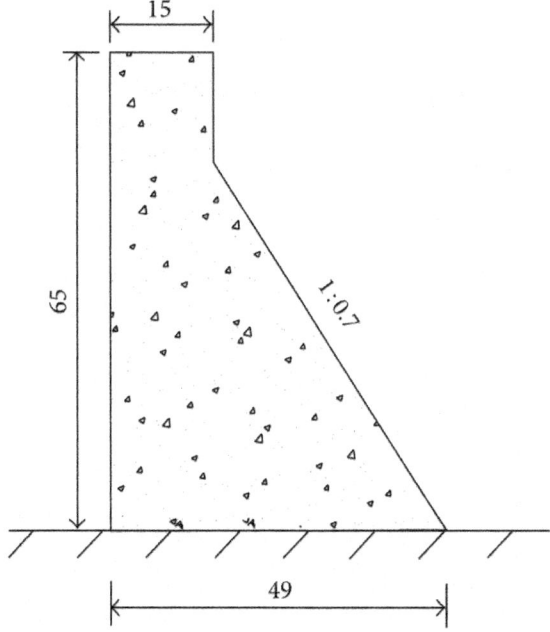

Figure 14: A concrete gravity dam.

The calculation is done using the mesh with the center nodes of edges of elements as shown in Figure 15. In this calculation, we do not consider the initial geostress field. The boundaries of the foundation are used the fixed constraint. The origin of coordinates is located at the bottom of the dam, and is 25 meters away from the dam heel.

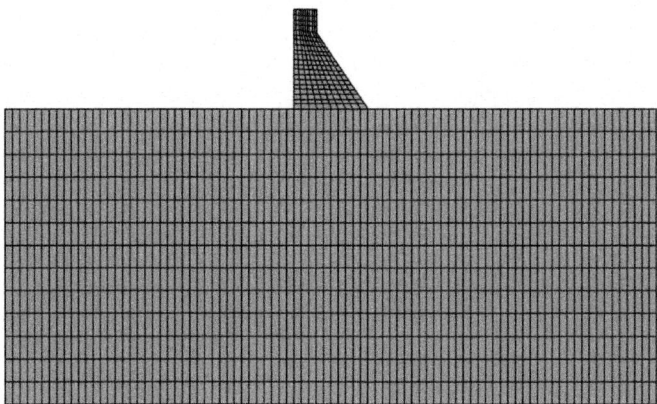

Figure 15: Meshes of the dam and its foundation (1344 elements, 2809 nodes).

The values of stress components and displacement components of the dam are plotted in Figures 16–25, respectively. Comparisons of the results from the conventional quadrilateral isoparametric element (Q4 model) and quadrilateral reduced integration element (Q4R model) are also given in Figures 16–25, respectively. The numerical results of the present model are consistent with those of the Q4 model and Q4R model and have shown good computational stability.

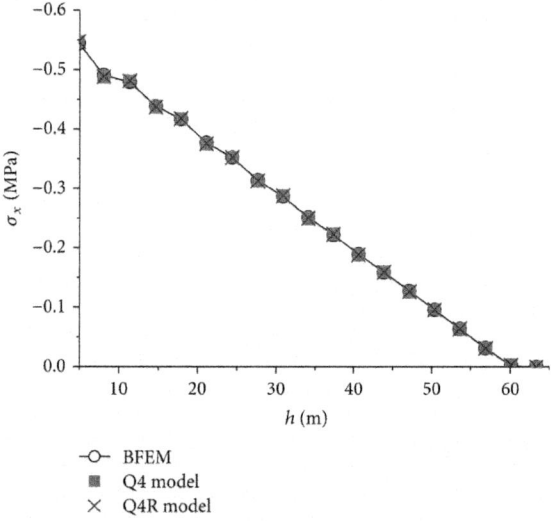

Figure 16: Figure 16: The $h\text{-}\sigma_x$ curves at the upstream face of the dam.

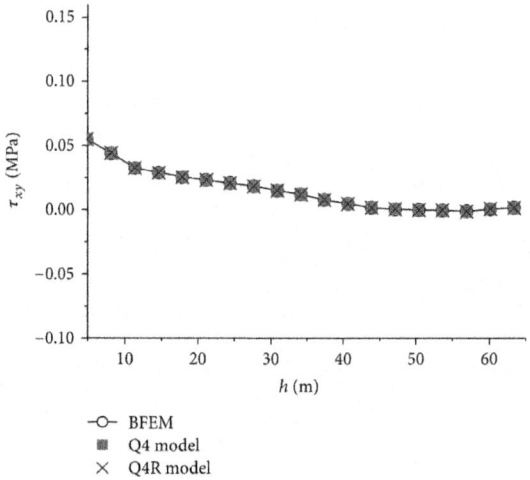

Figure 17: Figure 17: The $h\text{-}\tau_{xy}$ curves at the upstream face of the dam

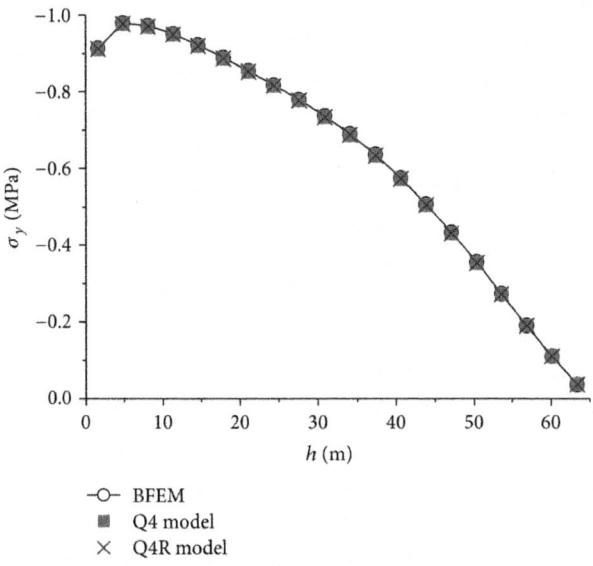

Figure 18: The h-σ_y curves at the upstream face of the dam.

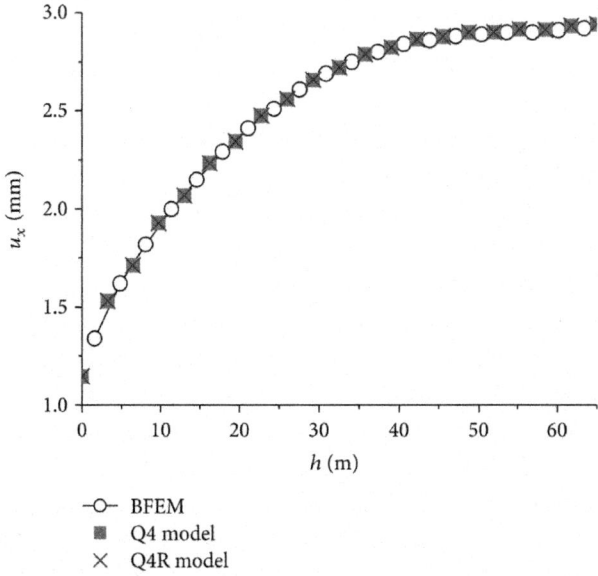

Figure 19: The h-u_x curves at the upstream face of the dam.

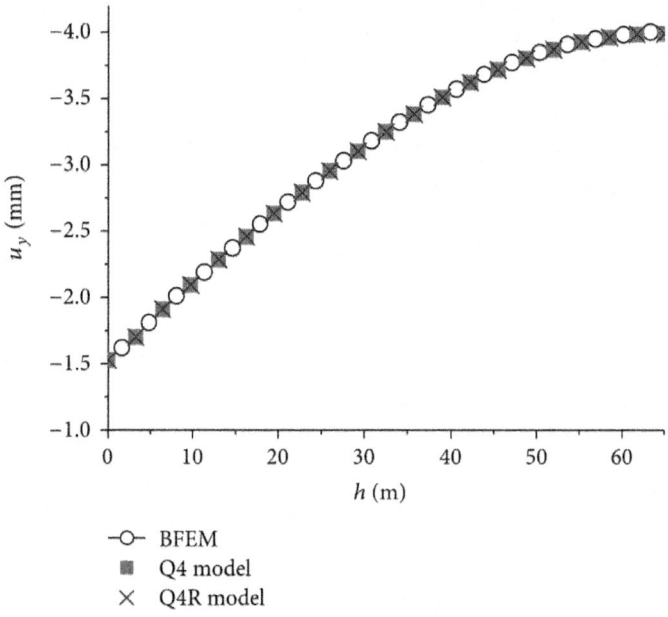

Figure 20: The h-u_y curves at the upstream face of the dam.

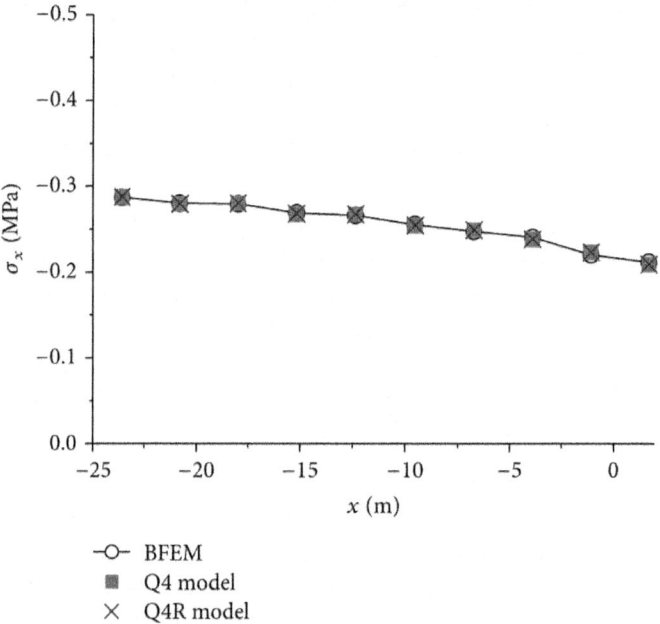

Figure 21: The h-σ_x curves at the half height of the dam.

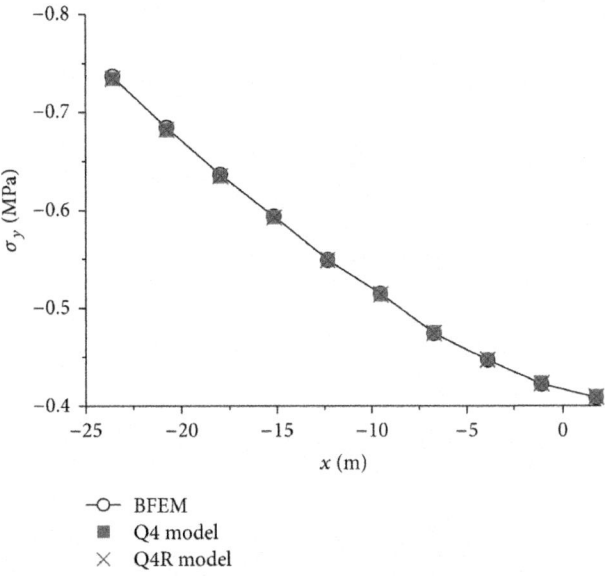

Figure 22: The h-σ_y curves at the half height of the dam

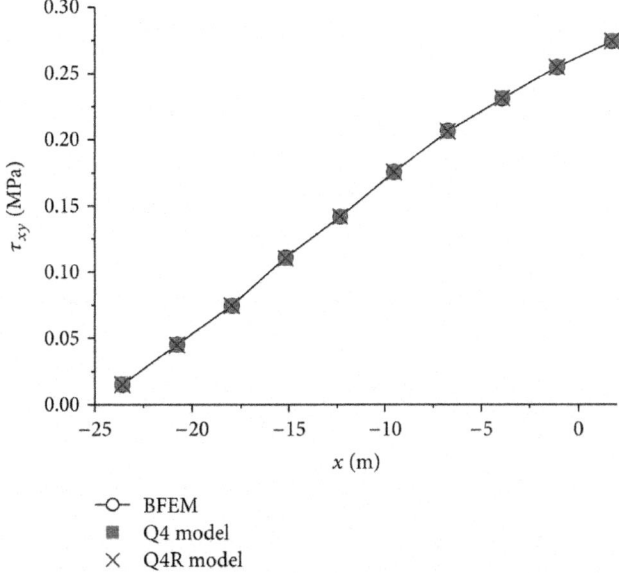

Figure 23: The h-τ_{xy} curves at the half height of the dam.

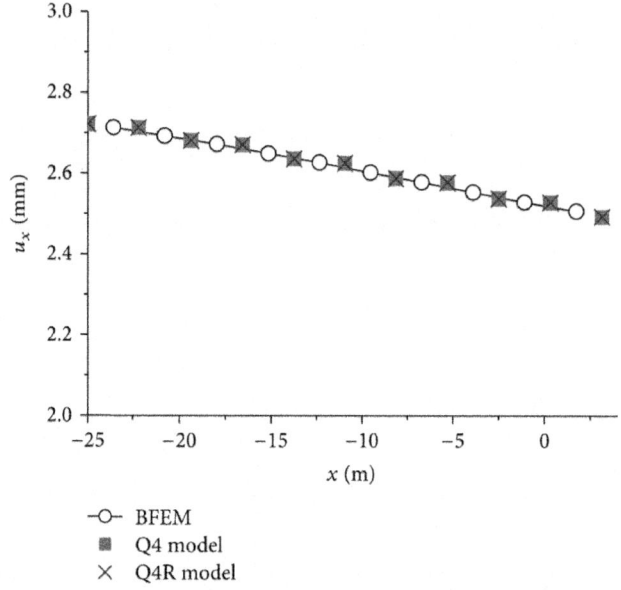

Figure 24: The h-u_x curves at the half height of the dam

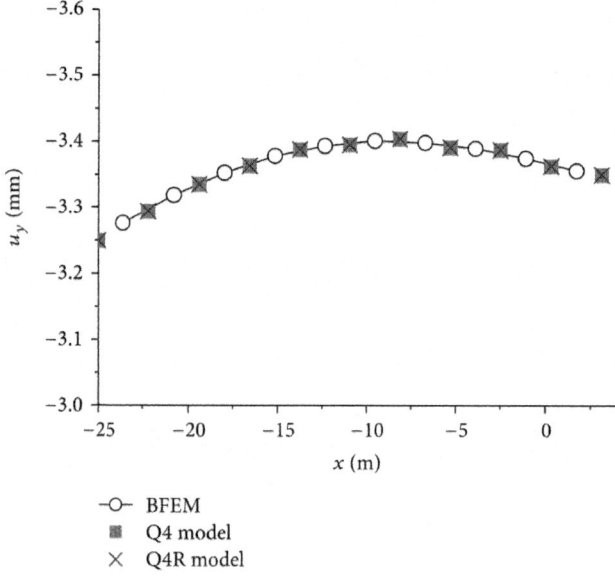

Figure 25: The h-u_y curves at the half height of the dam.

Example 5: Simulation and Analysis on the Horizontal Crack Propagation of Rock Block

Consider a rock block subjected by the horizontal thrust and vertical pressure shown in Figure 26. For the convenience of study, we do not consider the weight. And we use the dimensionless numerical analysis. Assuming elastic modulus $E=1$, Poisson ratio $v = 0.3$, tensile strength of a large number, and the uniform load $p_1 = 1$ and $p_2 = 1$. The calculation is considered into the plane stress problem and the dimensionless values.

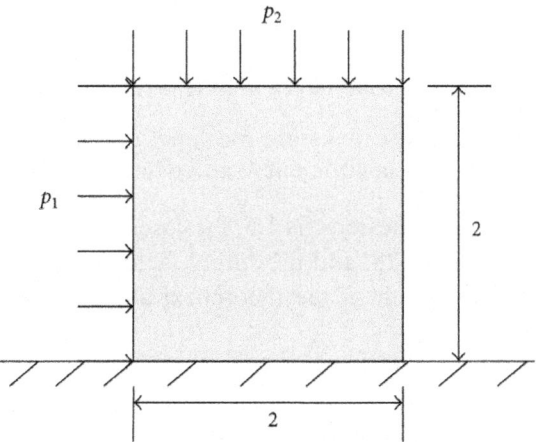

Figure 26: A rock block subjected by the horizontal thrust and vertical pressure.

The calculation is done using the mesh with the center nodes of edges of elements as shown in Figure 27.

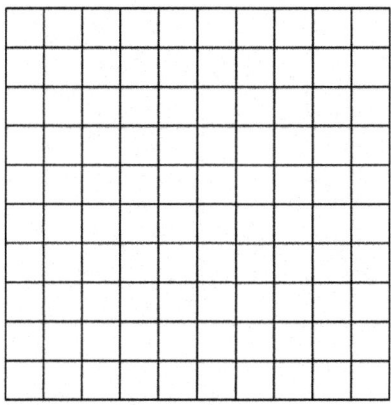

Figure 27: Meshes of the rock block.

In order to check whether the interface between the rock block and the ground will crack, we assume the friction coefficient of interface $f = 0.5$ and change the value of interface cohesion c. The results of calculation using the computer program of the nonlinear BFEM shown in Section 4 and the failure criteria in Section 4.1are as follows.

(1) When $c = 10$, there is no cracks.

(2) When $c = 2.8$, there is one element interface crack.

(3) When $c=2$, there are three elements' interfaces cracks.

(4) When $c = 1.5$, there are five elements' interfaces cracks.

(5) When $c = 1.1$, there are seven elements' interfaces cracks.

(6) When $c = 1.0$, the cracks are too long, and too little structural constraints have been insufficient to solve the equations

When the interface cohesion c is 1.5, the case of crack propagation of rock block is shown in Figure 28, and the safety factor of stability is $K=2$ which is consistent with the result of the theoretical analysis using the rigid limit equilibrium method.

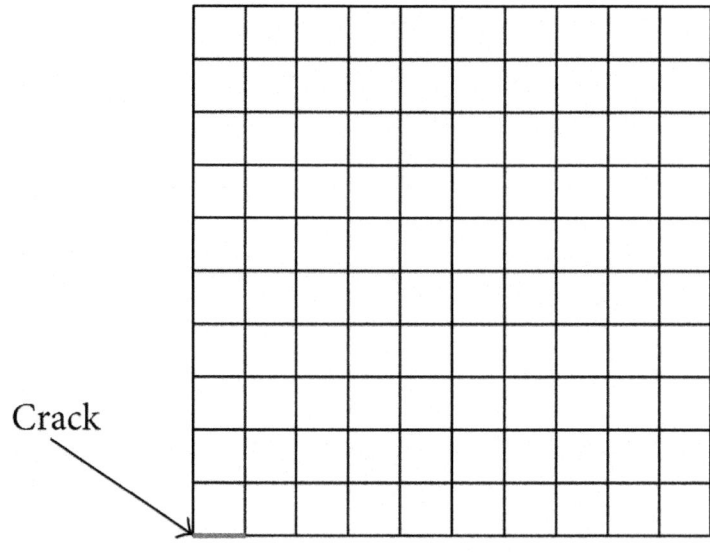

(a) In the first cycle calculation

Application of Base Force Element Method on Complementary ... 241

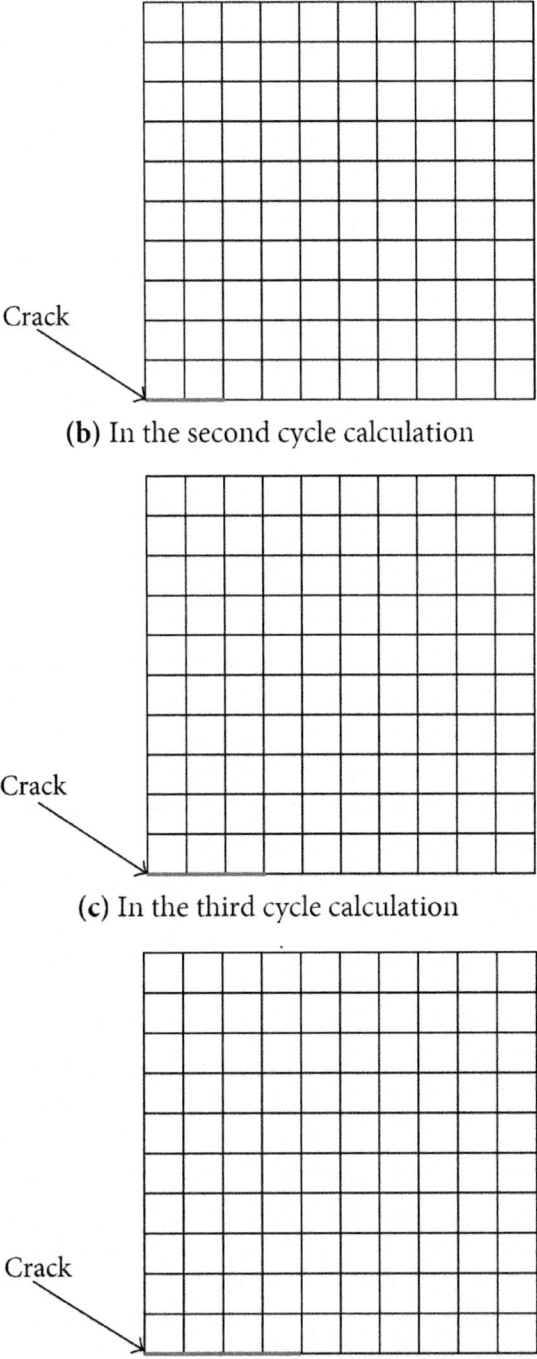

(b) In the second cycle calculation

(c) In the third cycle calculation

(d) In the fourth cycle calculation

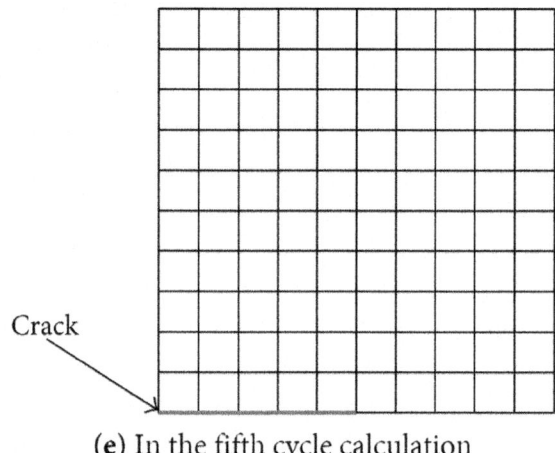

(e) In the fifth cycle calculation

Figure 28: Crack propagation path of the rock block.

Example 6: Analysis on the Crack Propagation of Concrete Gravity Dam

Consider a concrete gravity dam shown in Figure 14. And height of the dam is 65 m, bottom width is 49 m, the water level is 60 m, the elastic modulus of concrete $E = 15$ GPa, Poisson ratio of concrete $v = 0.2$, density of concrete is 2.45 t/m^3, density of water is 1 t/m^3, and acceleration of gravity $g = 9.8$m/s^2. The calculation is considered into the plane strain problem and is not considered the effect of rock foundation of the dam.

The calculation is done using the mesh with the center nodes of edges of elements as shown in Figure 29.

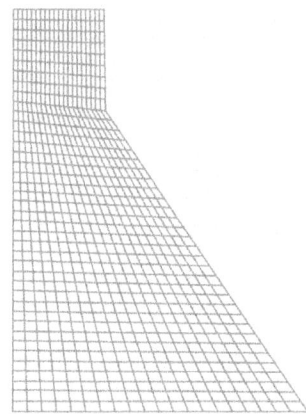

Figure 29: Mesh of gravity dam (800 elements, 1660 nodes).

No Initial Crack in Dam Foundation

We assume that the foundation surface of dam is weak structural interface and analyze the crack propagation and the safety factor by adjusting the value of the friction coefficient and the cohesion. The results of calculation using the computer program of the nonlinear BFEM shown in Section 4 and the failure criteria in Section 4.1 are as follows.

(1) When $c = 10^6$ Pa and $f = 0.95$, there is no cracks.

(2) When $c = 0.5 \times 10^6$ Pa and $f = 0.5$, there are two elements' interface cracks.

(3) When $c = 0.1 \times 10^6$ Pa and $f = 0.4$, there are eight elements' interfaces cracks.

The case of crack propagation of dam is shown in Figure 30. When $c = 10^6$ Pa and $f = 0.5$, the safety factor of stability $K = 4.0$ which is consistent with the result of the theoretical analysis. When $c = 0.5 \times 10^6$ Pa and $f = 0.5$, the safety factor of stability $K = 2.55$ which is consistent with the results of the theoretical analysis using the rigid limit equilibrium method.

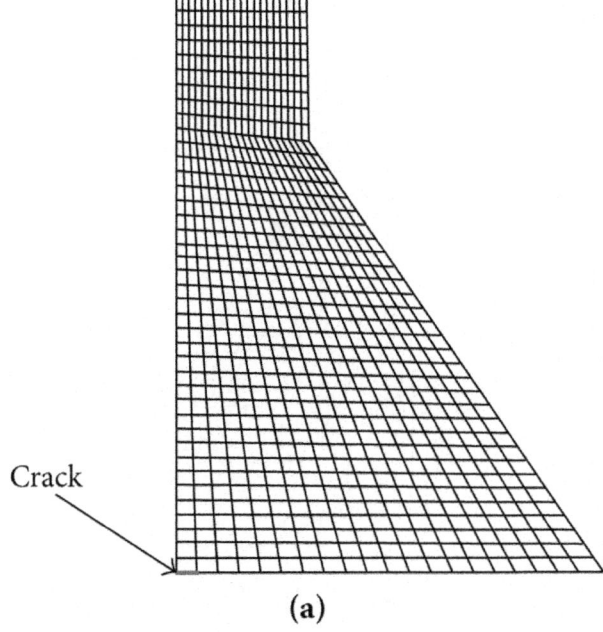

(a)

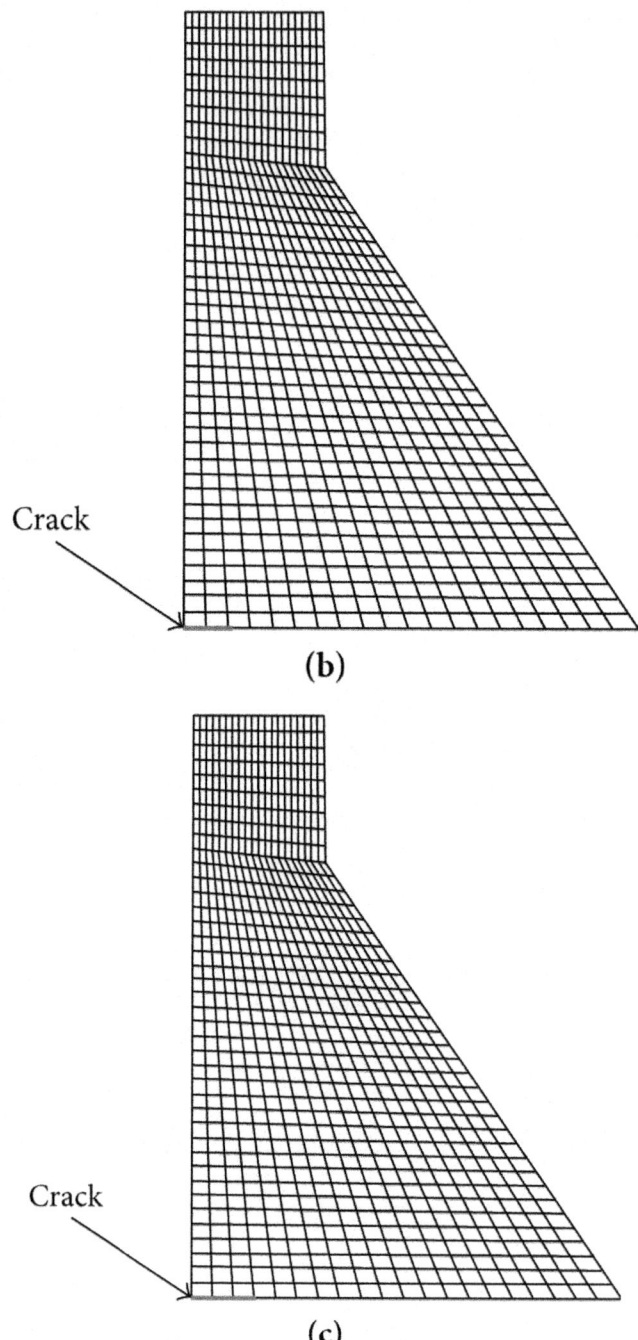

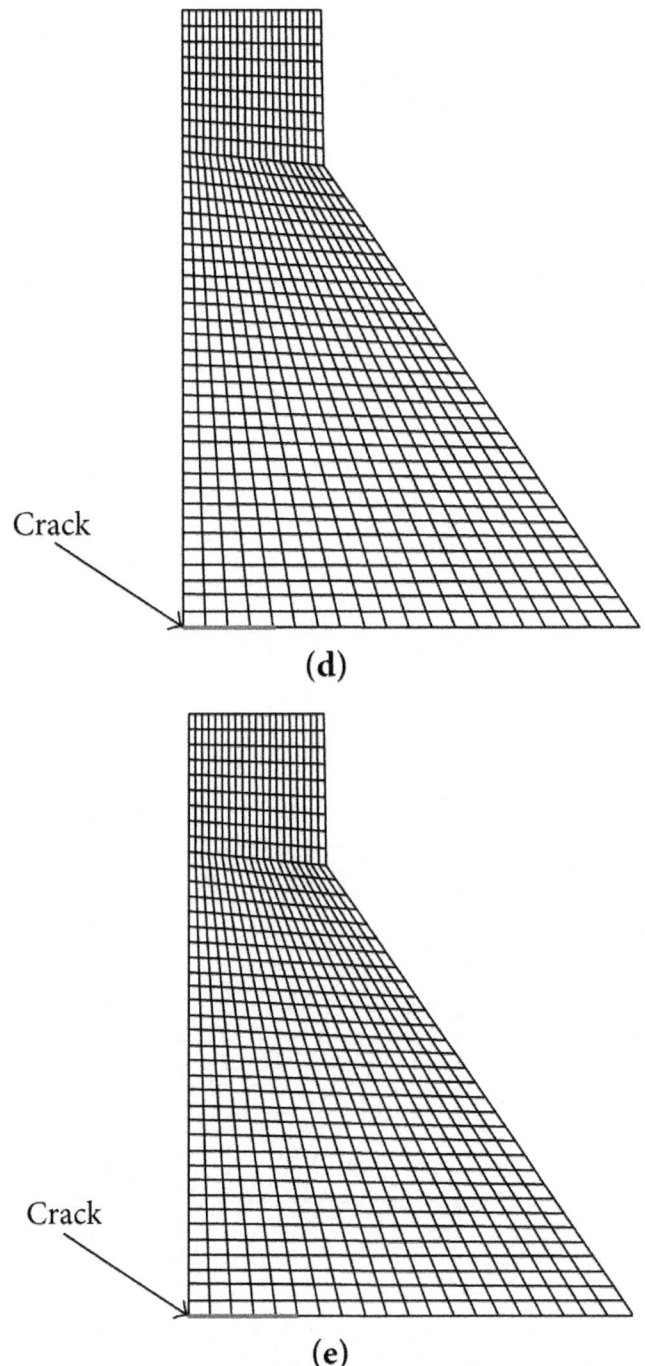

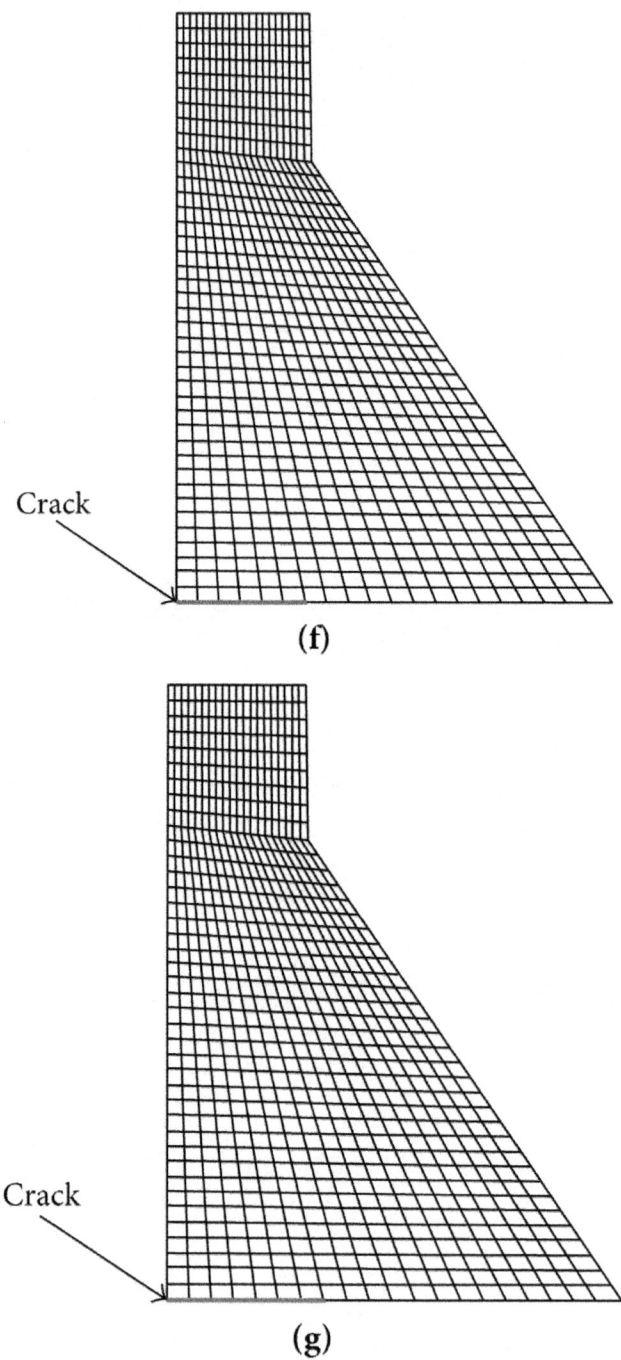

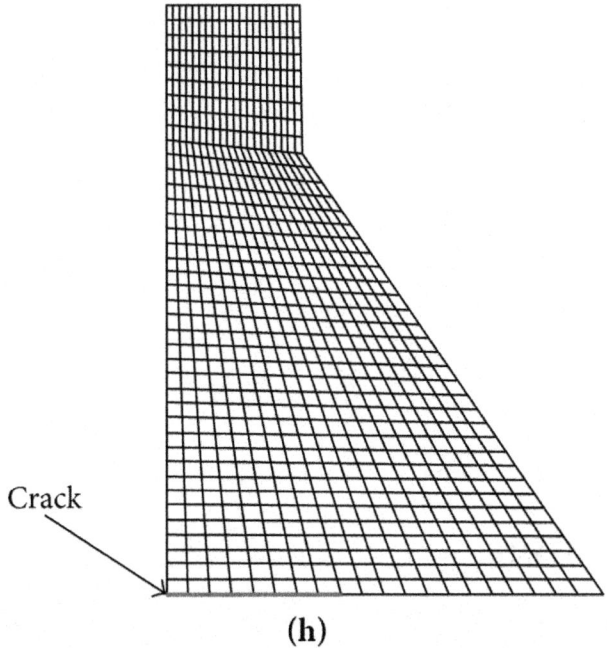

(h)

Figure 30: Crack propagation path of the gravity dam.

Existing an Initial Crack in Dam Foundation

There is an initial crack in the dam heel as shown in Figure 31. We assume that the dam foundation surface is weak structural interface and analyze the crack propagation and the safety factor by adjusting the value of the friction coefficient and the cohesion. The results of calculation using the computer program of the nonlinear BFEM shown in Section 4 and the failure criteria in Section 4.1 are as follows.

(1) When $c = 10^6$ Pa and $f = 0.95$, there is no cracks.

(2) When $c = 0.5 \times 10^6$ Pa and $f = 0.5$, there is one element interface cracks.

(3) When $c = 0.1 \times 10^6$ Pa and $f = 0.5$, there are three element' interface cracks.

(4) When $c = 0.1 \times 10^6$ Pa and $f = 0.4$, there are seven elements' interfaces cracks.

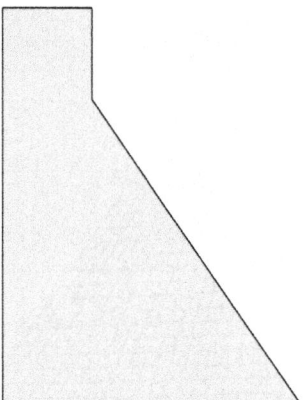

Figure 31: A gravity dam with a crack at the dam heel.

CONCLUSIONS

In this paper, the base force element method (BFEM) on complementary energy principle is used to analyze the rock mechanics problems. The methods to simulate the gravity of an element, the crack propagation and the safety factor of stability are proposed for the BFEM on complementary energy principle. The following conclusions can be drawn.

- The calculation results of the BFEM on complementary energy principle show that the numerical results of the present method coincide with the theoretical solution, the results of conventional quadrilateral isoparametric element (Q4 model) and quadrilateral reduced integration element (Q4R model). The correctness of the present method and its computer program is verified.
- The research results show that the BFEM on complementary energy principle has a good computational precision and stability is not sensitive to the effects on the aspect ratio of element and can be used for large-scale scientific and engineering computing.
- The results of the BFEM for crack propagation problems show that the nonlinear BFEM can solve the cracking problem and simulate the crack propagation of the interface in rock mechanics engineering.
- The BFEM on complementary energy principle was applied to analyze the stability of rock mass and dam, and the results of safety factor are consistent with the results of the theoretical solutions using the rigid limit equilibrium method. The research results show the present method can be easily used to calculate the safety factor in rock engineering.

- This paper researched only the cracking problems with horizontal crack in rock mass and calculated only the safety factor of a single slip channel in rock mass.
- The cracking problems of inclined cracks and the safety factor of multiple sliding channels in rock mass are studying, and the further research results will be published in the future.

ACKNOWLEDGMENTS

This work is supported by the National Natural Science Foundation of China, nos. 10972015 and 11172015 and the preexploration project of the Key Laboratory of Urban Security and Disaster Engineering, Ministry of Education, Beijing University of Technology, no. USDE201404.

REFERENCES

1. T. H. H. Pian, "Derivation of element stiffness matrices by assumed stress distributions," AIAA Journal, vol. 2, no. 7, pp. 1333–1336, 1964.
2. T. H. Pian and D. P. Chen, "Alternative ways for formulation of hybrid stress elements," International Journal for Numerical Methods in Engineering, vol. 18, no. 11, pp. 1679–1684, 1982.
3. T. H. H. Pian and K. Sumihara, "Rational approach for assumed stress finite elements," International Journal for Numerical Methods in Engineering, vol. 20, no. 9, pp. 1685–1695, 1984.
4. C. Zhang, D. Wang, J. Zhang, W. Feng, and Q. Huang, "On the equivalence of various hybrid finite elements and a new orthogonalization method for explicit element stiffness formulation," Finite Elements in Analysis and Design, vol. 43, no. 4, pp. 321–332, 2007.
5. B. Fraeijs de Veubeke, "Displacement and equilibrium models in the finite element method," in Stress Analysis, O. C. Zienkiewicz and G. S. Holister, Eds., pp. 145–197, John Wiley & Sons, New York, NY, USA, 1965.
6. B. F. de Veubeke, "A new variational principle for finite elastic displacements," International Journal of Engineering Science, vol. 10, no. 9, pp. 745–763, 1972.
7. R. L. Taylor and O. C. Zienkiewicz, "Complementary energy with penalty functions in finite element analysis," in Energy Methods in Finite Element Analysis, R. Glowinski, Ed., pp. 153–174, John Wiley & Sons, New York, NY, USA, 1979.
8. S. N. Patniak, "An integrated force method for discrete analysis,"

International Journal for Numerical Methods in Engineering, vol. 6, no. 2, pp. 237–251, 1973.

9. S. N. Patnaik, "The integrated force method versus the standard force method," Computers and Structures, vol. 22, no. 2, pp. 151–163, 1986.

10. S. N. Patnaik, "The variational energy formulation for the integrated force method," AIAA Journal, vol. 24, no. 1, pp. 129–137, 1986.

11. S. N. Patnaik, L. Berke, and R. H. Gallagher, "Integrated force method versus displacement method for finite element analysis," Computers and Structures, vol. 38, no. 4, pp. 377–407, 1991.

12. E. L. Wilson, R. L. Tayler, W. P. Doherty, and J. Ghaboussi, "Incompatible displacement models," inNumerical and Computational Methods in Structural Mechanics, S. J. Fenves, N. Perrone, A. R. Robinson, and W. C. Schnobrich, Eds., pp. 43–57, Academic Press, New York, NY, USA, 1973.

13. R. L. Taylor, P. J. Beresford, and E. L. Wilson, "A non-conforming element for stress analysis,"International Journal for Numerical Methods in Engineering, vol. 10, no. 6, pp. 1211–1219, 1976.

14. J. C. Simo and T. J. R. Hughes, "On the variational foundations of assumed strain methods," Journal of Applied Mechanics, vol. 53, no. 1, pp. 51–54, 1986.

15. J. C. Simo and M. S. Rifai, "Class of mixed assumed strain methods and the method of incompatible modes," International Journal for Numerical Methods in Engineering, vol. 29, no. 8, pp. 1595–1638, 1990.

16. R. H. Macneal, "Derivation of element stiffness matrices by assumed strain distributions," Nuclear Engineering and Design, vol. 70, no. 1, pp. 3–12, 1982.

17. T. Belytschko and L. P. Bindeman, "Assumed strain stabilization of the 4-node quadrilateral with 1-point quadrature for nonlinear problems," Computer Methods in Applied Mechanics and Engineering, vol. 88, no. 3, pp. 311–340, 1991.

18. R. Piltner and R. L. Taylor, "A quadrilateral mixed finite element with two enhanced strain modes,"International Journal for Numerical Methods in Engineering, vol. 38, no. 11, pp. 1783–1808, 1995.

19. R. Piltner and R. L. Taylor, "A systematic constructions of B-bar functions for linear and nonlinear mixed-enhanced finite elements for plane elasticity problems," International Journal for Numerical Methods in Engineering, vol. 44, no. 5, pp. 615–639, 1997.

20. T. J. R. Hughes, "Generalization of selective integration procedures to

anisotropic and nonlinear media,"International Journal for Numerical Methods in Engineering, vol. 15, no. 9, pp. 1413–1418, 1980.

21. T. Limin, C. Wanji, and L. Yingxi, "Formulation of quasi-conforming element and Hu-Washizu principle," Computers and Structures, vol. 19, no. 1-2, pp. 247–250, 1984.

22. L. Yu-qiu and H. Min-feng, "A generalized conforming isoparametric element," Applied Mathematics and Mechanics, vol. 9, no. 10, pp. 929–936, 1988.

23. G. R. Liu, T. Nguyen-Thoi, and K. Y. Lam, "A novel alpha finite element method (αFEM) for exact solution to mechanics problems using triangular and tetrahedral elements," Computer Methods in Applied Mechanics and Engineering, vol. 197, no. 45–48, pp. 3883–3897, 2008.

24. J. Chen, C.-J. Li, and W.-J. Chen, "A 17-node quadrilateral spline finite element using the triangular area coordinates," Applied Mathematics and Mechanics (English Edition), vol. 31, no. 1, pp. 125–134, 2010.

25. J. Chen, C.-J. Li, and W.-J. Chen, "A family of spline finite elements," Computers and Structures, vol. 88, no. 11-12, pp. 718–727, 2010.

26. S. Rajendran and K. M. Liew, "A novel unsymmetric 8-node plane element immune to mesh distortion under a quadratic displacement field," International Journal for Numerical Methods in Engineering, vol. 58, no. 11, pp. 1713–1748, 2003.

27. S. Rajendran, "A technique to develop mesh-distortion immune finite elements," Computer Methods in Applied Mechanics and Engineering, vol. 199, no. 17–20, pp. 1044–1063, 2010.

28. E. T. Ooi, S. Rajendran, and J. H. Yeo, "A 20-node hexahedron element with enhanced distortion tolerance," International Journal for Numerical Methods in Engineering, vol. 60, no. 15, pp. 2501–2530, 2004.

29. E. T. Ooi, S. Rajendran, and J. H. Yeo, "Remedies to rotational frame dependence and interpolation failure of US-QUAD8 element," Communications in Numerical Methods in Engineering, vol. 24, no. 11, pp. 1203–1217, 2008.

30. Y. Long, L. Juxuan, Z. Long, and C. Song, "Area co-ordinates used in quadrilateral elements,"Communications in Numerical Methods in Engineering, vol. 15, no. 8, pp. 533–543, 1999.

31. Y. Q. Long, S. Cen, and Z. F. Long, Advanced Finite Element Method in Structural Engineering, Springer/Tsinghua University Press, Berlin, Germany, 2009.

32. Z. F. Long, J. X. Li, S. Cen, and Y. Q. Long, "Some basic formulae for

area coordinates used in quadrilateral elements," Communications in Numerical Methods in Engineering, vol. 15, no. 12, pp. 841–852, 1999.

33. Z.-F. Long, S. Cen, L. Wang, X.-R. Fu, and Y.-Q. Long, "The third form of the quadrilateral area coordinate method (QACM-III): theory, application, and scheme of composite coordinate interpolation," Finite Elements in Analysis and Design, vol. 46, no. 10, pp. 805–818, 2010.

34. G. R. Liu, K. Y. Dai, and T. T. Nguyen, "A smoothed finite element method for mechanics problems,"Computational Mechanics, vol. 39, no. 6, pp. 859–877, 2007.

35. Y. Peng and Y. Liu, "Base force element method of complementary energy principle for large rotation problems," Acta Mechanica Sinica, vol. 25, no. 4, pp. 507–515, 2009.

36. Y. Liu and Y. Peng, "Base force element method (BFEM) on complementary energy principle for linear elasticity problem," Science China: Physics, Mechanics and Astronomy, vol. 54, no. 11, pp. 2025–2032, 2011.

37. Y. Peng, Z. Dong, B. Peng, and Y. Liu, "Base force element method (BFEM) on potential energy principle for elasticity problems," International Journal of Mechanics and Materials in Design, vol. 7, no. 3, pp. 245–251, 2011.

38. Y. Peng, Z. Dong, B. Peng, and N. Zong, "The application of 2D base force element method (BFEM) to geometrically non-linear analysis," International Journal of Non-Linear Mechanics, vol. 47, no. 3, pp. 153–161, 2012.

39. Y.-J. Peng, J.-W. Pu, B. Peng, and L.-J. Zhang, "Two-dimensional model of base force element method (BFEM) on complementary energy principle for geometrically nonlinear problems," Finite Elements in Analysis and Design, vol. 75, pp. 78–84, 2013.

40. Y. J. Peng, N. N. Zong, L. J. Zhang, and J. W. Pu, "Application of 2D base force element method with complementary energy principle for arbitrary meshes," Engineering Computations, vol. 31, no. 4, pp. 1–15, 2014.

41. Y. Peng, L. Zhang, J. Pu, and Q. Guo, "A two-dimensional base force element method using concave polygonal mesh," Engineering Analysis with Boundary Elements, vol. 42, pp. 45–50, 2014.

42. Y. Peng, Y. Liu, J. Pu, and L. Zhang, "Application of base force element method to mesomechanics analysis for recycled aggregate concrete," Mathematical Problems in Engineering, vol. 2013, Article ID 292801, 8 pages, 2013.

43. Y. Liu, Y. Peng, L. Zhang, and Q. Guo, "A 4-mid-node plane model of base force element method on complementary energy principle," Mathematical Problems in Engineering, vol. 2013, Article ID 706759, 8 pages, 2013.
44. C. Y. Dong and G. L. Zhang, "Boundary element analysis of three dimensional nanoscale inhomogeneities," International Journal of Solids and Structures, vol. 50, no. 1, pp. 201–208, 2013.
45. C. Y. Dong and E. Pan, "Boundary element analysis of nanoinhomogeneities of arbitrary shapes with surface and interface effects," Engineering Analysis with Boundary Elements, vol. 35, no. 8, pp. 996–1002, 2011.
46. S. S. Chen, Q. H. Li, Y. H. Liu, and Z. Q. Xue, "A meshless local natural neighbour interpolation method for analysis of two-dimensional piezoelectric structures," Engineering Analysis with Boundary Elements, vol. 37, no. 2, pp. 273–279, 2013.
47. S. Chen, Y. Liu, J. Li, and Z. Cen, "Performance of the MLPG method for static shakedown analysis for bounded kinematic hardening structures," European Journal of Mechanics, A/Solids, vol. 30, no. 2, pp. 183–194, 2011.
48. S. Cen, X.-R. Fu, and M.-J. Zhou, "8- and 12-node plane hybrid stress-function elements immune to severely distorted mesh containing elements with concave shapes," Computer Methods in Applied Mechanics and Engineering, vol. 200, no. 29–32, pp. 2321–2336, 2011.
49. S. Cen, G.-H. Zhou, and X.-R. Fu, "A shape-free 8-node plane element unsymmetric analytical trial function method," International Journal for Numerical Methods in Engineering, vol. 91, no. 2, pp. 158–185, 2012.
50. H. A. F. A. Santos, "Complementary-energy methods for geometrically non-linear structural models: an overview and recent developments in the analysis of frames," Archives of Computational Methods in Engineering, vol. 18, no. 4, pp. 405–440, 2011.
51. H. A. F. A. Santos and C. I. Almeida Paulo, "On a pure complementary energy principle and a force-based finite element formulation for non-linear elastic cables," International Journal of Non-Linear Mechanics, vol. 46, no. 2, pp. 395–406, 2011.
52. Y. C. Gao, "A new description of the stress state at a point with applications," Archive of Applied Mechanics, vol. 73, no. 3-4, pp. 171–183, 2003.
53. Y. C. Gao, "Asymptotic analysis of the nonlinear Boussinesq problem for a kind of incompressible rubber material (compression case)," Journal of Elasticity, vol. 64, no. 2-3, pp. 111–130, 2001.

54. Y. C. Gao and T. J. Gao, "Large deformation contact of a rubber notch with a rigid wedge," International Journal of Solids and Structures, vol. 37, no. 32, pp. 4319–4334, 2000.
55. Y. C. Gao and S. H. Chen, "Analysis of a rubber cone tensioned by a concentrated force," Mechanics Research Communications, vol. 28, no. 1, pp. 49–54, 2001.
56. Y.-C. Gao, M. Jin, and G.-S. Dui, "Stresses, singularities, and a complementary energy principle for large strain elasticity," Applied Mechanics Reviews, vol. 61, no. 3, Article ID 030801, 16 pages, 2008.

Chapter 10

THREE-DIMENSIONAL NUMERICAL MODEL OF HYDRAULIC FRACTURING IN FRACTURED ROCK MASSES

B. Damjanac[1], C. Detournay[1], P.A. Cundall[1] and Varun[1]

[1]Itasca Consulting Group, Inc., Minneapolis, Minnesota, USA

ABSTRACT

Conventional methods for simulation of hydraulic fracturing are based on assumptions of continuous, isotropic and homogeneous media. These assumptions are not valid for most rock mass formations, particularly shale gas reservoirs, as these typically consist of a large volume of naturally fractured rock in which propagation of a hydraulic fracture (HF) involves both fracturing of intact rock and opening or slip of pre-existing discontinuities (joints). The pre-existing joints can significantly affect the HF trajectory, the pressure required to propagate the fracture and also the leak-off from the fracture into the surrounding formation. None of these effects can be simulated using conventional methods.

HF Simulator is a new three-dimensional numerical code that can simulate propagation of hydraulic fracture in naturally fractured reservoirs, accounting for the interaction between the hydraulic fracture and pre-existing joints. In *HF Simulator*, fracture propagation occurs as a combination of intact-rock failure in tension, and slip and opening of joints. The code uses a lattice representation of brittle rock consisting of point masses (nodes) connected by springs. The pre-existing joints are derived from a user-specified discrete fracture network (DFN).

HF Simulator can model fluid injection or production from one or multiple boreholes each with one or multiple clusters. Non-steady, hydro-mechanically coupled fluid flow and pressure within the network of joint segments and the rock matrix are considered.

An outline of the code hydro-mechanical formulation is presented and examples are provided to illustrate the code capabilities.

INTRODUCTION

A new generation tool that uses the bonded particle model (BPM) [1] and the synthetic rock mass (SRM) concept [2] has been developed to model hydraulic fracture (HF) propagation in naturally fractured reservoirs (NFRs).

Most rock mass formations, and shale gas reservoirs in particular, consist of a large volume of fractured rock in which propagation of an HF involves both fracturing of intact rock and opening or slip of pre-existing discontinuities (joints). The pre-existing joints can significantly affect the HF trajectory, the pressure required to propagate the fracture, but also the leak-off from the fracture into the surrounding formation. None of these effects can be simulated using conventional hydraulic fracturing simulation methods, based on assumptions of continuous, isotropic and homogeneous media.

To address this challenge, a numerical approach called SRM method [2] has been developed recently based on the distinct element method. SRM method usually is realized as a bonded-particle assembly representing brittle rock containing multiple joints, each one consisting of a planar array of bonds that obey a special model, namely the smooth joint model (SJM). The SJM allows slip and separation at particle contacts, while respecting the given joint orientation rather than local contact orientations. Overall fracture of a synthetic rock mass depends on both fracture of intact material (bond breaks), as well as yield of joint segments.

Previous SRM models have used the general-purpose codes *PFC2D* and *PFC3D* [3,4], which employ assemblies of circular/spherical particles bonded together. Much greater efficiency can be realized if a "lattice," consisting of point masses (nodes) connected by springs, replaces the balls and contacts (respectively) of *PFC3D*. The lattice model still allows fracture through the breakage of springs along with joint slip, using a modified version of the SJM. The new 3D program, *HF Simulator* described in this paper, is based on such a lattice representation of brittle rock. *HF Simulator* overcomes all main limitations of the conventional methods for simulation of hydraulic fracturing in jointed rock masses and is computationally more efficient than *PFC*-based implementations of the SRM method.

The formulation of the code is described in this paper. The examples of code verification and application are also presented.

MODEL DESCRIPTION

Background: Synthetic Rock Mass Approach

Over past years, the SRM has been developed [2] as a more realistic representation of mechanical behavior of the fractured rock mass compared to conventional numerical models. The SRM consists of two components:

- the bonded particle model (BPM) of deformation and fracturing of intact rock, and
- the smooth joint model (SJM) of mechanical behavior of discontinuities.

The BPM, originally implemented in *PFC*, is created when the contacts between the particles (disks in 2D and spheres in 3D) are assigned certain bond strength (both in tension and shear). It was found that BPM quite well approximates mechanical behavior of the brittle rocks [1]. The elastic properties of the contacts (i.e., contact shear and normal stiffness) can be calibrated to match the desired elastic properties (e.g., Young's modulus and Poisson's ratio) of the assembly of the particles. Similarly, the tensile and shear contact strengths can be adjusted to match the macroscopic strengths under different loading conditions (e.g., direct tension, unconfined and confined compression).

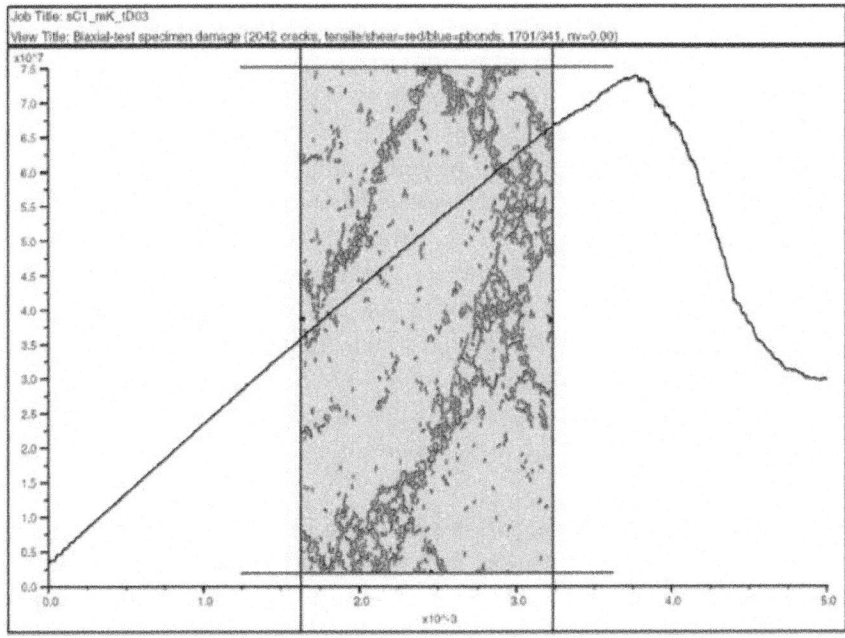

Figure 1: Example of unconfined compressive test using bonded particle model (BPM).

In the BPM, the contact behavior is perfectly brittle. Breakage of the bond, a function of the forces in the contact and the bond strength, corresponds to formation of a microcrack. An example of unconfined compression test conducted using *PFC2D* is illustrated in Figure 1, which shows recorded axial stress-strain response and the model configuration with generated microcracks. The shear microcracks are black; the tensile microcracks are red. Shown is the state when the sample is loaded beyond its peak strength. The stress-strain curve exhibits characteristics typical of brittle rock response. For the load levels less than ~80% of the peak strength, the stress-strain response is linearly elastic, with the slope of the line equal to the Young's modulus. Some microcracks, randomly distributed within the sample, start developing at the load levels greater than ~40% of the peak strength. Significant non-linearity develops as the load exceeds 80% of the peak strength. In this phase, the microcracks begin to coalesce, forming fractures on the scale of the sample. After the peak strength is reached, the material starts to soften (i.e., to lose the strength). At this stage, as shown in Figure 1, the failure mechanism and the "shear bands" are well developed. It is interesting that in the unconfined compression test, the majority of cracks are tensile (red lines in Figure 1). The "shear bands" on the scale of the sample are formed by coalescence of a large number of tensile microcracks.

In order to model a typical rock mass in the BPM, it is also necessary to represent pre-existing joints (discontinuities). A straightforward approach is to simply break or weaken the bonds (in the contacts between the particles) intersected by the pre-existing joints. The created discontinuity will have roughness with the amplitude and wavelength related to the resolution, or the particle size of the BPM. The mechanical behavior of discontinuities is very much affected by their roughness. The problem is that the selected particle size (or resolution) typically is not related to actual roughness of the pre-existing joints. The SJM overcomes this limitation. The contacts in the BPM model are oriented in the direction of the line connecting the centers of the particles involved in the contact. The SJM contacts are oriented perpendicular to the fracture plane irrespective of the relative position of the particles. Consequently, the particles can slide relative to each other in the plane of the fracture as if it is perfectly smooth.

The SRM and its components are shown in Figure 2. The BPM represents the intact rock, its deformation and damage. The pre-existing joints are represented explicitly, using the SJM. They can be treated deterministically, by specifying each discontinuity by its position and orientation as mapped in the field. However, typically, for practical reasons, it is not possible to treat the DFN deterministically. Instead, fracturing in the rock mass is characterized statistically. The synthetic DFNs that are statistically equivalent (i.e., fracture spacing, orientation and size) to fracturing of the rock mass are generated and imported into the SRM using SJM (Figure 2). Very often a reasonable compromise is to represent few dominant structures (faults) with their deterministic position and orientation and the rest of the fracturing in the rock mass (smaller structures) using a synthetic DFN.

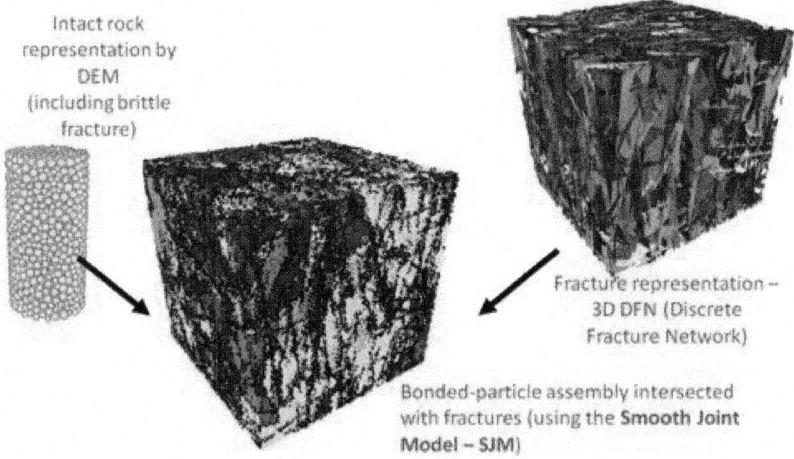

Figure 2: Synthetic rock mass (SRM).

One of the advantages of the SRM is that the components, the intact rock and the joints, can be mechanically characterized by standard laboratory tests. The mechanical response of the rock mass and the size effect are the model results, functions of the model size, DFN characteristics and mechanical properties of the components. Thus, it is not necessary to rely on empirical relations to estimate the rock mass properties and to account for the size effect considering the size of the samples tested in the laboratory and the scale of interest in the model.

The new code, *HF Simulator*, is based on implementation of the SRM in the lattice, which is a simplified, but also a computationally more efficient version of particle flow code (*PFC*). Despite simplifications, the lattice approach represents all physics important for simulation of hydraulic fracturing.

Lattice

The lattice is a quasi-random array of nodes (with given masses) in 3D connected by springs. It is formulated in small strain. The lattice nodes are connected by two springs, one representing the normal and the other shear contact stiffness. The springs represent elasticity of the rock mass. In *HF Simulator*, the calibration factors for spring stiffness are built-in and the user may specify typical macroscopic elastic properties as it is done for other conventional numerical models. The tensile and shear strengths of the springs control the macroscopic strength of the lattice. As for elastic constants, calibration factors are built-in for the strength parameters.

The model simulation is carried out by solving an equation of motions (three translations and three rotations) for all nodes in the model using an explicit numerical method. The following is the central difference equation for the translational degrees of freedom:

$$\dot{u}_i^{(t+\Delta t/2)} = \dot{u}_i^{(t-\Delta t/2)} + \Sigma F_i^{(t)} \Delta t / m$$
$$u_i^{(t+\Delta t)} = u_i^{(t)} + \dot{u}_i^{(t+\Delta t/2)} \Delta t \tag{1}$$

where $\dot{u}_i^{(t)}$ and $u_i^{(t)}$ are the velocity and position (respectively) of component i(i=1,3)i(i=1,3) at time t, ΣF_i is the sum of all force-components i, acting on the node of mass m, with time step Δt. The relative displacements of the nodes are used to calculate the force change in the springs:

$$F^N \leftarrow F^N + \dot{u}^N k^N \Delta t$$
$$F_i^S \leftarrow F_i^S + \dot{u}_i^S k^S \Delta t \tag{2}$$

where "N" denotes "normal," "S" denotes shear, k is spring stiffness and F is the spring force. If the force exceeds the calibrated spring strength, the spring breaks and the microcrack is formed. In other words, if $F^N > F^{Nmax}$, then $F^N = 0$, $F_i^S = 0$, and a "fracture flag" is set.

Fluid Flow

Fluid-flow model and hydro-mechanical coupling are essential parts of *HF Simulator*, as a code for simulation of hydraulic fracturing. The fluid flow occurs through the network of pipes that connect fluid elements, located at the centers of either broken springs or springs that represent pre-existing joints (i.e., springs intersected by the surfaces of pre-existing joints). (The code also can simulate the porous medium flow through unfractured blocks as a way to represent the leakoff. This capability is not discussed further in this paper.) The flow pipe network is dynamic and automatically updated by connecting newly formed microcracks to the existing flow network. The model uses the lubrication equation to approximate the flow within a fracture as a function of aperture. The flow rate along a pipe, from fluid node "A" to node "B," is calculated based on the following relation:

$$q = \beta k_r \frac{a^3}{12\mu}\left[p^A - p^B + \rho_w g\left(z^A - z^B\right)\right] \quad (3)$$

where a is hydraulic aperture, $\mu\mu$ is viscosity of the fluid, p^A and p^B are fluid pressures at nodes "A" and "B", respectively, z^A and z^B are elevations of nodes "A" and "B", respectively, and ρ_w is fluid density. The relative permeability, k_r, is a function of saturation, s:

$$k_r = s^2(3 - 2s) \quad (4)$$

Clearly, when the pipe is saturated, s=1s=1 and the relative permeability is 1. The dimensionless number $\beta\beta$ is a calibration parameter, a function of resolution, used to match conductivity of a pipe network to the conductivity of a joint represented by parallel plates with aperture aa. The calibrated relation between $\beta\beta$ and the resolution is built into the code.

Hydro-Mechanical Coupling

In *HF Simulator*, the mechanical and flow models are fully coupled.
- Fracture permeability depends on aperture, or on the deformation of the solid model.
- Fluid pressure affects both deformation and the strength of the solid model. The effective stress calculations are carried out.
- The deformation of the solid model affects the fluid pressures. In particular, the code can predict changes in fluid pressure under undrained conditions.

A new coupling scheme, in which the relaxation parameter is proportional to $K_R a/R$, where K_R is rock bulk modulus and R is the lattice resolution, is implemented in *HF Simulator*, allowing larger explicit time steps and faster simulation times compared to conventional methods that use fluid bulk modulus as a relaxation parameter.

VERIFICATION TEST: PENNY-SHAPED CRACK PROPAGATION IN MEDIUM WITH ZERO TOUGHNESS

The non-steady response of rock to injection of fluid depends on fracture toughness, the viscosity of the fluid and the rate of leak-off. In the case of zero fracture toughness and no leak-off, the response is viscosity-dominated, which corresponds to the "M-asymptote" identified by [5]. This condition is used for verification of *HF Simulator*.

In the simulated example, fluid is injected at a constant rate into a penny-shaped crack of low initial aperture (10^{-5}m). The crack has zero normal strength, and the in-situ stresses are also zero. Thus, the test conditions approximate those of the analytical solution for the no-lag case (i.e., no fluid pressure tension cut-off) provided by [5]. The injection rate is 0.01 m^3/s; the dynamic viscosity is 0.001 Pa×s. The mechanical properties of the rock are characterized by Young's modulus of 7×10^7Pa and Poisson's ratio of 0.22. Figure 3 provides a visualization of the state of the model at 10 s of elapsed time. Note that pressures are negative in the outer annulus of the flow disk.

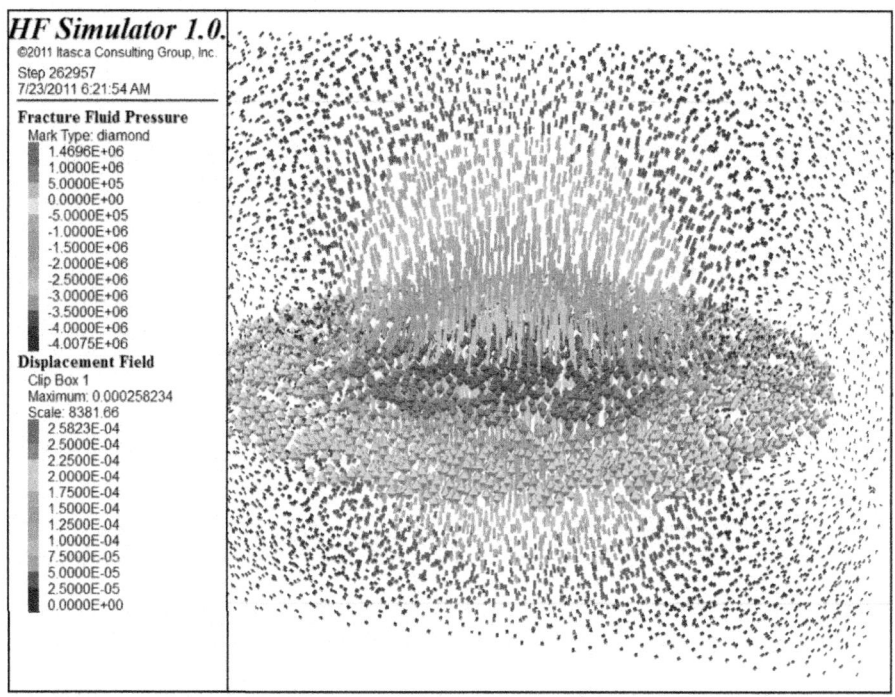

Figure 3: View of pressure (Pa) field (icons, colored according to magnitude) and cross-section of displacement (m) field (vectors, colored according to magnitude).

Figure 4 shows the aperture profiles at three times during the simulation — averaged numerical results (for 30 radial distances), together with asymptotic solutions (derived from the equations of [5]). Figure 5 shows the pressure profile at 10 s, together with the asymptotic solution. Note that there is a lack of match at small and large radial distances: at small distances, the numerical source is a finite volume, rather than a point source (which is assumed in the exact solution); at large distances, the finite initial aperture allows seepage (compared to zero seepage in the exact solution, which assumes zero initial aperture).

EXAMPLE APPLICATION

Two example problems are discussed in this section. Fracture propagation in a homogeneous (unfractured) and fractured media is analyzed. These two problems involve a horizontal borehole segment with two injection clusters with centers at 4.8 m distance (Figure 6). The model domain is 18 m × 18 m × 18 m, and the lattice resolution was set to 0.5 m. Fluid is injected into the clusters at rate of 0.01 m³/s. The assumed stress state is anisotropic with σ_{xx}=1MPa, σ_{yy}=12MPa and σ_{zz}=10MPa. The least principal stress is aligned with the horizontal section of the borehole. This stress state favors crack propagation in the direction normal to the horizontal section of the borehole. In order to initiate the fluid calculation, fluid-filled joints have been placed at the center of each cluster; these joints are slightly larger than the cluster size. The initial apertures in these joints have been set to 0.1 mm. Both example problems use this model configuration. The example shown on the left in Figure 6 simulates the response of an unfractured medium to fluid injection. Three discrete joints that interact with the induced fractures are introduced in the example on the right in Figure 6.

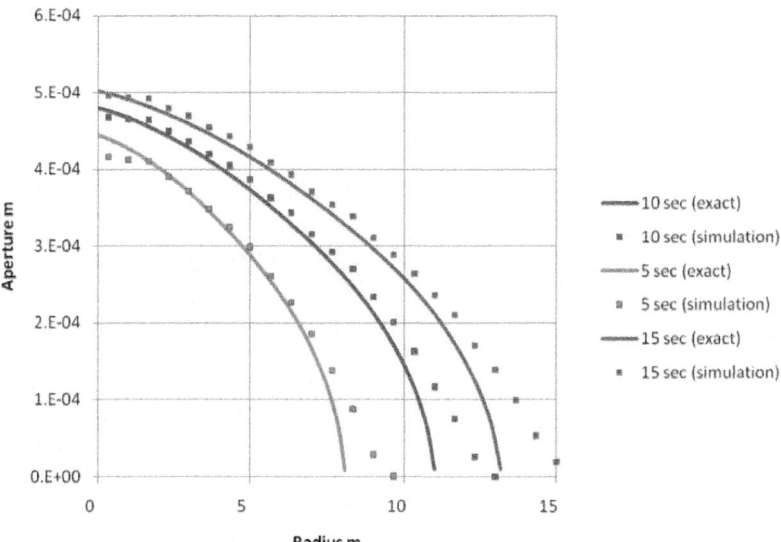

Figure 4: Aperture profiles for three times.

The induced microcracks in the homogeneous model after 15 s of injection are shown in Figure 7. The microcracks form two roughly circular (penny-shape) hydraulic fractures. In this example, the fractures are not parallel. There is a slight trend of fractures curving away from each other as a result of stress interaction.

In the second example, the HF propagation is clearly affected by the pre-existing joints, as shown inFigure 8. When the HF intersects the pre-existing joint, the fluid is diverted into the pre-existing joints. (In general case, the HF can cross or be diverted into the pre-existing joint, depending on a number of parameters, including stress state, strength and permeability of the pre-existing joint.) The propagation continues by reinitiation along the edges of pre-existing joints.

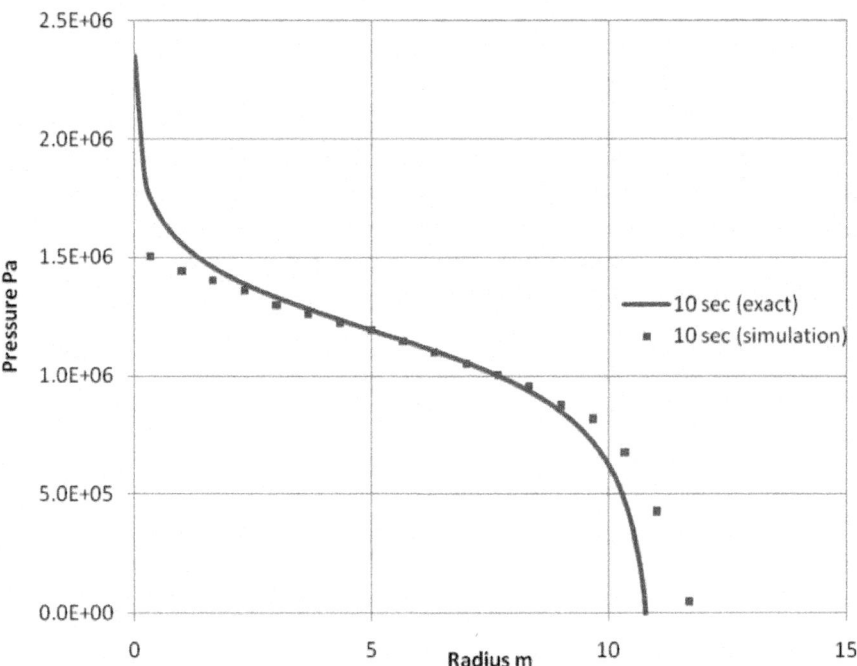

Figure 5: Pressure profile at 10 s.

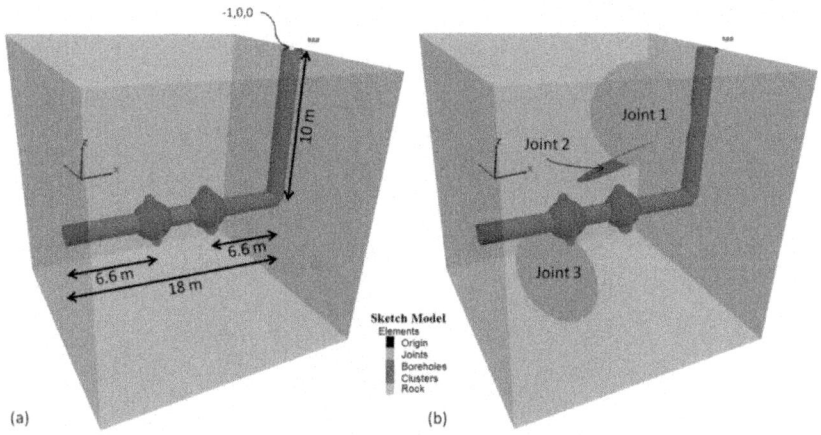

Figure 6: Geometry of two example problems.

CONCLUSION

HF Simulator is a powerful 3D simulator for hydraulic fracturing in jointed rock mass that allows the main mechanisms (nonlinear mechanical response, fluid flow in joints and coupled fluid-mechanical interaction) to be reproduced. The formulation of *HF Simulator* is based on a quasi-random lattice of nodes and springs.

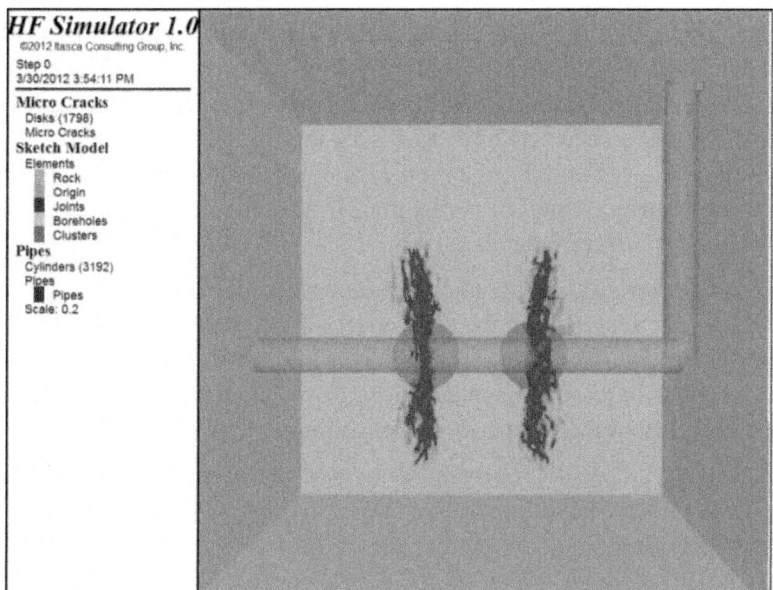

Figure 7: Hydraulic fractures generated in a homogeneous medium (dark blue disks are microcracks).

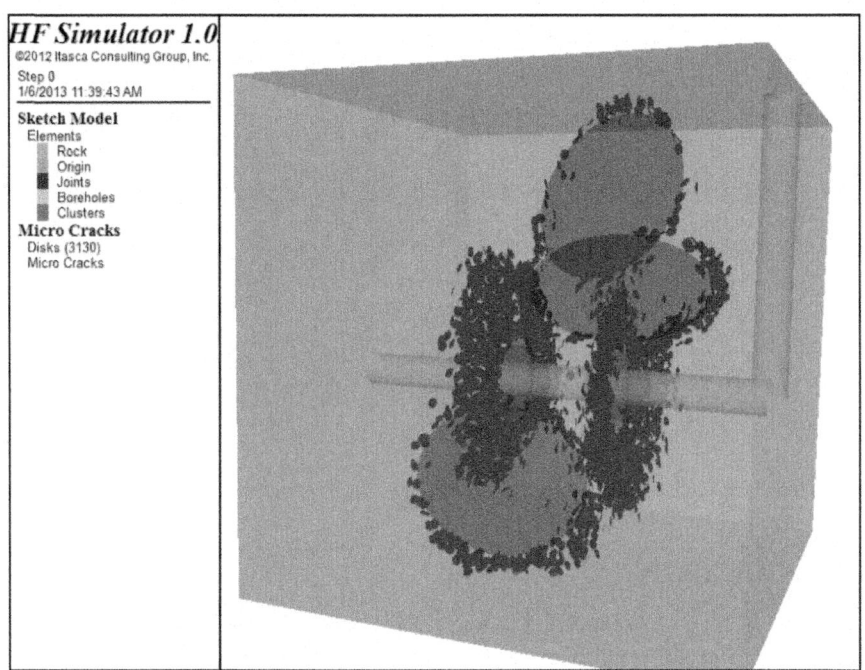

Figure 8: Hydraulic fractures generated in a medium with three pre-existing joints (blue disks are microcracks).

The springs between the nodes break when their strength (in tension) is exceeded. Breaking of the springs corresponds to the formation of microcracks, and microcracks may link to form macrofractures. The SJM (smooth joint model) is used to represent pre-existing joints in the model. Thus, the SJM allows simulation of sliding of a pre-existing joint in the model, unaffected by the apparent surface roughness resulting from lattice resolution and random arrangement of lattice nodes.

The model is fully coupled hydro-mechanically. There are several ways in which fluid interacts with the rock matrix. First, fluid pressures may induce opening or sliding of the fractures. Second, mechanical deformation of fractures causes changes in joint pressures. Third, the mechanical deformation changes the permeability of the rock mass as the joint apertures change.

The new code is a promising tool for simulation and understanding of complex processes, including propagation of HF and its interaction with DFN, during stimulation of unconventional reservoirs.

ACKNOWLEDGEMENTS

The development of the numerical code described in this paper was funded by BP America. The authors would like to thank BP America for their support. Matt Purvance, Jim Hazzard and Maurilio Torres of Itasca Consulting Group, Inc. are thanked for their valuable work on HF Simulator.

REFERENCES

1. D. O Potyondy, P. A Cundall, A Bonded-Particle Model of Rock. Int. J. Rock Mech. & Min. Sci., 41132913642004
2. M Pierce, Mas Ivars D., Cundall P.A., Potyondy D.O. "A Synthetic Rock Mass Model for Jointed Rock," in Rock Mechanics: Meeting Society's Challenges and Demands (1st Canada-U.S. Rock Mechanics Symposium, Vancouver, May 2007), 1Fundamentals, New Technologies & New Ideas, 341349E. Eberhardt et al., Eds. London: Taylor & Francis Group; 2007
3. Itasca Consulting GroupInc. PFC2D (Particle Flow Code in 2 Dimensions), Version 4.0. Minneapolis: Itasca; 2008
4. Itasca Consulting GroupInc. PFC3D (Particle Flow Code in 3 Dimensions), Version 4.0. Minneapolis: Itasca; 2008
5. A Peirce, E Detournay, An Implicit Set Method for Modeling Hydraulically Driven Fractures, Comput. Methods Appl. Mech. Engrg., 197285828852008

Chapter 11

ROUGHNESS RESEARCH OF CENTER PROFILE CURVE ON ROCK FRACTURE SURFACE BASED ON STATISTICAL METHOD

Xuezai Pan[1, 3], Zhigang Feng[2, 3], Guoxing Dai[3], and Hongguang Liu[4]

[1]School of Mathematics, Nanjing Normal University, Taizhou College, Taizhou, China
[2]State Key Laboratory of Coal Resources and Safe Mining, China University of Mining and Technology, Beijing, China
[3]Faculty of Science, Jiangsu University, Zhenjiang, China
[4]Faculty of Civil Engineering and Mechanics, Jiangsu University, Zhenjiang, China

ABSTRACT

In order to research roughness of rock fracture surfaces whether to depend on scale effect, Brazil discs were fractured under tensile and compression stresses in Brazil split test with MTS (Mechanics Test Systems) and a laser profilometer was used to scan rock fracture surfaces and coordinates datum of central profile were acquired. A figure of the central profile was plotted through the coordinates datum. A certain line segment length is regarded as a step length, which is called scale and the scale length is taken to connect pairs of closer peak points on the profile curve. The directional distribution of every scale's normal vector is analyzed by statistics and normal hypothesis test. Finally, some statistics of sample degrees datum are compared with other ones and reach a conclusion that roughness of center profile curve depends on scale effect. The distribution of degrees more and more approximates normal distribution along with increase of scale.

INTRODUCTION

Deformity and fracture of rock are involved in process of moving in earth crust, for example, earthquake, slide downhill, mud-stone flow and so on. In addition, rock fracture usually happens in rock project, for instance, explosion of rock, the project of tunnel, mining engineering etc. Studying rock fracture surfaces through morphology has been recognized by professionals since twenty-one century, because morphology of rock fracture surfaces implicates abundant information of rock fracture mechanics. Experts have discovered that rock fracture surface has characterization of roughness, irregularity and complexity. They attempt to depict the relation between complex morphology of rock fracture surfaces and roughness by all kinds of ways. For example, In order to assess the current state of rock masses and to predict the stability of jointed rock structures, the roughness of rock fracture surfaces has been studied to a higher level. Based on systematic experiments, Barton and Choubey in 1977 proposed a conceptual model to quantify the roughness of rock fracture surface [1]. They classified the roughness into ten categories and the Joint Roughness Coefficients (JRC) ranged from 0 to 20 [2-4]. Some investigators use fractal geometry and multifractal which have been developed since 1970's to describe rock fracture mechanism and have tried to establish the relationship between the fractal dimension and various mechanical parameters of rock fracture surfaces [5-8]. Some experts have indicated that the structural anisotropy of fracture surfaces in rocks greatly influences the mechanical behavior of rock joints under loading [9-11]. The following statement will study roughness of center profile curve on rock fracture surfaces from statistical view. Finally, three prospects will be put forward in the end of the paper.

EXPERIMENTAL METHOD AND ANALYSIS

Experimental Method

Firstly, a sort of special granites which were taken from Gansu province north mountain in China were used to experimental material, because the compactness of the rock material is relatively homogeneous. The granite material was made of the cylinder-shaped sample with rock drilling machine, and then the cylinder-shaped sample was cut into three Brazil discs samples with cutting off machine and buffing machine. The diameter and the height of the discs are equal to 112 mm and 28 mm respectively. Secondly, the rock discs were fractured under tensile and compression stresses in Brazil split test with MTS (Mechanics Test Systems). Loading speed was per minute 0.01 mm. When loading strength approximatively reaches 48 kN, the discs were fractured along vertical direction (refer with: Figure 1). Finally, according to rock mechanics principle, in indirect tensile stresses process of rock, the rock stress of the edge of disc is relatively centralized, so the edge of the disc was easily broken and a little stone chips fell. Whereas inner stress of the rock disc is relatively balanced [12-13], the inner of fracture rock has no stone chips fallen. So, 11 mm was removed from two ends of the rectangular fracture surfaces respectively. The length of center part of fracture surface is equal to 90 mm (refer with: Figure 2). The length of 90 mm is supposed to x axis direction. The center part of fracture surface was scanned by high-accuracy rock laser profilometer along x axis according to the way that interval of x axis is equal to 0.1 mm to acquired three dimension coordinates (x, y, z) of lattice.

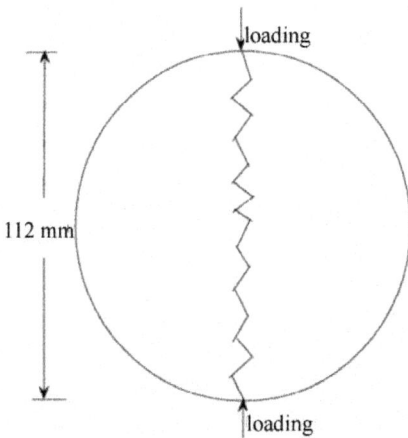

Figure 1: Indirect tensile diagrammatic sketch.

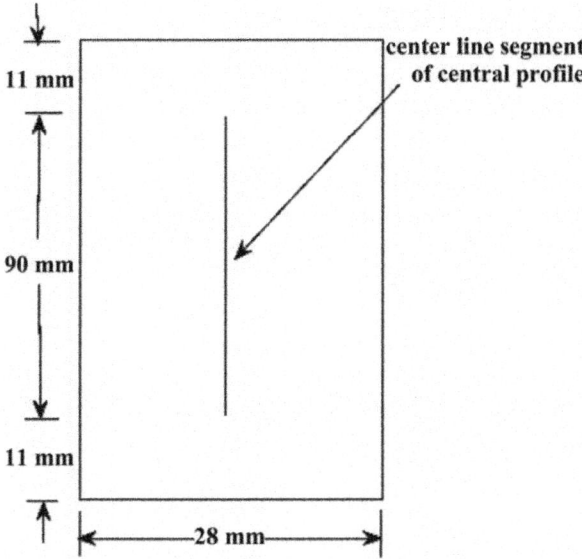

Figure 2: Center profile acquired datum.

Total 901 rope line segments were scanned through the above method, because the length of center part of fracture surface is 90 mm. Length of every line segment scanned is 28 mm, since the width of center part of rock fracture surface is equal to 28 mm (refer with: Figure 2).

Center Profile Curve Analysis on Rock Fracture Surface

The coordinates datum of the center segment of central profile curve on rock rectangle fracture surface (refer with: Figure 2) was extracted by computer procedure, and then the approximative two dimension curve figure of the profile was acquired by linear interpolating method (refer with: Figure 3). A unit length of step length is supposed to 0.1 mm. A step length was taken to connect pairs of points on the center profile [14-16]. Normal vectors' directional distribution of corresponding step length was considered and every angle of each normal vector departing from straight up vector is measured (refer with: Figure 4).The motivation for this approach is straightforward: for a perfectly straight profile the normal vectors will all be parallel and hence display zero dispersion, whereas the dispersion will increase as the profile departs from straight (i.e. becomes rougher). Suppose angle of straight up vector is zero degree. The degree of the angle in the direction skewing to left is negative, whereas that skewing to right is positive. Thus, these degree datum of angles can be obtained with computer procedure.

STATISTICAL METHOD

From statistical knowledge [17-20], mean and median of sample (refer with: Equation (1)) reflect concentrated tendency of sample datum. Range (refer with: Equation (2)), variance and sample standard deviation (refer with: Equation (3)) indicate the extent that sample datum depart from sample mean.

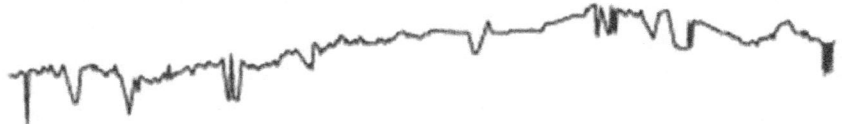

Figure 3: Linear interpolating figure of center profile.

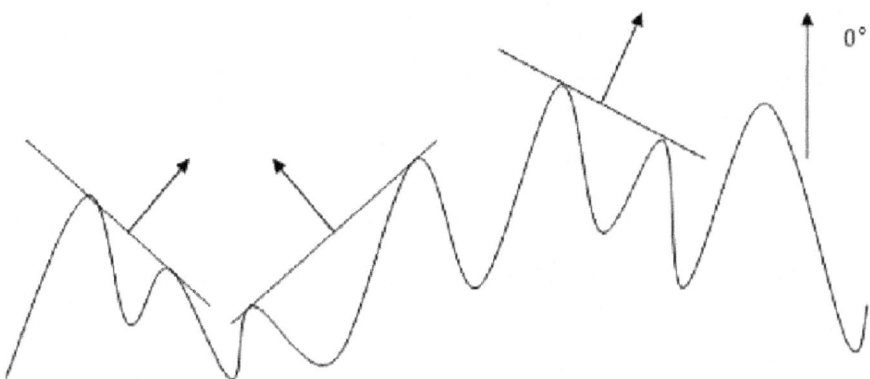

Figure 4: Orientation of normal vector to a step length connecting two points on a profile curve.

Skewness (refer with: Equation (4)) and kurtosis (refer with: Equation (5)) are such statistic describing the shape of sample datum. Skewness reflects dispersive symmetrical characterization of sample datum. Finally, kurtosis indicates the situation that sample datum deviate normal distribution.

$$\bar{X} = \frac{1}{n}\sum_{i=1}^{n} X_i \quad (1)$$

$$R = \max(X_i) - \min(X_i) \quad (2)$$

$$s = \sqrt{\frac{1}{n-1}\sum_{i=1}^{n}(X_i - \bar{X})^2} \quad (3)$$

$$g_1 = \frac{1}{s^3}\sum_{i=1}^{n}(X_i - \bar{X})^3 \quad (4)$$

$$g_2 = \frac{1}{s^4}\sum_{i=1}^{n}(X_i - \bar{X})^4 \quad (5)$$

where X_i denotes samples.

When $g_1 > 0$, the form is called right deviation, which illustrates the right datum of mean are more dispersive than those of the left datum; As $g_1 < 0$, the result is named left deviation, which illustrates the situation is opposite to that of right deviation. As g_1 approach zero, which is called impartiality, So, the distribution is regarded as symmetry. On the other hand, kurtosis of normal distribution is equal to 3. As $g_2 > 3$, there are a lot of datum departing from mean, whose shape of distributive curve is flatter than that of normal distribution accordingly; On the contrary, when $g_2 < 3$, the case is inverse to that of $g_2 > 3$. So, statistical method can be used to characterize roughness of profile curve subjected to fracture surfaces. The following discussion is concrete operation.

Suppose the step length is equal to 0.4 mm, 0.3 mm, 0.2 mm and 0.1 mm respectively, then the datum of angle variation are acquired by computer procedure corresponding to various step length. Furthermore, under the same scale, sample mean, median, range, variance, standard deviation, coefficient of skewness and kurtosis are computed respectively and frequency histogram [21-25] is drawn with computer program, which describes the distribution of orientation of normal vectors from the center profile curve. Frequency histogram under a step length is compared with that of other ones, which can show distributional differences each other. On the other hand, hypothesis test method is used to test whether the distribution of normal vectors obeys normal distribution or not and distribution function plots were drawn with computer program. For the degree datum input into computer, Jarque-Bera test is used to test these degree datum whether to obey normal distribution. Significance level α is supposed to 0.05. P is a probability value accepting original hypothesis. JBSTAT is test statistics value. CV is a threshold which can judge whether to refuse original hypothesis and H is test result. If H = 0, the distribution of the degree datum can be considered normal distribution; If H = 1, the distribution of the degree datum doesn't obey normal distribution. If $P < \alpha$, original hypothesis that the datum belong to normal distribution can be denied; If JBSTAT > CV, normal distributional original hypothesis can be negated. Every statistic in following tables is a mean value of corresponding statistic of three center profile curves datum under the same step length (i.e. the same scale), because there are three Brazil discs samples. The differences among the same statistic are compared within four tables under different scales.

1) If the step length is equal to 0.4 mm, the following Tables 1 and 2 indicate a statistical result.

In Table 1, unit of mean, median, range and standard deviation is degree, whereas other statistics have no unit, because they are only coefficients (below affinity).

Variables in Table 2 have no unit (below affinity). From coefficient of skewness -0.0013 (≈ 0), dispersive extent of datum with left side and right one deviating mean is almost comparative. Distribution of angles' degrees approximatively summits to normal distribution from kurtosis coefficient 3.1511 (≈ 3) and its frequency histogram is referred with Figure 5. From hypothesis test view, where H = 0, $P < \alpha$ and JBSTAT < CV, the normal distributional original hypothesis can be accepted. From normal probability plot shown in Figure 6, the absolute major points gather on the red straight line, which illustrates the normal distributional suppose can be accepted.

In Figure 5, i denotes positive integer and $1 \leq i \leq 12$, because frequency histogram consists of twelve columns.

2) If the step length is equal to 0.3 mm, the statistical datum result is shown in the Tables 3 and 4.

The extent of departing from sample mean increases; The skewness coefficient increases and is more than 0, which illustrates the right datum of mean is more dispersive than that of the left, but the dispersive extent is faint; Kurtosis coefficient is equal to 2.8296 (≈ 3), which illustrates distribution of angle datum approximate normal distribution. The frequency histogram reflects that the distribution of angle datum close to normal distribution (shown in Figure 7). From normal hypothesis test view, H = 0, P = 0.6492 and JBSTAT < CV indicate normal distributional hypothesis can be accepted with 64.92% probability. From normal probability plot shown in Figure 8, the absolute major points gather on the red straight line, which illustrates the normal distributional suppose can be accepted.

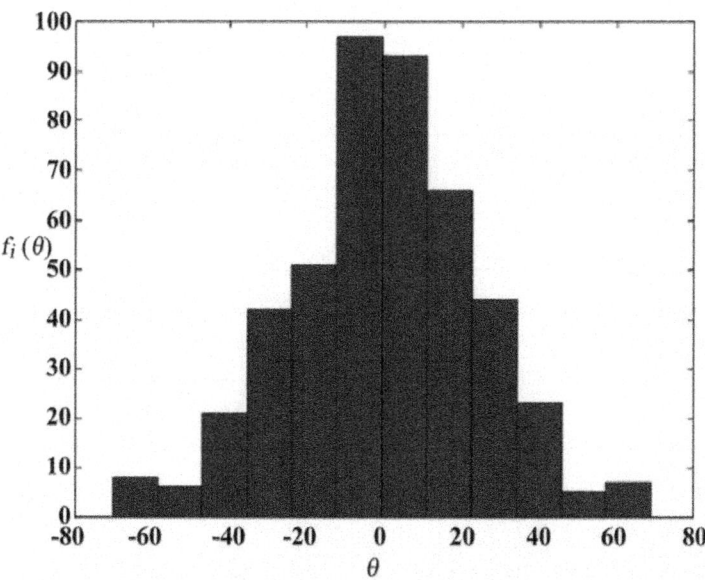

Figure 5: The histogram with the step length 0.4 mm (θ: degree; $f_i(\theta)$: frequency).

Table 1: Statistics of sample datum with the step length 0.4 mm

Mean	Median	Range	Variance	Standard deviation	Skewness	Kurtosis
−0.1333	0.0000	149.0381	603.3292	24.5628	−0.0013	3.1511

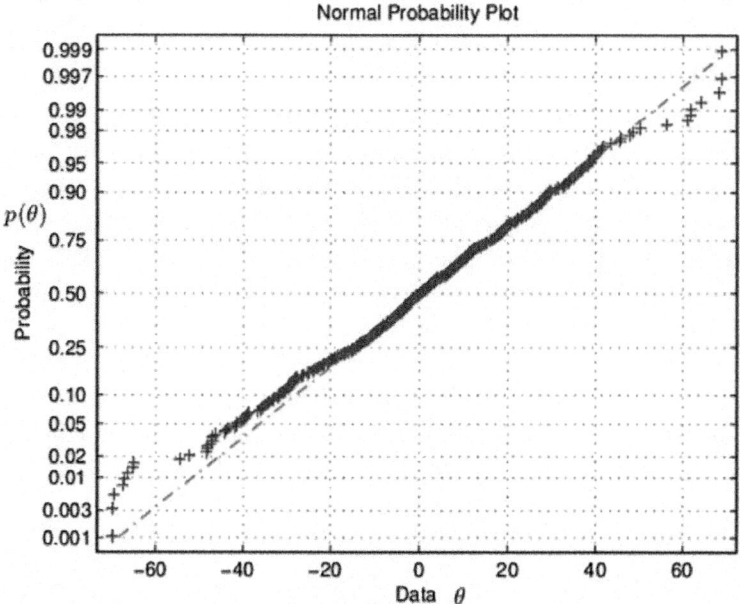

Figure 6: Normal probability plot with the step length 0.4 mm (θ: degree; p(θ): probability).

Table 2: Hypothesis test value with the step length 0.4 mm

H	P	JBSTAT	CV
0	0.2314	2.9269	5.9915

3) If the step length is equal to 0.2 mm, the statistical result is shown in the next Tables 5 and 6.

The sample mean and median increase; Variance and standard deviation increase furthermore; Range hardly change; Coefficient of skewness reduces, however the decrement is very little; coefficient of kurtosis decreases furthermore and reaches 1.9127, which illustrates the distribution of angle datum continues to deviate from normal distribution.

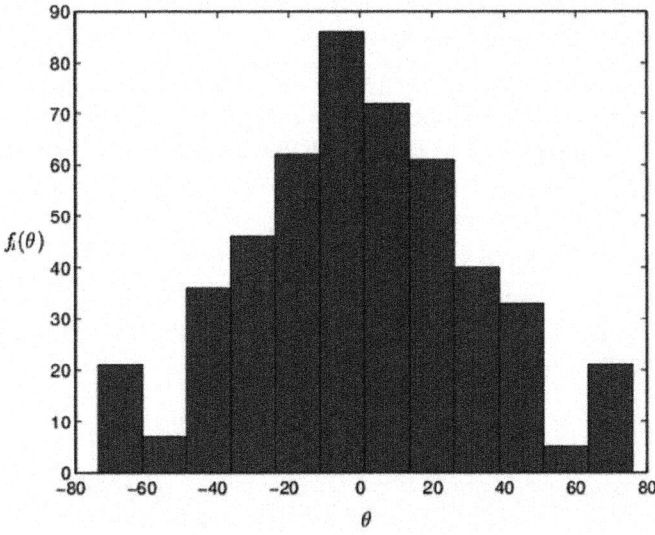

Figure 7: The histogram with the step length 0.3 mm (θ: degree; $f_i(\theta)$: frequency).

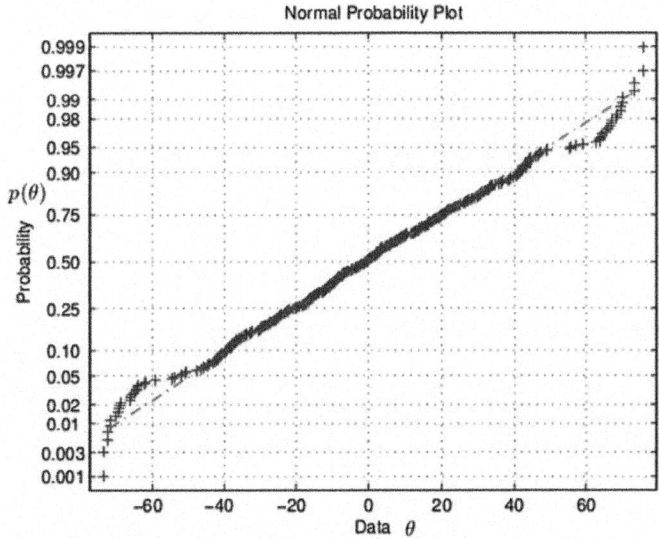

Figure 8: Normal probability plot with the step length 0.3 mm (θ: degree; p(θ): probability).

Frequency histogram is shown in Figure 9. From normal hypothesis test view, H = 1, the value of P and JBSTAT > CV indicate normal distributional hypothesis can be negated. Normal probability plot is referred with Figure 10.

278   Design Analysis in Rock Mechanics

4) If the step length is equal to 0.1 mm, the following Tables 7 and 8 indicate according statistical result.

The value of sample mean and median has a weak variation; Range still change rarely; Variance and standard deviation increase greatly; Coefficient of skewness continues to decrease, but it still fluctuates near 0, which still describes that two sides' datum of mean have the same dispersion characterization; Coefficient of kurtosis descends further and attains 1.6600, which indicates that the distribution of normal vector deviates from normal distribution; Frequency histogram is shown in Figure 11. From normal hypothesis test view, H = 1, the value of P = 0 and JBSTAT $\gg$ CV indicate normal distributional hypothesis can be negated completely.

Table 3: Statistics of sample datum with the step length 0.3 mm

Mean	Median	Range	Variance	Standard deviation	Skewness	Kurtosis
−0.2117	−0.8593	149.5869	990.3574	31.4699	0.0488	2.8296

Table 4: Hypothesis test value with the step length 0.3 mm

H	P	JBSTAT	CV
0	0.6492	0.8640	5.9915

Table 5: Statistics of sample datum with the step length 0.2 mm

Mean	Median	Range	Variance	Standard deviation	Skewness	Kurtosis
0.2166	0.0000	149.5887	1588.7498	39.8591	0.0081	1.9127

Table 6: Hypothesis test value with the step length 0.2 mm

H	P	JBSTAT	CV
1	2.5850e−011	48.7574	5.9915

Table 7: Statistics of sample datum with the step length 0.1 mm

Mean	Median	Range	Variance	Standard deviation	Skewness	Kurtosis
0.1318	0.2865	149.5898	1783.2761	42.2289	0.0072	1.6600

Table 8: Hypothesis test value with the step length 0.1 mm

H	P	JBSTAT	CV
1	0	112.9056	5.9915

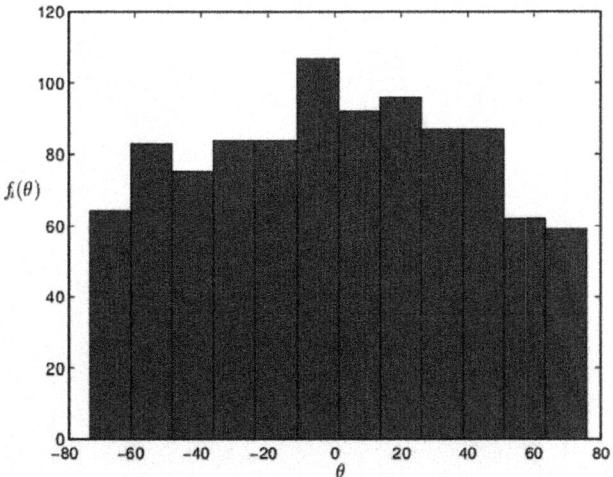

Figure 9: The histogram with the step length 0.2 mm (θ: degree; $f_i(\theta)$: frequency).

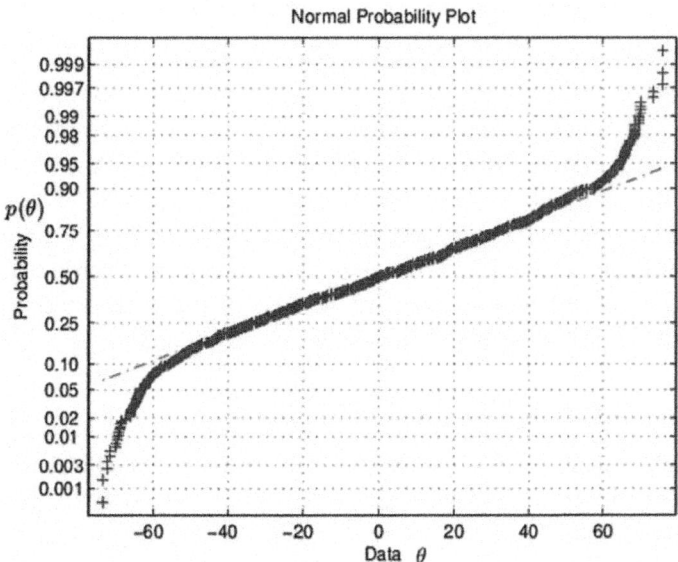

Figure 10: Normal probability plot with the step length 0.2 mm. (θ: degree; p(θ): probability).

Normal probability plot is referred with Figure 12. The absolute major points deviates from the red straight line.

In conclusion, the directional distribution of normal vectors is associated with the step length, which depends on scale effect.

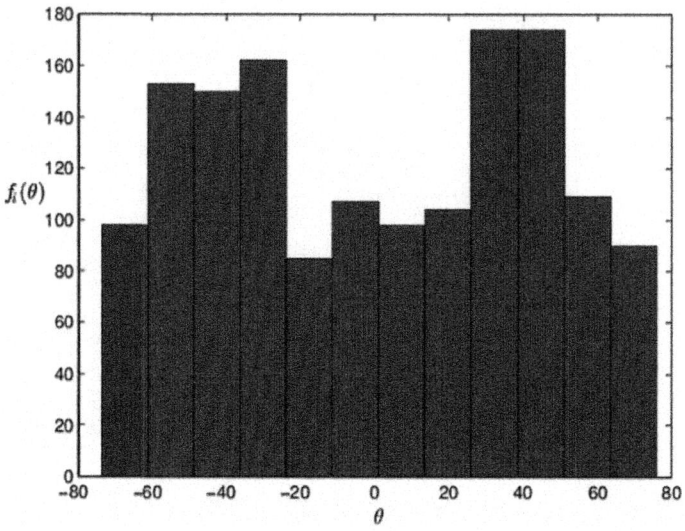

Figure 11: The histogram with the step length 0.1 mm (θ: degree; $f_i(θ)$: frequency).

The smaller the kurtosis is, the smaller the scale is, which illustrates the distribution of angle datum deviates from normal distribution continuously. The case that range of sample datum is close to 149 is discovered, which indicates that variable range of angle degrees is definite. And the dispersive extent between the left datum of sample mean and those of right is comparative, because skewness coefficient is almost equal to 0. In summary, the roughness of center profile curve could be supposed to standard state, when the distribution of angle datum belongs to standard normal distribution. Furthermore, the rougher center profile curve is, the smaller the step length is.

CONCLUSION AND PROSPECT

In generally, mean, median, range and skewness coefficient almost have no scale effect, however, variance and standard deviation will increase with decreasing of scale. Kurtosis coefficient will descent along with diminishing of scale. Accordingly, from normal hypothesis test view, H, P, JBSTAT have scale effect too. As a whole, H leaps from 0 to 1 along with decreasing of scale. P will decrease while scale drops. JBSTAT will raise as scale falls. That is to say, the rougher center profile curve is, the smaller the step length is. The ultimate purpose of researching morphology of rock fracture surfaces is that the information in process of rock fracture is acquired through methods of mathematical analysis. Successively, components of rock structure and limitation are discovered and further mechanics of rock fracture is posted. However, structure of rock and properties of mechanics exhibit nonlinear characterization. Rock fracture surface possesses fairly irregular and stochastic properties, hence the topic is developed so slowly that the results acquired through researching haven't been applied in forecasting and instructing practise of project. Therefore, the following three aspects of research will be evolved.

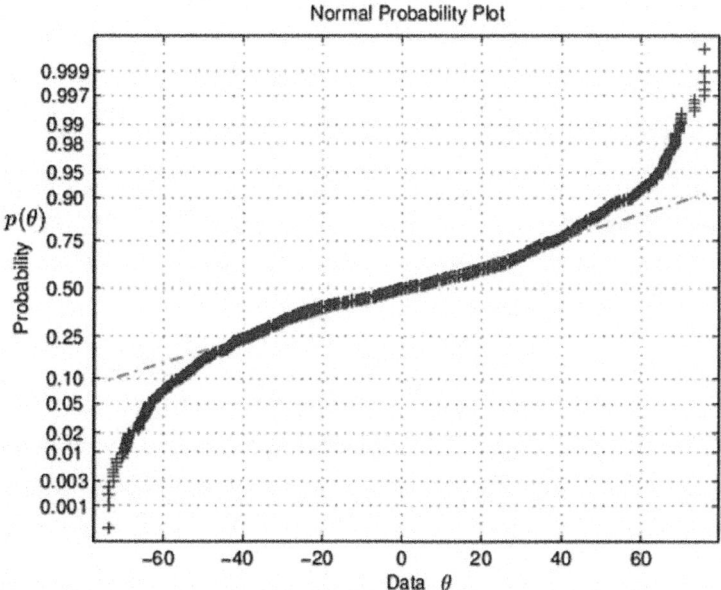

Figure 12: Normal probability plot with the step length 0.1 mm (θ: degree; p(θ): probability).

Firstly, advantages of approaches applied by experts now continue to be developed and perfected, and insufficiencies will be overcome extensively, which attempt to build the relationship between rock mechanics and topograph of rock fracture surfaces. Secondly, new approaches of studying rock fracture surfaces through morphology will be sought, which try to discover the mechanics of rock fracture. Finally, existing experimental results will be transformed into theoretical basis of guiding project practice possibly.

ACKNOWLEDGEMENTS

- Supported by the National Natural Science Foundation of China 51079064.
- Supported by State Key Laboratory of Coal Resources and Safe Mining, China University of Mining and Technology SKLCRSM10KFA02.
- Supported by Nanjing Normal University Taizhou College Q201234.

REFERENCES

1. N. Barton and V. Choubey, "The Shear Strength of Rock Jounts in Theory and Practice," Rock Mechanics, Vol. 1977, Vol. 10, No. 2, pp. 1-45. doi:10.1007/BF01261801
2. N. Barton, "Review of a New Shear Strength Criterion for Rock Joints," Engineering Geology, Vol. 7, No. 4, 1973, pp. 287-332. doi:10.1016/0013-7952(73)90013-6
3. R. Tse and D. M. Cruden, "Estimating Joint Roughness Coefficients," International Journal of Rock Mechanics and Mining Sciences & Geomechanics Abstracts, Vol. 16, No. 5, 1979, pp. 303-307. doi:10.1016/0148-9062(79)90241-9
4. N. H. Maerz, J. A. Franklin and C. P. Bennett, "Joint Roughness Measurement Using Shadow Profilometry," International Journal of Rock Mechanics and Mining Sciences & Geomechanics Abstracts, Vol. 27, No. 5, 1990, pp. 329-343. doi:10.1016/0148-9062(90)92708-M
5. K. Falconer and W. G. Yang, "Fractal Geometry Mathematical Foundations and Applications," 2nd Edition, Posts & Telecom Press, Beijing, 2007.
6. Y. H. Zhang, H. W. Zhou and H. P. Xie, "Improved Cubic Covering Method for Fractal Dimensions of a Fracture Surface of Rock," Chinese Journal of Rock Mechanics and Engineering, Vol. 24, No. 17, 2005, pp. 3192- 3196.
7. H. P. Xie, H. Q. Sun, Y. Ju, et al., "Study on Generation of Rock Surfaces by Using Fractal Interpolation," International Journal of Solid and Structure, Vol. 38, No. 32-33, 2001, pp. 5765-5787. doi:10.1016/S0020-7683(00)00390-5
8. H. P. Xie, J.-A. Wang and M. A. Kwa Niewski, "Multifractal Characterization of Rock Fracture Surfaces," International Journal of Rock Mechanics Mining Sciences, Vol. 36, No. 1, 1999, pp. 19-27. doi:10.1016/S0148-9062(98)00172-7
9. S. C. Bandis, A. C. Lumsden and N. R. Barton, "Fundamentals of Rock

Joint Deformation," International Journal of Rock Mechanics and Mining Sciences & Geomechanics Abstracts, Vol. 20, No. 6, 1983, pp. 249-268. doi:10.1016/0148-9062(83)90595-8

10. P. H. S. W. Kulatilake, G. Shou, T. H. Huang and R. M. Morgan, "New Peak Shear Strength Criteria for Anisotropic Rock Joints," International Journal of Rock Mechanics and Mining Sciences & Geomechanics Abstracts, Vol. 32, No. 7, 1995, pp. 673-697.doi:10.1016/0148-9062(95)00022-9

11. H. W. Zhou and H. Xie, "Anisotropic Characterization of Rock Fracture Surfaces Subjected to Profile Analysis," Physics Letters A, Vol. 325, No. 5-6, 2004, pp. 355-362.doi:10.1016/j.physleta.2004.04.006

12. W. Gao, "Mechanics of Rock," Peking University Press, Beijing, 2010.

13. Z. L. Fu, F. K. Xiao, Y. X. Liu and S. J. Chen, "Experiment Course on Rock Mechanics," Chemistry Engineering Press, Beijing, 2010.

14. V. Rasouli and J. P. Harrison, "Assessment of Rock Fracture Surface Roughness Using Riemannian Statistics of Linear Profiles," International Journal of Rock Mechanics Mining Sciences, Vol. 47, No. 6, 2010, pp. 940-948. doi:10.1016/j.ijrmms.2010.05.013

15. V. Rasouli and J. P. Harrison, "Scale Effect, Anisotropy and Directionality of Discontinuity Surface Roughness," Proceedings of the EUROCK Symposium, Aachen, 27-31 March 2000, pp. 751-756.

16. T. H. Wu and E. M. Ali, "Statistical Representation of Joint Roughness," International Journal of Rock Mechanics and Mining Science & Geomechanics Abstracts, Vol. 15, No. 5, 1978, pp. 259-262. doi:10.1016/0148-9062(78)90958-0

17. Z. S. Wei, "Probability and Statistics Tutorial," Higher Education Press, Beijing, 1983.

18. X. R. Chen, "Higher Statistics," China University of Science and Technology, Hefei, 1999

19. G. X. Liu, Z. F. He and J. L. Yang, "Probability and Statistics," Gansu Education, Lanzhou, 2002.

20. X. S. Liu, "Probability and Statistics," Sichuan University Press, Chengdu, 2009.

21. L. B. Wu and B. N. Li, "Mathematical Experiment and Modeling," Defense Industry Press, Changsha, 2007.

22. J. Zhao and Q. Dan, "Mathematical Modeling and Mathematical Experiment," Higher Education Press, Beijing, 2000.

23. H. G. Zhang, "Practical Tutorial of MATLAB/SIMULINK," Posts & Telecom Press, Beijing, 2009.
24. D. X. Zhang and L. S. Zhao, "Teaching of Language Procedure Design of MATLAB," China Railway Press, Beijing, 2010.
25. P. Wu, "Technology and Application of Effective Procedure of MATLAB: Analysis of 25 Cases," Beihang University Press, Beijing, 2010.

CITATION

CHAPTER 1
M. Maleki, M. Mahyar and K. Meshkabadi, "Design of Overall Slope Angle and Analysis of Rock Slope Stability of Chadormalu Mine Using Empirical and Numerical Methods," Engineering, Vol. 3 No. 9, 2011, pp. 965-971. doi:10.4236/eng.2011.39119.

CHAPTER 2
Li, Y.; Zhou, H.; Zhu, W.; Li, S.; Liu, J. Numerical Study on Crack Propagation in Brittle Jointed Rock Mass Influenced by Fracture Water Pressure. Materials 2015, 8, 3364-3376.

CHAPTER 3
Xu Q, Chen J, Li J, Zhao C, Yuan C (2015) Study on the Constitutive Model for Jointed Rock Mass. PLoS ONE 10(4): e0121850. doi:10.1371/journal.pone.0121850.

CHAPTER 4
Xiaojun Zhou, Jinghe Wang, Bentao Lin, "Study on calculation of rock pressure for ultra-shallow tunnel in poor surrounding rock and its tunneling procedure," J. Mod. Transport. (2014) 22(1):1–11 DOI 10.1007/s40534-013-0025-8.

CHAPTER 5
Meng-Chia Weng, "A Generalized Plasticity-Based Model for Sandstone Considering Time-Dependent Behavior and Wetting Deterioration," Rock Mech Rock Eng (2014) 47:1197–1209 DOI 10.1007/s00603-013-0466-8.

CHAPTER 6

P. Samui and T. Sitharam, "Application of Geostatistical Models for Estimating Spatial Variability of Rock Depth,"Engineering, Vol. 3 No. 9, 2011, pp. 886-894. doi: 10.4236/eng.2011.39108.

CHAPTER 7

M. Cano, R. Tomás, "Proposal of a New Parameter for the Weathering Characterization of Carbonate Flysch-Like Rock Masses: The Potential Degradation Index (PDI)," Rock Mech Rock Eng DOI 10.1007/s00603-016-0915-2.

CHAPTER 8

Shih-Meng Hsu, Hung-Chieh Lo, Shue-Yeong Chi and Cheng-Yu Ku (2011). Rock Mass Hydraulic Conductivity Estimated by Two Empirical Models, Developments in Hydraulic Conductivity Research, Dr. Oagile Dikinya (Ed.), ISBN: 978-953-307-470-2, InTech, DOI: 10.5772/15669.

CHAPTER 9

Yijiang Peng, Qing Guo, Zhaofeng Zhang, and Yanyan Shan, "Application of Base Force Element Method on Complementary Energy Principle to Rock Mechanics Problems," Mathematical Problems in Engineering, vol. 2015, Article ID 292809, 16 pages, 2015. doi:10.1155/2015/292809.

CHAPTER 10

B. Damjanac, C. Detournay, P.A. Cundall and Varun (2013). Three-Dimensional Numerical Model of Hydraulic Fracturing in Fractured Rock Masses, Effective and Sustainable Hydraulic Fracturing, Dr. Rob Jeffrey (Ed.), ISBN: 978-953-51-1137-5, InTech, DOI: 10.5772/56313.

CHAPTER 11

X. Pan, Z. Feng, G. Dai and H. Liu, "Roughness Research of Center Profile Curve on Rock Fracture Surface Based on Statistical Method," Geomaterials, Vol. 3 No. 2, 2013, pp. 47-53. doi: 10.4236/gm.2013.32006.

INDEX

A

American Society for testing and Materials (ASTM) 140, 150, 171
Analyses of stability 5

B

base force element method (BFEM) 209, 210, 248, 252
Basic Quality (BQ) 72
Biot's coefficient 15
Bonded particle model (BPM) 256, 257
boundary element method (BEM) 12

C

calcarenites 142, 143, 145, 146, 147, 159, 167, 168, 169, 170
calculation method 209
cohesion 29, 30, 31, 32, 36, 37, 55

D

depth index (DI) 176, 177, 179, 203
dilatancy 86, 87, 88, 92, 93
diluvial silty subclay 72, 78
discrete element method (DEM) 12

Discrete fracture network (DFN) 255
displacement discontinuity method (DDM) 12

E

elasticity 83, 84, 86, 110
eluvial subclay 72, 78
equilibrium method 240, 243, 248

F

Factor of Safety (FOS) 1
finite element (FE) 46
finite element method (FEM) 12, 209
friction coefficient 30, 36, 37

G

Geographic Information System (GIS) 117
Geological Strength Index (GSI) 3
gouge content designation (GCD) 176, 177, 179, 203
gravity 214, 215, 216, 217, 218, 221, 222, 226, 227, 232, 233, 242, 247, 248
ground Reduced levels (GRL) 117

H

heat-pulse flowmeter (HPFM) 192
Hydraulic fracture (HF) 255, 256
hydrogeologic system 191
hyperelasticity 86, 87
hyperelasticity theory 86

I

International Society for Rock Mechanics (ISRM) 140

J

Jarque-Bera test 274
Joint Roughness Coefficients (JRC) 270

K

kriging 115, 116, 120, 121, 122, 123, 124, 125, 126, 127, 128, 129, 130, 131, 132, 133, 134

L

Lagrangian analysis of continua 11, 12, 16, 17
limestones 142, 143, 145, 146, 159, 160, 167, 168, 169
lithologies 137, 138, 140, 141, 143, 144, 145, 146, 147, 148, 149, 150, 151, 152, 166, 169
lithology permeability index (LPI) 176, 177, 179, 203

M

microcracks 12, 27
microfabric 139, 167
Morh-Coulomb model 30, 31
MTS (Mechanics Test Systems) 269, 270

N

Naturally fractured reservoirs (NFRs) 256
Numerical analysis 6

numerical analysis methods 210

P

Particle flow code (PFC) 259
plastic modulus 85, 86, 90, 91, 92, 93, 104, 106
Poisson ratio 221, 225, 232, 239, 242
Potential Degradation Index (PDI) 137, 152, 159, 286
profilometer 269, 271

R

recycled aggregate concrete (RAC) 211
rockbolt 80
rock foundation 232, 242
Rock Mass Rating (RMR) 3, 4
rock mass stability 209
rock quality designation (RQD) 176, 177, 178, 179, 203
Rock Quality Designation (RQD) 72, 177

S

safety factor 209, 221, 240, 243, 247, 248, 249
semivariogram 116, 120, 121, 122, 124, 125, 126, 127, 133, 134
Slake Durability Test (SDT) 149
Slope Rock Mass Rating (SRMR) 3, 4, 6
Smooth joint model (SJM) 256, 257
Synthetic rock mass (SRM) 256

T

transmissivity 191

V

viscoplastic flows 84

X

X-ray diffractograms 148